Trigonometry

SECOND EDITION

JUDITH A. BEECHER
Indiana University Purdue University Indianapolis

JUDITH A. PENNA
Indiana University Purdue University Indianapolis

MARVIN L. BITTINGER
Indiana University Purdue University Indianapolis

PEARSON
Addison
Wesley

Y0-CAY-901

Boston San Francisco New York
London Toronto Sydney Tokyo Singapore Madrid
Mexico City Munich Paris Cape Town Hong Kong Montreal

Publisher:	Greg Tobin
Executive Editor:	Anne Kelly
Executive Project Manager:	Kari Heen
Assistant Editors:	Joanna Doxey and Ashley O'Shaughnessy
Production Manager:	Ron Hampton
Cover Design:	Christina Gleason
Digital Assets Manager:	Marianne Groth
Media Producer:	Christine Stavrou
Software Development:	Mary Durnwald, TestGen; Bob Carroll, MathXL
Executive Marketing Manager:	Becky Anderson
Senior Author Support/ Technology Specialist:	Joseph K. Vetere
Senior Prepress Supervisor:	Caroline Fell
Manufacturing Manager:	Evelyn M. Beaton
Senior Media Buyer:	Ginny Michaud
Art and Text Design:	The Davis Group, Inc.
Editorial and Production Services:	Martha K. Morong/Quadrata, Inc.
Composition:	Pre-Press PMG
Illustrations:	William Melvin and Network Graphics

Many of the designations used by manufacturers and sellers to distinguish their products are claimed as trademarks. Where those designations appear in this book, and Addison-Wesley was aware of a trademark claim, the designations have been printed in initial caps or all caps.

Photo Credits: 1, Doug Berry/telluridestock.com 7, Brand X Pictures/ Getty Images 16, © Royalty-Free/Corbis 18, Doug Berry/telluridestock.com 20 (**top**), © Royalty-Free/Corbis 20 (**bottom**), © Reuters/CORBIS 51, © 2006 The Children's Museum of Indianapolis 56, Courtesy of Goss Graphics Systems 86, © Royalty-Free/Corbis 93, PhotoDisc 94, © Royalty-Free/Corbis 101, 114, © John and Lisa Merrill/Corbis 157, © Michael S. Yamashita/CORBIS 165, DigitalVision/Getty 170, ThinkStock/Getty Images 176, © Royalty-Free/Corbis 177, © Neil Rabinowitz/CORBIS 182, DigitalVision/Getty 239, © Tom Stewart/CORBIS

ISBN-13: 978-0-321-53630-3/ISBN-10: 0-321-53630-4

2 3 4 5 6 7 8 9 10–DOW–11 10

Contents

2 Trigonometric Identities, Inverse Functions, and Equations 101

3 Applications of Trigonometry 165

The Trigonometric Functions

APPLICATION

In Telluride, Colorado, there is a free gondola ride that provides a spectacular view of the town and the surrounding mountains. The gondolas that begin in the town at an elevation of 8725 ft travel 5750 ft to Station St. Sophia, whose altitude is 10,550 ft. They then continue 3913 ft to Mountain Village, whose elevation is 9500 ft. **(a)** What is the angle of elevation from the town to Station St. Sophia? **(b)** What is the angle of depression from Station St. Sophia to Mountain Village?

This problem appears as Example 5 in Section 1.2.

1.1 Trigonometric Functions of Acute Angles

✦ Determine the six trigonometric ratios for a given acute angle of a right triangle.

✦ Determine the trigonometric function values of 30°, 45°, and 60°.

✦ Using a calculator, find function values for any acute angle, and given a function value of an acute angle, find the angle.

✦ Given the function values of an acute angle, find the function values of its complement.

✦ The Trigonometric Ratios

We begin our study of trigonometry by considering right triangles and acute angles measured in degrees. An **acute angle** is an angle with measure greater than 0° and less than 90°. Greek letters such as α (alpha), β (beta), γ (gamma), θ (theta), and ϕ (phi) are often used to denote an angle. Consider a right triangle with one of its acute angles labeled θ. The side opposite the right angle is called the **hypotenuse.** The other sides of the triangle are referenced by their position relative to the acute angle θ. One side is opposite θ and one is adjacent to θ.

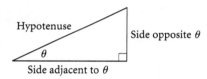

The *lengths* of the sides of the triangle are used to define the six trigonometric ratios:

sine (sin),	cosecant (csc),
cosine (cos),	secant (sec),
tangent (tan),	cotangent (cot).

The **sine of θ** is the *length* of the side opposite θ divided by the *length* of the hypotenuse (see Fig. 1):

$$\sin \theta = \frac{\text{length of side opposite } \theta}{\text{length of hypotenuse}}.$$

The ratio depends on the measure of angle θ and thus is a function of θ. The notation sin θ actually means sin (θ), where sin, or sine, is the name of the function.

The **cosine of θ** is the *length* of the side adjacent to θ divided by the *length* of the hypotenuse (see Fig. 2):

$$\cos \theta = \frac{\text{length of side adjacent to } \theta}{\text{length of hypotenuse}}.$$

The six trigonometric ratios, or trigonometric functions, are defined as follows.

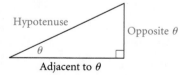

Figure 1

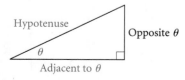

Figure 2

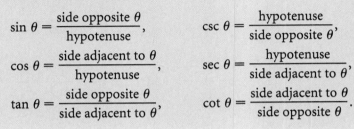

Trigonometric Function Values of an Acute Angle θ

Let θ be an acute angle of a right triangle. Then the six trigonometric functions of θ are as follows:

$$\sin \theta = \frac{\text{side opposite } \theta}{\text{hypotenuse}}, \qquad \csc \theta = \frac{\text{hypotenuse}}{\text{side opposite } \theta},$$

$$\cos \theta = \frac{\text{side adjacent to } \theta}{\text{hypotenuse}}, \qquad \sec \theta = \frac{\text{hypotenuse}}{\text{side adjacent to } \theta},$$

$$\tan \theta = \frac{\text{side opposite } \theta}{\text{side adjacent to } \theta}, \qquad \cot \theta = \frac{\text{side adjacent to } \theta}{\text{side opposite } \theta}.$$

EXAMPLE 1 In the right triangle shown at left, find the six trigonometric function values of **(a)** θ and **(b)** α.

Solution We use the definitions.

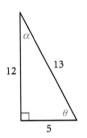

a) $\sin \theta = \dfrac{\text{opp}}{\text{hyp}} = \dfrac{12}{13},$ $\qquad \csc \theta = \dfrac{\text{hyp}}{\text{opp}} = \dfrac{13}{12},$

$\cos \theta = \dfrac{\text{adj}}{\text{hyp}} = \dfrac{5}{13},$ $\qquad \sec \theta = \dfrac{\text{hyp}}{\text{adj}} = \dfrac{13}{5},$

$\tan \theta = \dfrac{\text{opp}}{\text{adj}} = \dfrac{12}{5},$ $\qquad \cot \theta = \dfrac{\text{adj}}{\text{opp}} = \dfrac{5}{12}$

The references to opposite, adjacent, and hypotenuse are relative to *θ*.

b) $\sin \alpha = \dfrac{\text{opp}}{\text{hyp}} = \dfrac{5}{13},$ $\qquad \csc \alpha = \dfrac{\text{hyp}}{\text{opp}} = \dfrac{13}{5},$

$\cos \alpha = \dfrac{\text{adj}}{\text{hyp}} = \dfrac{12}{13},$ $\qquad \sec \alpha = \dfrac{\text{hyp}}{\text{adj}} = \dfrac{13}{12},$

$\tan \alpha = \dfrac{\text{opp}}{\text{adj}} = \dfrac{5}{12},$ $\qquad \cot \alpha = \dfrac{\text{adj}}{\text{opp}} = \dfrac{12}{5}$

The references to opposite, adjacent, and hypotenuse are relative to *α*.

▶ **Now Try Exercise 1.**

In Example 1(a), we note that the value of $\sin \theta$, $\frac{12}{13}$, is the reciprocal of $\frac{13}{12}$, the value of $\csc \theta$. Likewise, we see the same reciprocal relationship between the values of $\cos \theta$ and $\sec \theta$ and between the values of $\tan \theta$ and $\cot \theta$. For any angle, the cosecant, secant, and cotangent values are the reciprocals of the sine, cosine, and tangent function values, respectively.

Reciprocal Functions

$$\csc \theta = \frac{1}{\sin \theta}, \qquad \sec \theta = \frac{1}{\cos \theta}, \qquad \cot \theta = \frac{1}{\tan \theta}$$

If we know the values of the sine, cosine, and tangent functions of an angle, we can use these reciprocal relationships to find the values of the cosecant, secant, and cotangent functions of that angle.

EXAMPLE 2   Given that $\sin \phi = \frac{4}{5}$, $\cos \phi = \frac{3}{5}$, and $\tan \phi = \frac{4}{3}$, find $\csc \phi$, $\sec \phi$, and $\cot \phi$.

Solution   Using the reciprocal relationships, we have

$$\csc \phi = \frac{1}{\sin \phi} = \frac{1}{\frac{4}{5}} = \frac{5}{4}, \qquad \sec \phi = \frac{1}{\cos \phi} = \frac{1}{\frac{3}{5}} = \frac{5}{3},$$

and $\cot \phi = \dfrac{1}{\tan \phi} = \dfrac{1}{\frac{4}{3}} = \dfrac{3}{4}.$

▶ **Now Try Exercise 7.**

Triangles are said to be **similar** if their corresponding angles have the *same* measure. In similar triangles, the lengths of corresponding sides are in the same ratio. The right triangles shown below are similar. Note that the corresponding angles are equal and the length of each side of the second triangle is four times the length of the corresponding side of the first triangle.

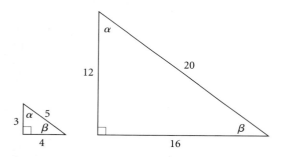

Let's observe the sine, cosine, and tangent values of β in each triangle. Can we expect corresponding function values to be the same?

FIRST TRIANGLE	SECOND TRIANGLE
$\sin \beta = \dfrac{3}{5}$	$\sin \beta = \dfrac{12}{20} = \dfrac{3}{5}$
$\cos \beta = \dfrac{4}{5}$	$\cos \beta = \dfrac{16}{20} = \dfrac{4}{5}$
$\tan \beta = \dfrac{3}{4}$	$\tan \beta = \dfrac{12}{16} = \dfrac{3}{4}$

For the two triangles, the corresponding values of $\sin \beta$, $\cos \beta$, and $\tan \beta$ are the same. The lengths of the sides are proportional—thus the *ratios* are the same. This must be the case because in order for the sine, cosine, and tangent to be functions, there must be only one output (the ratio) for each input (the angle β).

> The trigonometric function values of θ depend only on the measure of the angle, *not* on the size of the triangle.

✦ The Six Functions Related

We can find the other five trigonometric function values of an acute angle when one of the function-value ratios is known.

EXAMPLE 3 If $\sin \beta = \frac{6}{7}$ and β is an acute angle, find the other five trigonometric function values of β.

Solution We know from the definition of the sine function that the ratio

$$\frac{6}{7} \quad \text{is} \quad \frac{\text{opp}}{\text{hyp}}.$$

Using this information, let's consider a right triangle in which the hypotenuse has length 7 and the side opposite β has length 6. To find the length of the side adjacent to β, we recall the *Pythagorean theorem*:

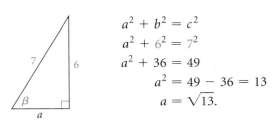

$$a^2 + b^2 = c^2$$
$$a^2 + 6^2 = 7^2$$
$$a^2 + 36 = 49$$
$$a^2 = 49 - 36 = 13$$
$$a = \sqrt{13}.$$

We now use the lengths of the three sides to find the other five ratios:

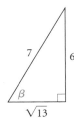

$$\sin \beta = \frac{6}{7}, \qquad\qquad \csc \beta = \frac{7}{6},$$

$$\cos \beta = \frac{\sqrt{13}}{7}, \qquad\qquad \sec \beta = \frac{7}{\sqrt{13}}, \quad \text{or} \quad \frac{7\sqrt{13}}{13},$$

$$\tan \beta = \frac{6}{\sqrt{13}}, \quad \text{or} \quad \frac{6\sqrt{13}}{13}, \qquad \cot \beta = \frac{\sqrt{13}}{6}.$$

▶ Now Try Exercise 9.

✦ Function Values of 30°, 45°, and 60°

In Examples 1 and 3, we found the trigonometric function values of an acute angle of a right triangle when the lengths of the three sides were known. In most situations, we are asked to find the function values when the measure of the acute angle is given. For certain special angles such as 30°, 45°, and 60°, which are frequently seen in applications, we can use geometry to determine the function values.

A right triangle with a 45° angle actually has two 45° angles. Thus the triangle is *isosceles*, and the legs are the same length. Let's consider such a triangle whose legs have length 1. Then we can find the length of its hypotenuse, c, using the Pythagorean theorem as follows:

$$1^2 + 1^2 = c^2, \quad \text{or} \quad c^2 = 2, \quad \text{or} \quad c = \sqrt{2}.$$

Such a triangle is shown below. From this diagram, we can easily determine the trigonometric function values of 45°.

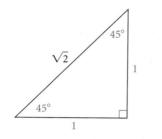

$$\sin 45° = \frac{\text{opp}}{\text{hyp}} = \frac{1}{\sqrt{2}} = \frac{\sqrt{2}}{2} \approx 0.7071,$$

$$\cos 45° = \frac{\text{adj}}{\text{hyp}} = \frac{1}{\sqrt{2}} = \frac{\sqrt{2}}{2} \approx 0.7071,$$

$$\tan 45° = \frac{\text{opp}}{\text{adj}} = \frac{1}{1} = 1$$

It is sufficient to find only the function values of the sine, cosine, and tangent, since the others are their reciprocals.

It is also possible to determine the function values of 30° and 60°. A right triangle with 30° and 60° acute angles is half of an equilateral triangle, as shown in the following figure. Thus if we choose an equilateral triangle whose sides have length 2 and take half of it, we obtain a right triangle that has a hypotenuse of length 2 and a leg of length 1. The other leg has length a, which can be found as follows:

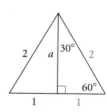

$$a^2 + 1^2 = 2^2$$
$$a^2 + 1 = 4$$
$$a^2 = 3$$
$$a = \sqrt{3}.$$

We can now determine the function values of 30° and 60°:

$$\sin 30° = \frac{1}{2} = 0.5, \qquad\qquad \sin 60° = \frac{\sqrt{3}}{2} \approx 0.8660,$$

$$\cos 30° = \frac{\sqrt{3}}{2} \approx 0.8660, \qquad \cos 60° = \frac{1}{2} = 0.5,$$

$$\tan 30° = \frac{1}{\sqrt{3}} = \frac{\sqrt{3}}{3} \approx 0.5774, \qquad \tan 60° = \frac{\sqrt{3}}{1} = \sqrt{3} \approx 1.7321.$$

Since we will often use the function values of 30°, 45°, and 60°, either the triangles that yield them or the values themselves should be memorized.

	30°	45°	60°
sin	1/2	$\sqrt{2}/2$	$\sqrt{3}/2$
cos	$\sqrt{3}/2$	$\sqrt{2}/2$	1/2
tan	$\sqrt{3}/3$	1	$\sqrt{3}$

Let's now use what we have learned about trigonometric functions of special angles to solve problems. We will consider such applications in greater detail in Section 1.2.

EXAMPLE 4   *Height of a Hot-Air Balloon.*   As a hot-air balloon began to rise, the ground crew drove 1.2 mi to an observation station. The initial observation from the station estimated the angle between the ground and the line of sight to the balloon to be 30°. Approximately how high was the balloon at that point? (We are assuming that the wind velocity was low and that the balloon rose vertically for the first few minutes.)

Solution   We begin with a drawing of the situation. We know the measure of an acute angle and the length of its adjacent side.

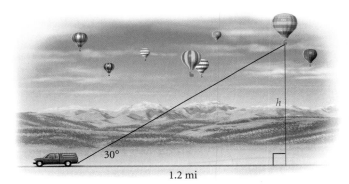

Since we want to determine the length of the opposite side, we can use the tangent ratio, or the cotangent ratio. Here we use the tangent ratio:

$$\tan 30° = \frac{\text{opp}}{\text{adj}} = \frac{h}{1.2}$$

$$1.2 \tan 30° = h$$

$$1.2\left(\frac{\sqrt{3}}{3}\right) = h \qquad \text{Substituting; } \tan 30° = \frac{\sqrt{3}}{3}$$

$$0.7 \approx h.$$

The balloon is approximately 0.7 mi, or 3696 ft, high.

▶ Now Try Exercise 29.

◆ Function Values of Any Acute Angle

Historically, the measure of an angle has been expressed in degrees, minutes, and seconds. One minute, denoted $1'$, is such that $60' = 1°$, or $1' = \frac{1}{60} \cdot (1°)$. One second, denoted $1''$, is such that $60'' = 1'$, or $1'' = \frac{1}{60} \cdot (1')$. Then 61 degrees, 27 minutes, 4 seconds could be written as $61°27'4''$. This **D°M'S'' form** was common before the widespread use of scientific calculators. Now the preferred notation is to express fractional parts of degrees in **decimal degree form.** Although the D°M'S'' notation is still widely used in navigation, we will most often use the decimal form in this text.

Most scientific calculators can convert D°M'S'' notation to decimal degree notation and vice versa. Procedures among calculators vary.

EXAMPLE 5 Convert $5°42'30''$ to decimal degree notation.

Solution We enter $5°42'30''$. The calculator gives us

$$5°42'30'' \approx 5.71°,$$

rounded to the nearest hundredth of a degree.

Without a calculator, we can convert as follows:

$$5°42'30'' = 5° + 42' + 30''$$
$$= 5° + 42' + \frac{30\,'}{60} \qquad 1'' = \frac{1\,'}{60}; 30'' = \frac{30\,'}{60}$$
$$= 5° + 42.5' \qquad \frac{30\,'}{60} = 0.5'$$
$$= 5° + \frac{42.5\,°}{60} \qquad 1' = \frac{1\,°}{60}; 42.5' = \frac{42.5\,°}{60}$$
$$\approx 5.71°. \qquad \frac{42.5\,°}{60} \approx 0.71°$$

▶ Now Try Exercise 37.

EXAMPLE 6 Convert $72.18°$ to D°M'S'' notation.

Solution On a calculator, we enter 72.18. The result is

$$72.18° = 72°10'48''.$$

Without a calculator, we can convert as follows:

$$72.18° = 72° + 0.18 \times 1°$$
$$= 72° + 0.18 \times 60' \qquad 1° = 60'$$
$$= 72° + 10.8'$$
$$= 72° + 10' + 0.8 \times 1'$$
$$= 72° + 10' + 0.8 \times 60'' \qquad 1' = 60''$$
$$= 72° + 10' + 48''$$
$$= 72°10'48''.$$

▶ Now Try Exercise 45.

So far we have measured angles using degrees. Another useful unit for angle measure is the radian, which we will study in Section 1.4. Calculators

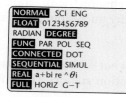

work with either degrees or radians. Be sure to use whichever mode is appropriate. In this section, we use the degree mode.

Keep in mind the difference between an exact answer and an approximation. For example,

$$\sin 60° = \frac{\sqrt{3}}{2}.$$ **This is exact!**

But using a calculator, you get an answer like

$$\sin 60° \approx 0.8660254038.$$ **This is an approximation!**

Calculators generally provide values only of the sine, cosine, and tangent functions. You can find values of the cosecant, secant, and cotangent by taking reciprocals of the sine, cosine, and tangent functions, respectively.

EXAMPLE 7   Find the trigonometric function value, rounded to four decimal places, of each of the following.

a) $\tan 29.7°$ 　　　　　 b) $\sec 48°$ 　　　　　 c) $\sin 84°10'39''$

Solution

a) We check to be sure that the calculator is in DEGREE mode. The function value is

$$\tan 29.7° \approx 0.5703899297$$
$$\approx 0.5704.$$ Rounded to four decimal places

b) The secant function value can be found by taking the reciprocal of the cosine function value:

$$\sec 48° = \frac{1}{\cos 48°} \approx 1.49447655 \approx 1.4945.$$

c) We enter $\sin 84°10'39''$. The result is

$$\sin 84°10'39'' \approx 0.9948409474 \approx 0.9948.$$

▷ Now Try Exercises 61 and 69.

We can use a calculator to find an angle for which we know a trigonometric function value.

EXAMPLE 8   Find the acute angle, to the nearest tenth of a degree, whose sine value is approximately 0.20113.

Solution   The quickest way to find the angle with a calculator is to use an inverse function key. First check to be sure that your calculator is in DEGREE mode. Usually two keys must be pressed in sequence. For this example, if we press

2ND **SIN** .20113 **ENTER** ,

we find that the acute angle whose sine is 0.20113 is approximately 11.60304613°, or 11.6°.

▷ Now Try Exercise 75.

25 ft

θ

6.5 ft

EXAMPLE 9 *Ladder Safety.* A paint crew has purchased new 30-ft extension ladders. The manufacturer states that the safest placement on a wall is to extend the ladder to 25 ft and to position the base 6.5 ft from the wall (*Source*: R. D. Werner Co., Inc.). What angle does the ladder make with the ground in this position?

Solution We make a drawing and then use the most convenient trigonometric function. Because we know the length of the side adjacent to θ and the length of the hypotenuse, we choose the cosine function.

From the definition of the cosine function, we have

$$\cos \theta = \frac{\text{adj}}{\text{hyp}} = \frac{6.5 \text{ ft}}{25 \text{ ft}} = 0.26.$$

Using a calculator, we find the acute angle whose cosine is 0.26:

$$\theta \approx 74.92993786°.$$ Pressing **2ND** **COS** 0.26 **ENTER**

Thus when the ladder is in its safest position, it makes an angle of about 75° with the ground. ◀

✦ Cofunctions and Complements

We recall that two angles are **complementary** whenever the sum of their measures is 90°. Each is the complement of the other. In a right triangle, the acute angles are complementary, since the sum of all three angle measures is 180° and the right angle accounts for 90° of this total. Thus if one acute angle of a right triangle is θ, the other is 90° − θ.

The six trigonometric function values of each of the acute angles in the triangle below are listed at the right. Note that 53° and 37° are complementary angles since 53° + 37° = 90°.

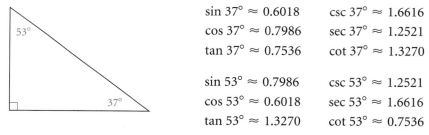

sin 37° ≈ 0.6018	csc 37° ≈ 1.6616
cos 37° ≈ 0.7986	sec 37° ≈ 1.2521
tan 37° ≈ 0.7536	cot 37° ≈ 1.3270
sin 53° ≈ 0.7986	csc 53° ≈ 1.2521
cos 53° ≈ 0.6018	sec 53° ≈ 1.6616
tan 53° ≈ 1.3270	cot 53° ≈ 0.7536

Try this with the acute, complementary angles 20.3° and 69.7° as well. What pattern do you observe? Look for this same pattern in Example 1 earlier in this section.

Note that the sine of an angle is also the cosine of the angle's complement. Similarly, the tangent of an angle is the cotangent of the angle's complement, and the secant of an angle is the cosecant of the angle's complement. These pairs of functions are called **cofunctions.** A list of cofunction identities follows.

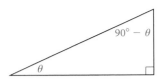

> **Cofunction Identities**
>
> $\sin \theta = \cos (90° - \theta),$ $\qquad \cos \theta = \sin (90° - \theta),$
>
> $\tan \theta = \cot (90° - \theta),$ $\qquad \cot \theta = \tan (90° - \theta),$
>
> $\sec \theta = \csc (90° - \theta),$ $\qquad \csc \theta = \sec (90° - \theta)$

EXAMPLE 10   Given that $\sin 18° \approx 0.3090$, $\cos 18° \approx 0.9511$, and $\tan 18° \approx 0.3249$, find the six trigonometric function values of $72°$.

Solution   Using reciprocal relationships, we know that

$$\csc 18° = \frac{1}{\sin 18°} \approx 3.2361,$$

$$\sec 18° = \frac{1}{\cos 18°} \approx 1.0515,$$

and $\quad \cot 18° = \frac{1}{\tan 18°} \approx 3.0777.$

Since $72°$ and $18°$ are complementary, we have

$$\sin 72° = \cos 18° \approx 0.9511, \qquad \cos 72° = \sin 18° \approx 0.3090,$$
$$\tan 72° = \cot 18° \approx 3.0777, \qquad \cot 72° = \tan 18° \approx 0.3249,$$
$$\sec 72° = \csc 18° \approx 3.2361, \qquad \csc 72° = \sec 18° \approx 1.0515.$$

▶ Now Try Exercise 97.

◆ 1.1   EXERCISE SET

In Exercises 1–6, find the six trigonometric function values of the specified angle.

1.

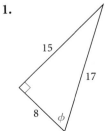

2.

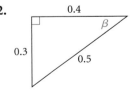

3.

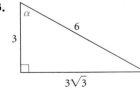

4.

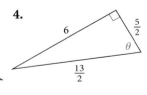

5.

6.

7. Given that $\sin \alpha = \dfrac{\sqrt{5}}{3}$, $\cos \alpha = \dfrac{2}{3}$, and

$\tan \alpha = \dfrac{\sqrt{5}}{2}$, find csc α, sec α, and cot α.

8. Given that $\sin \beta = \dfrac{2\sqrt{2}}{3}$, $\cos \beta = \dfrac{1}{3}$, and

$\tan \beta = 2\sqrt{2}$, find csc β, sec β, and cot β.

Given a function value of an acute angle, find the other five trigonometric function values.

9. $\sin \theta = \frac{24}{25}$

10. $\cos \sigma = 0.7$

11. $\tan \phi = 2$

12. $\cot \theta = \frac{1}{3}$

13. $\csc \theta = 1.5$

14. $\sec \beta = \sqrt{17}$

15. $\cos \beta = \dfrac{\sqrt{5}}{5}$

16. $\sin \sigma = \frac{10}{11}$

Find the exact function value.

17. cos 45°

18. tan 30°

19. sec 60°

20. sin 45°

21. cot 60°

22. csc 45°

23. sin 30°

24. cos 60°

25. tan 45°

26. sec 30°

27. csc 30°

28. tan 60°

29. *Distance across a River.* Find the distance a across the river.

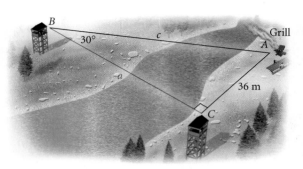

30. *Distance between Bases.* A baseball diamond is actually a square 90 ft on a side. If a line is drawn from third base to first base, then a right triangle QPH is formed, where $\angle QPH$ is 45°. Using a trigonometric function, find the distance from third base to first base.

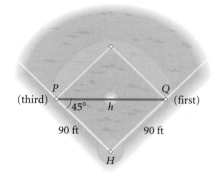

Convert to decimal degree notation. Round to two decimal places.

31. 9°43′

32. 52°15′

33. 35°50″

34. 64°53′

35. 3°2′

36. 19°47′23″

37. 49°38′46″

38. 76°11′34″

39. 15′5″

40. 68°2″

41. 5°53″

42. 44′10″

Convert to degrees, minutes, and seconds. Round to the nearest second.

43. 17.6°

44. 20.14°

45. 83.025°

46. 67.84°

47. 11.75°

48. 29.8°

49. 47.8268°

50. 0.253°

51. 0.9°

52. 30.2505°

53. 39.45°

54. 2.4°

Find the function value. Round to four decimal places.

55. cos 51°

56. cot 17°

57. tan 4°13′

58. sin 26.1°

59. sec 38.43°

60. cos 74°10′40″

61. cos 40.35°

62. csc 45.2°

63. sin 69°

64. tan 63°48′

65. tan 85.4°

66. cos 4°

67. csc 89.5°

68. sec 35.28°

69. cot 30°25′6″

70. sin 59.2°

Find the acute angle θ, to the nearest tenth of a degree, for the given function value.

71. $\sin \theta = 0.5125$

72. $\tan \theta = 2.032$

73. $\tan \theta = 0.2226$

74. $\cos \theta = 0.3842$

75. $\sin \theta = 0.9022$

76. $\tan \theta = 3.056$

77. $\cos \theta = 0.6879$

78. $\sin \theta = 0.4005$

79. $\cot \theta = 2.127$

$\left(Hint: \tan \theta = \dfrac{1}{\cot \theta}.\right)$

80. $\csc \theta = 1.147$

81. $\sec \theta = 1.279$

82. $\cot \theta = 1.351$

Find the exact acute angle θ for the given function value.

83. $\sin \theta = \dfrac{\sqrt{2}}{2}$

84. $\cot \theta = \dfrac{\sqrt{3}}{3}$

85. $\cos \theta = \dfrac{1}{2}$

86. $\sin \theta = \dfrac{1}{2}$

87. $\tan \theta = 1$

88. $\cos \theta = \dfrac{\sqrt{3}}{2}$

89. $\csc \theta = \dfrac{2\sqrt{3}}{3}$

90. $\tan \theta = \sqrt{3}$

91. $\cot \theta = \sqrt{3}$

92. $\sec \theta = \sqrt{2}$

Use the cofunction and reciprocal identities to complete each of the following.

93. $\cos 20° = \underline{\quad} 70° = \dfrac{1}{\underline{\quad} 20°}$

94. $\sin 64° = \underline{\quad} 26° = \dfrac{1}{\underline{\quad} 64°}$

95. $\tan 52° = \cot \underline{\quad} = \dfrac{1}{\underline{\quad} 52°}$

96. $\sec 13° = \csc \underline{\quad} = \dfrac{1}{\underline{\quad} 13°}$

97. Given that

sin 65° ≈ 0.9063,	cos 65° ≈ 0.4226,
tan 65° ≈ 2.1445,	cot 65° ≈ 0.4663,
sec 65° ≈ 2.3662,	csc 65° ≈ 1.1034,

find the six function values of 25°.

98. Given that

sin 8° ≈ 0.1392,	cos 8° ≈ 0.9903,
tan 8° ≈ 0.1405,	cot 8° ≈ 7.1154,
sec 8° ≈ 1.0098,	csc 8° ≈ 7.1853,

find the six function values of 82°.

99. Given that sin 71°10′5″ ≈ 0.9465, cos 71°10′5″ ≈ 0.3228, and tan 71°10′5″ ≈ 2.9321, find the six function values of 18°49′55″.

100. Given that sin 38.7° ≈ 0.6252, cos 38.7° ≈ 0.7804, and tan 38.7° ≈ 0.8012, find the six function values of 51.3°.

101. Given that sin 82° = *p*, cos 82° = *q*, and tan 82° = *r*, find the six function values of 8° in terms of *p*, *q*, and *r*.

Technology Connection

102. Using the TABLE feature, scroll through a table of values to find the acute angle *θ* in each of Exercises 71–80, to the nearest tenth of a degree, for the given function value.

Collaborative Discussion and Writing

103. Explain why it is not necessary to memorize the function values for both 30° and 60°.

104. Explain the difference between reciprocal functions and cofunctions.

Skill Maintenance

Graph the function.

105. $f(x) = e^{x/2}$

106. $f(x) = 2^{-x}$

107. $h(x) = \ln x$

108. $g(x) = \log_2 x$

Solve.

109. $5^x = 625$

110. $e^t = 10,000$

111. $\log_7 x = 3$

112. $\log(3x + 1) - \log(x - 1) = 2$

Synthesis

113. Given that $\sec \beta = 1.5304$, find $\sin(90° - \beta)$.

114. Find the six trigonometric function values of α.

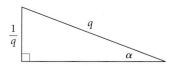

115. Show that the area of this right triangle is
$$\tfrac{1}{2} bc \sin A.$$

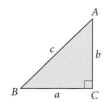

116. Show that the area of this triangle is
$$\tfrac{1}{2} ab \sin \theta.$$

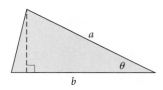

1.2

Applications of Right Triangles

◆ Solve right triangles.

◆ Solve applied problems involving right triangles and trigonometric functions.

◆ Solving Right Triangles

Now that we can find function values for any acute angle, it is possible to *solve* right triangles. To **solve** a triangle means to find the lengths of *all* sides and the measures of *all* angles.

EXAMPLE 1 In $\triangle ABC$ (shown on the following page), find a, b, and B, where a and b represent lengths of sides and B represents the measure of $\angle B$. Here we use standard lettering for naming the sides and angles of a right triangle: Side a is opposite angle A, side b is opposite angle B, where a and b are the legs, and side c, the hypotenuse, is opposite angle C, the right angle.

Solution   In △*ABC*, we know three of the measures:

$$A = 61.7°, \qquad a = ?,$$
$$B = ?, \qquad b = ?,$$
$$C = 90°, \qquad c = 106.2.$$

Since the sum of the angle measures of any triangle is 180° and $C = 90°$, the sum of A and B is 90°. Thus,

$$B = 90° - A = 90° - 61.7° = 28.3°.$$

We are given an acute angle and the hypotenuse. This suggests that we can use the sine and cosine ratios to find a and b, respectively:

$$\sin 61.7° = \frac{\text{opp}}{\text{hyp}} = \frac{a}{106.2} \quad \text{and} \quad \cos 61.7° = \frac{\text{adj}}{\text{hyp}} = \frac{b}{106.2}.$$

Solving for a and b, we get

$$a = 106.2 \sin 61.7° \quad \text{and} \quad b = 106.2 \cos 61.7°$$
$$a \approx 93.5 \qquad\qquad\qquad b \approx 50.3.$$

Thus,

$$A = 61.7°, \qquad a \approx 93.5,$$
$$B = 28.3°, \qquad b \approx 50.3,$$
$$C = 90°, \qquad c = 106.2.$$

▶ **Now Try Exercise 1.**

EXAMPLE 2   In △*DEF* (shown at left), find D and F. Then find d.

Solution   In △*DEF*, we know three of the measures:

$$D = ?, \qquad d = ?,$$
$$E = 90°, \qquad e = 23,$$
$$F = ?, \qquad f = 13.$$

We know the side adjacent to D and the hypotenuse. This suggests the use of the cosine ratio:

$$\cos D = \frac{\text{adj}}{\text{hyp}} = \frac{13}{23}.$$

We now find the angle whose cosine is $\frac{13}{23}$. To the nearest hundredth of a degree,

$$D \approx 55.58°. \qquad \text{Pressing } \boxed{\text{2ND}} \boxed{\text{COS}} (13/23) \boxed{\text{ENTER}}$$

Since the sum of D and F is 90°, we can find F by subtracting:

$$F = 90° - D \approx 90° - 55.58° \approx 34.42°.$$

We could use the Pythagorean theorem to find d, but we will use a trigonometric function here. We could use $\cos F$, $\sin D$, or the tangent or cotangent ratios for either D or F. Let's use $\tan D$:

$$\tan D = \frac{\text{opp}}{\text{adj}} = \frac{d}{13}, \quad \text{or} \quad \tan 55.58° \approx \frac{d}{13}.$$

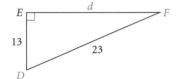

Then
$$d \approx 13 \tan 55.58° \approx 19.$$

The six measures are

$D \approx 55.58°,$ $d \approx 19,$

$E = 90°,$ $e = 23,$

$F \approx 34.42°,$ $f = 13.$

▶ **Now Try Exercise 5.**

◆ Applications

Right triangles can be used to model and solve many applied problems in the real world.

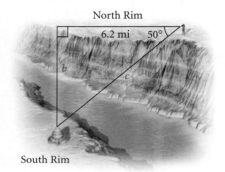

North Rim

6.2 mi   50°

b

c

South Rim

EXAMPLE 3 *Hiking at the Grand Canyon.* A backpacker hiking east along the North Rim of the Grand Canyon notices an unusual rock formation directly across the canyon. She decides to continue watching the landmark while hiking along the rim. In 2 hr, she has gone 6.2 mi due east and the landmark is still visible but at approximately a 50° angle to the North Rim. (See the figure at left.)

a) How many miles is she from the rock formation?

b) How far is it across the canyon from her starting point?

Solution

a) We know the side adjacent to the 50° angle and want to find the hypotenuse. We can use the cosine function:

$$\cos 50° = \frac{6.2 \text{ mi}}{c}$$

$$c = \frac{6.2 \text{ mi}}{\cos 50°} \approx 9.6 \text{ mi}.$$

After hiking 6.2 mi, she is approximately 9.6 mi from the rock formation.

b) We know the side adjacent to the 50° angle and want to find the opposite side. We can use the tangent function:

$$\tan 50° = \frac{b}{6.2 \text{ mi}}$$

$$b = 6.2 \text{ mi} \cdot \tan 50° \approx 7.4 \text{ mi}.$$

Thus it is approximately 7.4 mi across the canyon from her starting point.

▶ **Now Try Exercise 19.**

EXAMPLE 4 *Rafters for a House.* House framers can use trigonometric functions to determine the lengths of rafters for a house. They first choose the pitch of the roof, or the ratio of the rise over the run. Then using a triangle with that ratio, they calculate the length of the rafter needed for the house. José is constructing rafters for a roof with a 10/12 pitch on a house that is 42 ft wide. Find the length x of the rafter of the house to the nearest tenth of a foot.

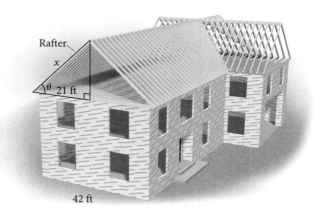

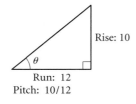

Solution   We first find the angle θ that the rafter makes with the side wall. We know the rise, 10, and the run, 12, so we can use the tangent function to determine the angle that corresponds to the pitch of 10/12:

$$\tan \theta = \frac{10}{12} \approx 0.8333.$$

Thus, $\theta \approx 39.8°$. Since trigonometric function values of θ depend only on the measure of the angle and not on the size of the triangle, the angle for the rafter is also 39.8°.

To determine the length x of the rafter, we can use the cosine function. (See the figure at left.) Note that the width of the house is 42 ft, and a leg of this triangle is half that length, 21 ft.

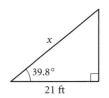

$$\cos 39.8° = \frac{21 \text{ ft}}{x}$$

$$x \cos 39.8° = 21 \text{ ft} \qquad \text{Multiplying by } x$$

$$x = \frac{21 \text{ ft}}{\cos 39.8°} \qquad \text{Dividing by } \cos 39.8°$$

$$x \approx 27.3 \text{ ft}$$

The length of the rafter for this house is approximately 27.3 ft.

▶ Now Try Exercise 31.

Many applications with right triangles involve an *angle of elevation* or an *angle of depression*. The angle between the horizontal and a line of sight above the horizontal is called an **angle of elevation.** The angle between the horizontal and a line of sight below the horizontal is called an **angle of depression.** For example, suppose that you are looking straight ahead and then you move your eyes up to look at an approaching airplane. The angle that your eyes pass through is an angle of elevation. If the pilot of the plane is looking forward and then looks down, the pilot's eyes pass through an angle of depression.

EXAMPLE 5   *Gondola Aerial Lift.*   In Telluride, Colorado, there is a free gondola ride that provides a spectacular view of the town and the surrounding mountains. The gondolas that begin in the town at an elevation of 8725 ft travel 5750 ft to Station St. Sophia, whose altitude is 10,550 ft. They then continue 3913 ft to Mountain Village, whose elevation is 9500 ft.

a) What is the angle of elevation from the town to Station St. Sophia?

b) What is the angle of depression from Station St. Sophia to Mountain Village?

Solution   We begin by labeling a drawing with the given information.

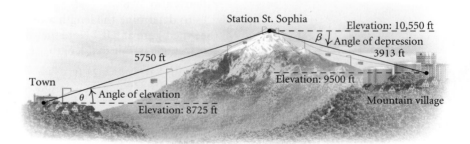

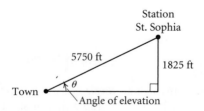

a) The difference in the elevation of Station St. Sophia and the elevation of the town is 10,550 ft − 8725 ft, or 1825 ft. This measure is the length of the side opposite the angle of elevation, θ, in the right triangle shown at left. Since we know the side opposite θ and the hypotenuse, we can find θ by using the sine function. We first find $\sin \theta$:

$$\sin \theta = \frac{1825 \text{ ft}}{5750 \text{ ft}} \approx 0.3174.$$

Using a calculator, we find that

$$\theta \approx 18.5°. \quad \text{Pressing } \boxed{\text{2ND}} \ \boxed{\text{SIN}} \ 0.3174 \ \boxed{\text{ENTER}}$$

Thus the angle of elevation from the town to Station St. Sophia is approximately 18.5°.

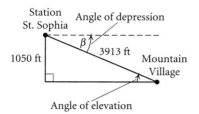

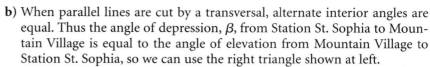

b) When parallel lines are cut by a transversal, alternate interior angles are equal. Thus the angle of depression, β, from Station St. Sophia to Mountain Village is equal to the angle of elevation from Mountain Village to Station St. Sophia, so we can use the right triangle shown at left.

The difference in the elevation of Station St. Sophia and the elevation of Mountain Village is 10,550 ft − 9500 ft, or 1050 ft. Since we know the side opposite the angle of elevation and the hypotenuse, we can again use the sine function:

$$\sin \beta = \frac{1050 \text{ ft}}{3913 \text{ ft}} \approx 0.2683.$$

Using a calculator, we find that

$$\beta = 15.6°.$$

The angle of depression from Station St. Sophia to Mountain Village is approximately 15.6°. ▶ Now Try Exercise 17.

EXAMPLE 6 *Cloud Height.* To measure cloud height at night, a vertical beam of light is directed on a spot on the cloud. From a point 135 ft away from the light source, the angle of elevation to the spot is found to be 67.35°. Find the height of the cloud.

Solution From the figure, we have

$$\tan 67.35° = \frac{h}{135 \text{ ft}}$$

$$h = 135 \text{ ft} \cdot \tan 67.35° \approx 324 \text{ ft}.$$

The height of the cloud is about 324 ft. ▶ Now Try Exercise 21.

Some applications of trigonometry involve the concept of direction, or bearing. In this text we present two ways of giving direction, the first below and the second in Exercise Set 1.3.

Bearing: First-Type

One method of giving direction, or bearing, involves reference to a north–south line using an acute angle. For example, N55°W means 55° west of north and S67°E means 67° east of south.

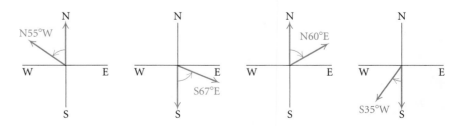

EXAMPLE 7 *Distance to a Forest Fire.* A forest ranger at point *A* sights a fire directly south. A second ranger at point *B*, 7.5 mi east, sights the same fire at a bearing of S27°23′W. How far from *A* is the fire?

Solution We first find the complement of 27°23′:

$$B = 90° - 27°23'$$ Angle *B* is opposite side *d* in the right triangle.

$$= 62°37'$$

$$\approx 62.62°.$$

From the figure shown above, we see that the desired distance *d* is part of a right triangle. We have

$$\frac{d}{7.5 \text{ mi}} \approx \tan 62.62°$$

$$d \approx 7.5 \text{ mi } \tan 62.62° \approx 14.5 \text{ mi.}$$

The forest ranger at point *A* is about 14.5 mi from the fire.

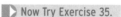 Now Try Exercise 35.

EXAMPLE 8 *U.S. Cellular Field.* In U.S. Cellular Field, the home of the Chicago White Sox baseball team, the first row of seats in the upper deck is farther away from home plate than the last row of seats in the original Comiskey Park. Although there is no obstructed view in the U.S. Cellular Field, some of the fans still complain about the present distance from home plate to the upper deck of seats. (*Source*: *Chicago Tribune*, September 19, 1993) From a seat in the last row of the upper deck directly behind the batter, the angle of depression to home plate is 29.9°, and the angle of depression to the pitcher's mound is 24.2°. Find (**a**) the viewing distance to home plate and (**b**) the viewing distance to the pitcher's mound.

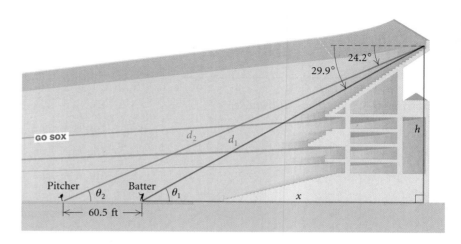

Solution From geometry we know that $\theta_1 = 29.9°$ and $\theta_2 = 24.2°$. The standard distance from home plate to the pitcher's mound is 60.5 ft. In the drawing, we let d_1 be the viewing distance to home plate, d_2 the viewing distance to the pitcher's mound, h the elevation of the last row, and x the horizontal distance from the batter to a point directly below the seat in the last row of the upper deck.

We begin by determining the distance x. We use the tangent function with $\theta_1 = 29.9°$ and $\theta_2 = 24.2°$:

$$\tan 29.9° = \frac{h}{x} \qquad \text{and} \quad \tan 24.2° = \frac{h}{x + 60.5}$$

or $\qquad h = x \tan 29.9° \quad$ and $\qquad h = (x + 60.5) \tan 24.2°.$

Then substituting $x \tan 29.9°$ for h in the second equation, we obtain

$$x \tan 29.9° = (x + 60.5) \tan 24.2°.$$

Solving for x, we get

$$x \tan 29.9° = x \tan 24.2° + 60.5 \tan 24.2°$$
$$x \tan 29.9° - x \tan 24.2° = x \tan 24.2° + 60.5 \tan 24.2° - x \tan 24.2°$$
$$x(\tan 29.9° - \tan 24.2°) = 60.5 \tan 24.2°$$
$$x = \frac{60.5 \tan 24.2°}{\tan 29.9° - \tan 24.2°}$$
$$x \approx 216.5.$$

We can then find d_1 and d_2 using the cosine function:

$$\cos 29.9° = \frac{216.5}{d_1} \qquad \text{and} \quad \cos 24.2° = \frac{216.5 + 60.5}{d_2}$$

or $\qquad d_1 = \frac{216.5}{\cos 29.9°} \quad$ and $\qquad d_2 = \frac{277}{\cos 24.2°}$

$$d_1 \approx 249.7 \qquad\qquad\qquad d_2 \approx 303.7.$$

The distance to home plate is about 250 ft (In the original Comiskey Park, the distance to home plate was only 150 ft.), and the distance to the pitcher's mound is about 304 ft.

1.2 EXERCISE SET

In Exercises 1–6, solve the right triangle.

1.

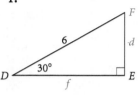

2.

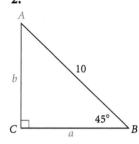

3.

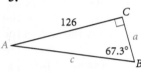

4.

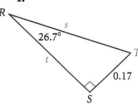

5.

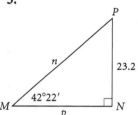

6.

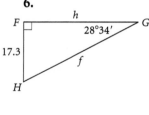

In Exercises 7–16, solve the right triangle. (Standard lettering has been used.)

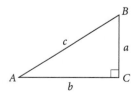

7. $A = 87°43'$, $a = 9.73$

8. $a = 12.5$, $b = 18.3$

9. $b = 100$, $c = 450$

10. $B = 56.5°$, $c = 0.0447$

11. $A = 47.58°$, $c = 48.3$

12. $B = 20.6°$, $a = 7.5$

13. $A = 35°$, $b = 40$

14. $B = 69.3°$, $b = 93.4$

15. $b = 1.86$, $c = 4.02$

16. $a = 10.2$, $c = 20.4$

17. *Aerial Photography.* An aerial photographer who photographs farm properties for a real estate company has determined from experience that the best photo is taken at a height of approximately 475 ft and a distance of 850 ft from the farmhouse. What is the angle of depression from the plane to the house?

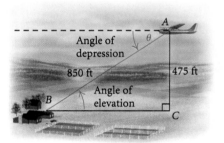

18. *Memorial Flag Case.* A tradition in the United States is to drape an American flag over the casket of a deceased U.S. armed forces veteran. At the burial, the flag is removed, folded into a triangle, and presented to the family. The folded flag will fit in an isosceles right triangle case, as shown below. The inside dimension across the bottom is $21\frac{1}{2}$ in. (*Source:* Bruce Kieffer, *Woodworker's Journal,* August 2006). Using trigonometric functions, find the length x and round the answer to the nearest tenth of an inch.

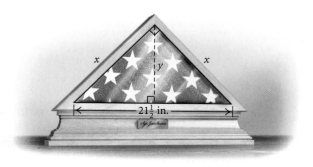

19. *Safety Line to Raft.* Each spring, Bryan uses his vacation time to ready his lake property for the summer. He wants to run a new safety line from point *B* on the shore to the corner of the anchored diving raft. The current safety line, which runs perpendicular to the shore line to point *A*, is 40 ft long. He estimates the angle from *B* to the corner of the raft to be 50°. Approximately how much rope does he need for the new safety line if he allows 5 ft of rope at each end to fasten the rope?

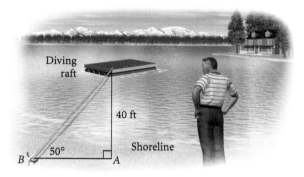

20. *Enclosing an Area.* Alicia is enclosing a triangular area in a corner of her fenced rectangular backyard for her Labrador retriever. In order for a certain tree to be included in this pen, one side needs to be 14.5 ft and make a 53° angle with the new side. How long is the new side?

21. *Height of a Tree.* A supervisor must train a new team of loggers to estimate the heights of trees. As an example, she walks off 40 ft from the base of a tree and estimates the angle of elevation to the tree's peak to be 70°. Approximately how tall is the tree?

22. *Easel Display.* A marketing group is designing an easel to display posters advertising their

newest products. They want the easel to be 6 ft tall and the back of it to fit flush against a wall. For optimal eye contact, the best angle between the front and back legs of the easel is 23°. How far from the wall should the front legs be placed in order to obtain this angle?

23. *Golden Gate Bridge.* The Golden Gate Bridge has two main towers of equal height that support the two main cables. A visitor on a tour ship passing through San Francisco Bay views the top of one of the towers and estimates the angle of elevation to be 30°. After sailing 670 ft closer, he estimates the angle of elevation to this same tower to be 50°. Approximate the height of the tower to the nearest foot.

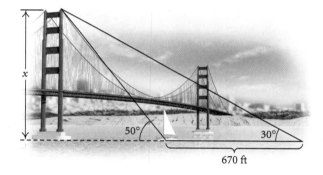

24. *Sand Dunes National Park.* While visiting the Sand Dunes National Park in Colorado, Cole approximated the angle of elevation to the top of a sand dune to be 20°. After walking 800 ft closer, he guessed that the angle of elevation had increased by 15°. Approximately how tall is the dune he was observing?

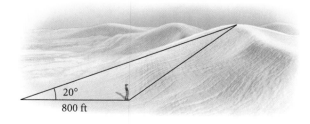

25. *Inscribed Pentagon.* A regular pentagon is inscribed in a circle of radius 15.8 cm. Find the perimeter of the pentagon.

26. *Height of a Weather Balloon.*   A weather balloon is directly west of two observing stations that are 10 mi apart. The angles of elevation of the balloon from the two stations are 17.6° and 78.2°. How high is the balloon?

27. *Height of a Kite.*   For a science fair project, a group of students tested different materials used to construct kites. Their instructor provided an instrument that accurately measures the angle of elevation. In one of the tests, the angle of elevation was 63.4° with 670 ft of string out. Assuming the string was taut, how high was the kite?

28. *Height of a Building.*   A window washer on a ladder looks at a nearby building 100 ft away, noting that the angle of elevation to the top of the building is 18.7° and the angle of depression to the bottom of the building is 6.5°. How tall is the nearby building?

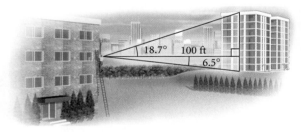

29. *Quilt Design.*   Nancy is designing a quilt that she will enter in the quilt competition at the State Fair. The quilt consists of twelve identical squares with 4 rows of 3 squares each. Each square is to have a regular octagon inscribed in a circle, as shown in the figure. Each side of the octagon is to be 7 in. long. Find the radius of the circumscribed circle and the dimensions of the quilt. Round the answers to the nearest hundredth of an inch.

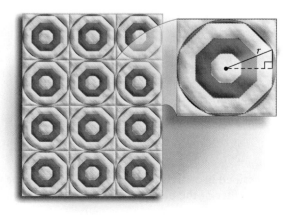

30. *Rafters for a House.*   Blaise, an architect for luxury homes, is designing a house that is 46 ft wide with a roof whose pitch is 11/12. Determine the length of the rafters needed for this house. Round the answer to the nearest tenth of a foot.

31. *Rafters for a Medical Office.*   The pitch of the roof for a medical office needs to be 5/12. If the building is 33 ft wide, how long must the rafters be?

32. *Angle of Elevation.*   What is the angle of elevation of the sun when a 35-ft mast casts a 20-ft shadow?

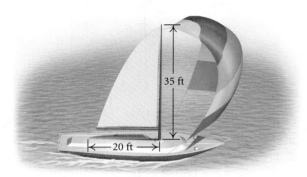

33. *Distance between Towns.*   From a hot-air balloon 2 km high, the angles of depression to two towns in line with the balloon are 81.2° and 13.5°. How far apart are the towns?

34. *Distance from a Lighthouse.*   From the top of a lighthouse 55 ft above sea level, the angle of depression to a small boat is 11.3°. How far from the foot of the lighthouse is the boat?

35. *Lightning Detection.*   In extremely large forests, it is not cost-effective to position forest rangers in towers or to use small aircraft to continually watch for fires. Since lightning is a frequent cause of fire, lightning detectors are now commonly used instead. These devices not only give a bearing on the location but also measure the intensity of the lightning. A detector at point *Q* is situated 15 mi west of a central fire station at point *R*. The bearing from *Q* to where lightning hits due south of *R* is S37.6°E. How far is the hit from point *R*?

36. *Length of an Antenna.*   A vertical antenna is mounted atop a 50-ft pole. From a point on level ground 75 ft from the base of the pole, the antenna subtends an angle of 10.5°. Find the length of the antenna.

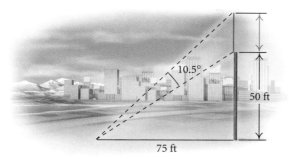

37. *Lobster Boat.*   A lobster boat is situated due west of a lighthouse. A barge is 12 km south of the lobster boat. From the barge, the bearing to the lighthouse is N63°20′E. How far is the lobster boat from the lighthouse?

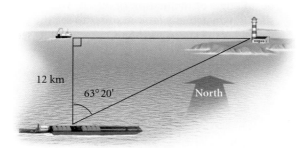

Collaborative Discussion and Writing

38. Explain in your own words five ways in which length *c* can be determined in this triangle. Which way seems the most efficient?

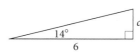

39. In this section, the trigonometric functions have been defined as functions of acute angles. Thus the set of angles whose measures are greater than 0° and less than 90° is the domain for each function. What appear to be the ranges for the sine, the cosine, and the tangent functions given this domain?

Skill Maintenance

Find the distance between the points.

40. $(-9, 3)$ and $(0, 0)$

41. $(8, -2)$ and $(-6, -4)$

42. Convert to a logarithmic equation: $e^4 = t$.

43. Convert to an exponential equation: $\log 0.001 = -3$.

Synthesis

44. Find *a*, to the nearest tenth.

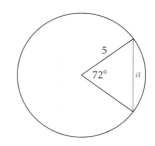

45. Find *h*, to the nearest tenth.

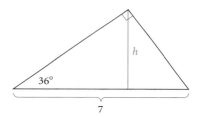

46. *Diameter of a Pipe.*   A V-gauge is used to find the diameter of a pipe. The advantage of such a device is that it is rugged, it is accurate, and it has no moving parts to break down. In the figure, the measure of angle *AVB* is 54°. A pipe is placed in the V-shaped slot and the distance *VP* is used to estimate the diameter. The line *VP* is calibrated by listing as its units the corresponding diameters. This, in effect, establishes a function between *VP* and *d*.

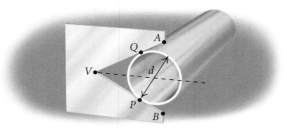

a) Suppose that the diameter of a pipe is 2 cm. What is the distance *VP*?

b) Suppose that the distance *VP* is 3.93 cm. What is the diameter of the pipe?

c) Find a formula for *d* in terms of *VP*.

d) Find a formula for *VP* in terms of *d*.

47. *Construction of Picnic Pavilions.* A construction company is mass-producing picnic pavilions for national parks, as shown in the figure. The rafter ends are to be sawed in such a way that they will be vertical when in place. The front is 8 ft high, the back is $6\frac{1}{2}$ ft high, and the distance between the front and back is 8 ft. At what angle should the rafters be cut?

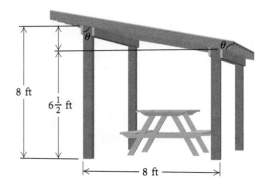

48. *Measuring the Radius of the Earth.* One way to measure the radius of the Earth is to climb to the top of a mountain whose height above sea level is known and measure the angle between a vertical line to the center of the Earth from the top of the mountain and a line drawn from the top of the mountain to the horizon, as shown in the figure. The height of Mt. Shasta in California is 14,162 ft. From the top of Mt. Shasta, one can see the horizon on the Pacific Ocean. The angle formed between a line to the horizon and the vertical is found to be 87°53′.

Use this information to estimate the radius of the Earth, in miles.

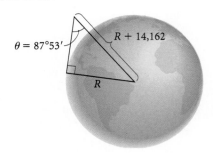

49. *Sound of an Airplane.* It is common experience to hear the sound of a low-flying airplane and look at the wrong place in the sky to see the plane. Suppose that a plane is traveling directly at you at a speed of 200 mph and an altitude of 3000 ft, and you hear the sound at what seems to be an angle of inclination of 20°. At what angle θ should you actually look in order to see the plane? Consider the speed of sound to be 1100 ft/sec.

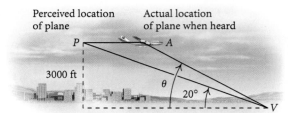

1.3 Trigonometric Functions of Any Angle

◆ Find angles that are coterminal with a given angle and find the complement and the supplement of a given angle.

◆ Determine the six trigonometric function values for any angle in standard position when the coordinates of a point on the terminal side are given.

◆ Find the function values for any angle whose terminal side lies on an axis.

◆ Find the function values for an angle whose terminal side makes an angle of 30°, 45°, or 60° with the *x*-axis.

◆ Use a calculator to find function values and angles.

◆ Angles, Rotations, and Degree Measure

An *angle* is a familiar figure in the world around us.

An **angle** is the union of two rays with a common endpoint called the **vertex.** In trigonometry, we often think of an angle as a **rotation.** To do so, think of locating a ray along the positive *x*-axis with its endpoint at the origin. This ray is called the **initial side** of the angle. Though we leave that ray fixed, think of making a copy of it and rotating it. A rotation *counterclockwise* is a **positive rotation,** and a rotation *clockwise* is a **negative rotation.** The ray at the end of the rotation is called the **terminal side** of the angle. The angle formed is said to be in **standard position.**

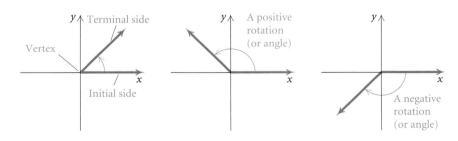

The measure of an angle or rotation may be given in degrees. The Babylonians developed the idea of dividing the circumference of a circle into 360 equal parts, or degrees. If we let the measure of one of these parts be 1°, then one complete positive revolution or rotation has a measure of 360°. One half of a revolution has a measure of 180°, one fourth of a revolution has a measure of 90°, and so on. We can also speak of an angle of measure 60°, 135°, 330°, or 420°. The terminal sides of these angles lie in quadrants I, II, IV, and I, respectively. The negative rotations −30°, −110°, and −225° represent angles with terminal sides in quadrants IV, III, and II, respectively.

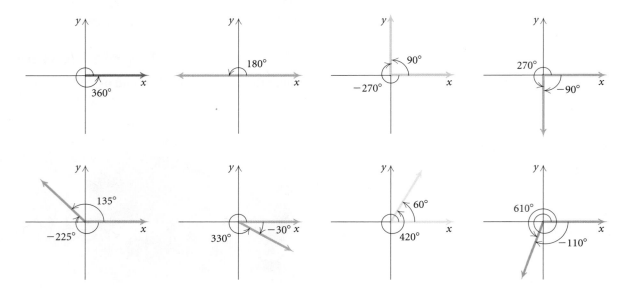

If two or more angles have the same terminal side, the angles are said to be **coterminal.** To find angles coterminal with a given angle, we add or subtract multiples of 360°. For example, 420°, shown above, has the same terminal side as 60°, since 420° = 360° + 60°. Thus we say that angles of measure 60° and 420° are coterminal. The negative rotation that measures −300° is also coterminal with 60° because 60° − 360° = −300°. The set of all angles coterminal with 60° can be expressed as 60° + n · 360°, where n is an integer. Other examples of coterminal angles shown above are 90° and −270°, −90° and 270°, 135° and −225°, −30° and 330°, and −110° and 610°.

EXAMPLE 1 Find two positive and two negative angles that are coterminal with (**a**) 51° and (**b**) −7°.

Solution

a) We add and subtract multiples of 360°. Many answers are possible.

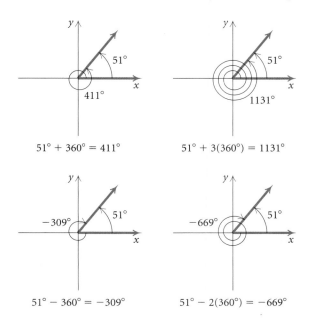

$$51° + 360° = 411°$$ $$51° + 3(360°) = 1131°$$

$$51° − 360° = −309°$$ $$51° − 2(360°) = −669°$$

Thus angles of measure 411°, 1131°, −309°, and −669° are coterminal with 51°.

b) We have the following:

$$-7° + 360° = 353°, \qquad -7° + 2(360°) = 713°,$$
$$-7° - 360° = -367°, \qquad -7° - 10(360°) = -3607°.$$

Thus angles of measure 353°, 713°, −367°, and −3607° are coterminal with −7°.

▶ Now Try Exercise 13.

Angles can be classified by their measures, as seen in the following figure.

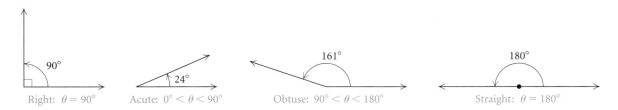

Right: $\theta = 90°$ Acute: $0° < \theta < 90°$ Obtuse: $90° < \theta < 180°$ Straight: $\theta = 180°$

Recall that two acute angles are **complementary** if their sum is 90°. For example, angles that measure 10° and 80° are complementary because $10° + 80° = 90°$. Two positive angles are **supplementary** if their sum is

180°. For example, angles that measure 45° and 135° are supplementary because 45° + 135° = 180°.

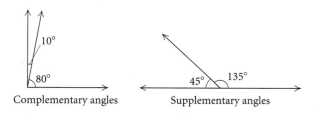

Complementary angles   Supplementary angles

EXAMPLE 2   Find the complement and the supplement of 71.46°.

Solution   We have

$$90° - 71.46° = 18.54°,$$
$$180° - 71.46° = 108.54°.$$

Thus the complement of 71.46° is 18.54° and the supplement is 108.54°.

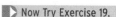
▶ Now Try Exercise 19.

✦ Trigonometric Functions of Angles or Rotations

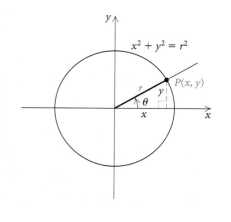

Many applied problems in trigonometry involve the use of angles that are not acute. Thus we need to extend the domains of the trigonometric functions defined in Section 1.1 to angles, or rotations, of *any* size. To do this, we first consider a right triangle with one vertex at the origin of a coordinate system and one vertex *on the positive x-axis*. (See the figure at left.) The other vertex is at *P*, a point on the circle whose center is at the origin and whose radius *r* is the length of the hypotenuse of the triangle. This triangle is a **reference triangle** for angle θ, which is in standard position. Note that *y* is the length of the side opposite θ and *x* is the length of the side adjacent to θ.

Recalling the definitions in Section 1.1, we note that three of the trigonometric functions of angle θ are defined as follows:

$$\sin \theta = \frac{\text{opp}}{\text{hyp}} = \frac{y}{r}, \qquad \cos \theta = \frac{\text{adj}}{\text{hyp}} = \frac{x}{r}, \qquad \tan \theta = \frac{\text{opp}}{\text{adj}} = \frac{y}{x}.$$

Since *x* and *y* are the coordinates of the point *P* and the length of the radius is the length of the hypotenuse, we can also define these functions as follows:

$$\sin \theta = \frac{y\text{-coordinate}}{\text{radius}},$$

$$\cos \theta = \frac{x\text{-coordinate}}{\text{radius}},$$

$$\tan \theta = \frac{y\text{-coordinate}}{x\text{-coordinate}}.$$

We will use these definitions for functions of angles of any measure. The following figures show angles whose terminal sides lie in quadrants II, III, and IV.

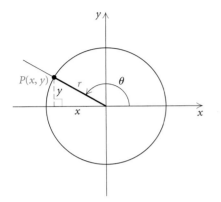

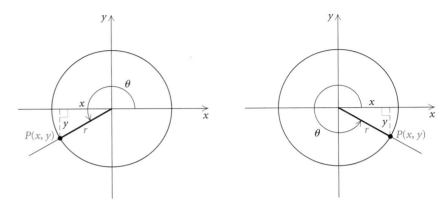

A reference triangle can be drawn for angles in any quadrant, as shown. Note that the angle is in standard position; that is, it is always measured from the positive half of the x-axis. The point $P(x, y)$ is a point, other than the vertex, on the terminal side of the angle. Each of its two coordinates may be positive, negative, or zero, depending on the location of the terminal side. *The length of the radius, which is also the length of the hypotenuse of the reference triangle, is always considered positive.* $\left(\text{Note that } x^2 + y^2 = r^2, \text{ or } r = \sqrt{x^2 + y^2}.\right)$ Regardless of the location of P, we have the following definitions.

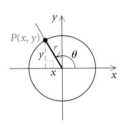

Trigonometric Functions of Any Angle θ

Suppose that $P(x, y)$ is any point other than the vertex on the terminal side of any angle θ in standard position, and r is the radius, or distance from the origin to $P(x, y)$. Then the trigonometric functions are defined as follows:

$$\sin \theta = \frac{y\text{-coordinate}}{\text{radius}} = \frac{y}{r}, \qquad \csc \theta = \frac{\text{radius}}{y\text{-coordinate}} = \frac{r}{y},$$

$$\cos \theta = \frac{x\text{-coordinate}}{\text{radius}} = \frac{x}{r}, \qquad \sec \theta = \frac{\text{radius}}{x\text{-coordinate}} = \frac{r}{x},$$

$$\tan \theta = \frac{y\text{-coordinate}}{x\text{-coordinate}} = \frac{y}{x}, \qquad \cot \theta = \frac{x\text{-coordinate}}{y\text{-coordinate}} = \frac{x}{y}.$$

Values of the trigonometric functions can be positive, negative, or zero, depending on where the terminal side of the angle lies. The length of the radius is always positive. Thus the signs of the function values depend only on the coordinates of the point P on the terminal side of the angle. In the first quadrant, all function values are positive because both coordinates are positive. In the second quadrant, first coordinates are negative and second coordinates are positive; thus only the sine and the cosecant values are positive. Similarly, we can determine the signs of the function values in the third and fourth quadrants. *Because of the reciprocal relationships, we need learn only the signs for the sine, cosine, and tangent functions.*

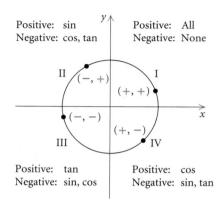

Positive: sin
Negative: cos, tan

Positive: All
Negative: None

Positive: tan
Negative: sin, cos

Positive: cos
Negative: sin, tan

EXAMPLE 3   Find the six trigonometric function values for each angle shown.

a)

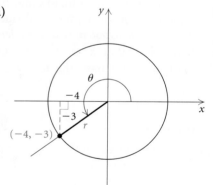

b)

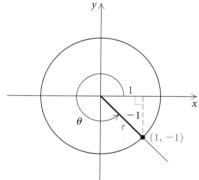

c)

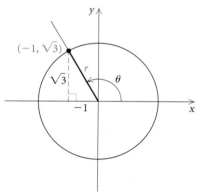

Solution

a) We first determine r, the distance from the origin $(0,0)$ to the point $(-4, -3)$. The distance between $(0,0)$ and any point (x, y) on the terminal side of the angle is

$$r = \sqrt{(x - 0)^2 + (y - 0)^2}$$
$$= \sqrt{x^2 + y^2}.$$

Substituting -4 for x and -3 for y, we find

$$r = \sqrt{(-4)^2 + (-3)^2}$$
$$= \sqrt{16 + 9} = \sqrt{25} = 5.$$

Using the definitions of the trigonometric functions, we can now find the function values of θ. We substitute -4 for x, -3 for y, and 5 for r:

$$\sin\theta = \frac{y}{r} = \frac{-3}{5} = -\frac{3}{5}, \qquad \csc\theta = \frac{r}{y} = \frac{5}{-3} = -\frac{5}{3},$$

$$\cos\theta = \frac{x}{r} = \frac{-4}{5} = -\frac{4}{5}, \qquad \sec\theta = \frac{r}{x} = \frac{5}{-4} = -\frac{5}{4},$$

$$\tan\theta = \frac{y}{x} = \frac{-3}{-4} = \frac{3}{4}, \qquad \cot\theta = \frac{x}{y} = \frac{-4}{-3} = \frac{4}{3}.$$

As expected, the tangent and the cotangent values are positive and the other four are negative. This is true for all angles in quadrant III.

b) We first determine r, the distance from the origin to the point $(1, -1)$:

$$r = \sqrt{1^2 + (-1)^2} = \sqrt{1 + 1} = \sqrt{2}.$$

Substituting 1 for x, -1 for y, and $\sqrt{2}$ for r, we find

$$\sin\theta = \frac{y}{r} = \frac{-1}{\sqrt{2}} = -\frac{\sqrt{2}}{2}, \qquad \csc\theta = \frac{r}{y} = \frac{\sqrt{2}}{-1} = -\sqrt{2},$$

$$\cos\theta = \frac{x}{r} = \frac{1}{\sqrt{2}} = \frac{\sqrt{2}}{2}, \qquad \sec\theta = \frac{r}{x} = \frac{\sqrt{2}}{1} = \sqrt{2},$$

$$\tan\theta = \frac{y}{x} = \frac{-1}{1} = -1, \qquad \cot\theta = \frac{x}{y} = \frac{1}{-1} = -1.$$

c) We determine r, the distance from the origin to the point $\left(-1, \sqrt{3}\right)$:

$$r = \sqrt{(-1)^2 + \left(\sqrt{3}\right)^2} = \sqrt{1 + 3} = \sqrt{4} = 2.$$

Substituting -1 for x, $\sqrt{3}$ for y, and 2 for r, we find the trigonometric function values of θ are

$$\sin\theta = \frac{\sqrt{3}}{2}, \qquad \csc\theta = \frac{2}{\sqrt{3}} = \frac{2\sqrt{3}}{3},$$

$$\cos\theta = \frac{-1}{2} = -\frac{1}{2}, \qquad \sec\theta = \frac{2}{-1} = -2,$$

$$\tan\theta = \frac{\sqrt{3}}{-1} = -\sqrt{3}, \qquad \cot\theta = \frac{-1}{\sqrt{3}} = -\frac{\sqrt{3}}{3}.$$

▶ **Now Try Exercise 27.**

Any point other than the origin on the terminal side of an angle in standard position can be used to determine the trigonometric function values of that angle. The function values are the same regardless of which point is used. To illustrate this, let's consider an angle θ in standard position whose terminal side lies on the line $y = -\frac{1}{2}x$. We can determine two

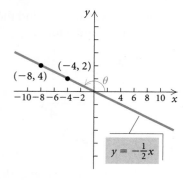

$$y = -\frac{1}{2}x$$

second-quadrant solutions of the equation, find the length r for each point, and then compare the sine, cosine, and tangent function values using each point.

If $x = -4$, then $y = -\frac{1}{2}(-4) = 2$.

If $x = -8$, then $y = -\frac{1}{2}(-8) = 4$.

For $(-4, 2)$, $r = \sqrt{(-4)^2 + 2^2} = \sqrt{20} = 2\sqrt{5}$.

For $(-8, 4)$, $r = \sqrt{(-8)^2 + 4^2} = \sqrt{80} = 4\sqrt{5}$.

Using $(-4, 2)$ and $r = 2\sqrt{5}$, we find that

$$\sin \theta = \frac{2}{2\sqrt{5}} = \frac{1}{\sqrt{5}} = \frac{\sqrt{5}}{5}, \qquad \cos \theta = \frac{-4}{2\sqrt{5}} = \frac{-2}{\sqrt{5}} = -\frac{2\sqrt{5}}{5},$$

and $\tan \theta = \dfrac{2}{-4} = -\dfrac{1}{2}$.

Using $(-8, 4)$ and $r = 4\sqrt{5}$, we find that

$$\sin \theta = \frac{4}{4\sqrt{5}} = \frac{1}{\sqrt{5}} = \frac{\sqrt{5}}{5}, \qquad \cos \theta = \frac{-8}{4\sqrt{5}} = \frac{-2}{\sqrt{5}} = -\frac{2\sqrt{5}}{5},$$

and $\tan \theta = \dfrac{4}{-8} = -\dfrac{1}{2}$.

We see that the function values are the same using either point. Any point other than the origin on the terminal side of an angle can be used to determine the trigonometric function values.

> The trigonometric function values of θ depend only on the angle, not on the choice of the point on the terminal side that is used to compute them.

✦ The Six Functions Related

When we know one of the function values of an angle, we can find the other five if we know the quadrant in which the terminal side lies. The procedure is to sketch a reference triangle in the appropriate quadrant, use the Pythagorean theorem as needed to find the lengths of its sides, and then find the ratios of the sides.

EXAMPLE 4   Given that $\tan \theta = -\frac{2}{3}$ and θ is in the second quadrant, find the other function values.

Solution   We first sketch a second-quadrant angle. Since

$$\tan \theta = \frac{y}{x} = -\frac{2}{3} = \frac{2}{-3}, \qquad \text{Expressing } -\frac{2}{3} \text{ as } \frac{2}{-3} \text{ since } \theta \text{ is in quadrant II}$$

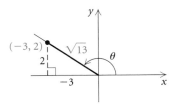

we make the legs lengths 2 and 3. The hypotenuse must then have length $\sqrt{2^2 + 3^2}$, or $\sqrt{13}$. Now we read off the appropriate ratios:

$$\sin \theta = \frac{2}{\sqrt{13}}, \quad \text{or} \quad \frac{2\sqrt{13}}{13}, \qquad \csc \theta = \frac{\sqrt{13}}{2},$$

$$\cos \theta = -\frac{3}{\sqrt{13}}, \quad \text{or} \quad -\frac{3\sqrt{13}}{13}, \qquad \sec \theta = -\frac{\sqrt{13}}{3},$$

$$\tan \theta = -\frac{2}{3}, \qquad \cot \theta = -\frac{3}{2}.$$

▶ **Now Try Exercise 33.**

◆ Terminal Side on an Axis

An angle whose terminal side falls on one of the axes is a **quadrantal angle.** One of the coordinates of any point on that side is 0. The definitions of the trigonometric functions still apply, but in some cases, function values will not be defined because a denominator will be 0.

EXAMPLE 5 Find the sine, cosine, and tangent values for 90°, 180°, 270°, and 360°.

Solution We first make a drawing of each angle in standard position and label a point on the terminal side. Since the function values are the same for all points on the terminal side, we choose $(0, 1)$, $(-1, 0)$, $(0, -1)$, and $(1, 0)$ for convenience. Note that $r = 1$ for each choice.

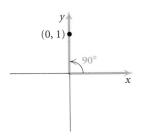 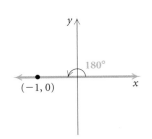 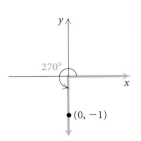 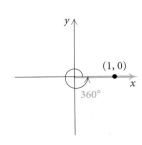

Then by the definitions we get

$$\sin 90° = \frac{1}{1} = 1, \qquad \sin 180° = \frac{0}{1} = 0, \qquad \sin 270° = \frac{-1}{1} = -1, \qquad \sin 360° = \frac{0}{1} = 0,$$

$$\cos 90° = \frac{0}{1} = 0, \qquad \cos 180° = \frac{-1}{1} = -1, \qquad \cos 270° = \frac{0}{1} = 0, \qquad \cos 360° = \frac{1}{1} = 1,$$

$$\tan 90° = \frac{1}{0}, \quad \text{Not defined} \qquad \tan 180° = \frac{0}{-1} = 0, \qquad \tan 270° = \frac{-1}{0}, \quad \text{Not defined} \qquad \tan 360° = \frac{0}{1} = 0.$$

◀

In Example 5, all the values can be found using a calculator, but you will find that it is convenient to be able to compute them mentally. It is also helpful to note that coterminal angles have the same function values. For example, 0° and 360° are coterminal; thus, sin 0° = 0, cos 0° = 1, and tan 0° = 0.

TECHNOLOGY ··················
CONNECTION
Trigonometric values can always
be checked using a calculator.
When the value is undefined,
the calculator will display an
ERROR message.

ERR: DIVIDE BY 0
1: Quit
2: Goto

EXAMPLE 6 Find each of the following.

a) $\sin(-90°)$ **b)** $\csc 540°$

Solution

a) We note that $-90°$ is coterminal with $270°$. Thus,

$$\sin(-90°) = \sin 270° = \frac{-1}{1} = -1.$$

b) Since $540° = 180° + 360°$, $540°$ and $180°$ are coterminal. Thus,

$$\csc 540° = \csc 180° = \frac{1}{\sin 180°} = \frac{1}{0}, \quad \text{which is not defined.}$$

▶ **Now Try Exercises 43 and 53.**

✦ Reference Angles: 30°, 45°, and 60°

We can also mentally determine trigonometric function values whenever the terminal side makes a 30°, 45°, or 60° angle with the *x*-axis. Consider, for example, an angle of 150°. The terminal side makes a 30° angle with the *x*-axis, since $180° - 150° = 30°$.

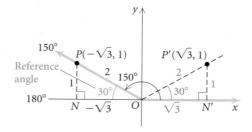

As the figure shows, $\triangle ONP$ is congruent to $\triangle ON'P'$; therefore, the ratios of the sides of the two triangles are the same. Thus the trigonometric function values are the same except perhaps for the sign. We could determine the function values directly from $\triangle ONP$, but this is not necessary. If we remember that in quadrant II, the sine is positive and the cosine and the tangent are negative, we can simply use the function values of 30° that we already know and prefix the appropriate sign. Thus,

$$\sin 150° = \sin 30° = \frac{1}{2},$$

$$\cos 150° = -\cos 30° = -\frac{\sqrt{3}}{2},$$

and $\quad \tan 150° = -\tan 30° = -\frac{1}{\sqrt{3}}, \quad \text{or} \quad -\frac{\sqrt{3}}{3}.$

Triangle ONP is the reference triangle and the acute angle $\angle NOP$ is called a *reference angle*.

> ### Reference Angle
> The **reference angle** for an angle is the acute angle formed by the terminal side of the angle and the *x*-axis.

EXAMPLE 7 Find the sine, cosine, and tangent function values for each of the following.

a) 225° **b)** −780°

Solution

a) We draw a figure showing the terminal side of a 225° angle. The reference angle is 225° − 180°, or 45°.

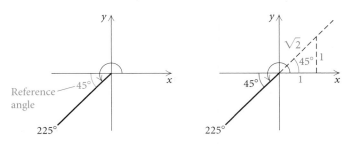

Recall from Section 1.1 that sin 45° = $\sqrt{2}/2$, cos 45° = $\sqrt{2}/2$, and tan 45° = 1. Also note that in the third quadrant, the sine and the cosine are negative and the tangent is positive. Thus we have

$$\sin 225° = -\frac{\sqrt{2}}{2}, \quad \cos 225° = -\frac{\sqrt{2}}{2}, \quad \text{and} \quad \tan 225° = 1.$$

b) We draw a figure showing the terminal side of a −780° angle. Since −780° + 2(360°) = −60°, we know that −780° and −60° are coterminal.

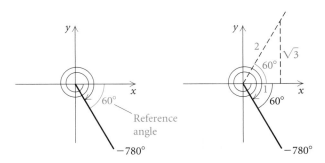

The reference angle for −60° is the acute angle formed by the terminal side of the angle and the *x*-axis. Thus the reference angle for −60° is 60°. We know that since −780° is a fourth-quadrant angle, the cosine is positive and the sine and the tangent are negative. Recalling that $\sin 60° = \sqrt{3}/2$, $\cos 60° = 1/2$, and $\tan 60° = \sqrt{3}$, we have

$$\sin(-780°) = -\frac{\sqrt{3}}{2}, \qquad \cos(-780°) = \frac{1}{2},$$

and $\tan(-780°) = -\sqrt{3}.$ ▶ Now Try Exercises 45 and 49.

**TECHNOLOGY
CONNECTION**

To find trigonometric function values of angles measured in degrees, we set the calculator in DEGREE mode. In the windows below, parts (a)–(f) of Example 8 are shown.

```
cos(112)
              -.3746065934
1/cos(500)
              -1.305407289
tan(-83.4)
               -8.64274761
```

```
1/sin(351.75)
              -6.968999424
cos(2400)
                       -.5
sin(175°40'9")
                .0755153443
```

✦ Function Values for Any Angle

When the terminal side of an angle falls on one of the axes or makes a 30°, 45°, or 60° angle with the x-axis, we can find exact function values without the use of a calculator. But this group is only a small subset of *all* angles. Using a calculator, we can approximate the trigonometric function values of *any* angle. In fact, we can approximate or find exact function values of all angles without using a reference angle.

EXAMPLE 8   Find each of the following function values using a calculator and round the answer to four decimal places, where appropriate.

a) $\cos 112°$ b) $\sec 500°$
c) $\tan(-83.4°)$ d) $\csc 351.75°$
e) $\cos 2400°$ f) $\sin 175°40'9''$
g) $\cot(-135°)$

Solution   Using a calculator set in DEGREE mode, we find the values.

a) $\cos 112° \approx -0.3746$

b) $\sec 500° = \dfrac{1}{\cos 500°} \approx -1.3054$

c) $\tan(-83.4°) \approx -8.6427$

d) $\csc 351.75° = \dfrac{1}{\sin 351.75°} \approx -6.9690$

e) $\cos 2400° = -0.5$

f) $\sin 175°40'9'' \approx 0.0755$

g) $\cot(-135°) = \dfrac{1}{\tan(-135°)} = 1$ ▶ Now Try Exercises 87 and 93.

In many applications, we have a trigonometric function value and want to find the measure of a corresponding angle. When only acute angles are considered, there is only one angle for each trigonometric function value. This is not the case when we extend the domain of the trigonometric functions to the set of *all* angles. For a given function value, there is an infinite number of angles that have that function value. There can be two such angles for each value in the range from 0° to 360°. To determine a unique answer in the interval $(0°, 360°)$, the quadrant in which the terminal side lies must be specified.

The calculator gives the reference angle as an output for each function value that is entered as an input. Knowing the reference angle and the quadrant in which the terminal side lies, we can find the specified angle.

EXAMPLE 9 Given the function value and the quadrant restriction, find θ.

a) $\sin \theta = 0.2812, \ 90° < \theta < 180°$

b) $\cot \theta = -0.1611, \ 270° < \theta < 360°$

Solution

a) We first sketch the angle in the second quadrant. We use the calculator to find the acute angle (reference angle) whose sine is 0.2812. The reference angle is approximately 16.33°. We find the angle θ by subtracting 16.33° from 180°:

$$180° - 16.33° = 163.67°.$$

Thus, $\theta \approx 163.67°$.

b) We begin by sketching the angle in the fourth quadrant. Because the tangent and cotangent values are reciprocals, we know that

$$\tan \theta \approx \frac{1}{-0.1611} \approx -6.2073.$$

We use the calculator to find the acute angle (reference angle) whose tangent is 6.2073, ignoring the fact that $\tan \theta$ is negative. The reference angle is approximately 80.85°. We find angle θ by subtracting 80.85° from 360°:

$$360° - 80.85° = 279.15°.$$

Thus, $\theta \approx 279.15°$.

▶ **Now Try Exercise 99.**

1.3 EXERCISE SET

For angles of the following measures, state in which quadrant the terminal side lies. It helps to sketch the angle in standard position.

1. 187°

2. −14.3°

3. 245°15′

4. −120°

5. 800°

6. 1075°

7. −460.5°

8. 315°

9. −912°

10. 13°15′60″

11. 537°

12. −345.14°

Find two positive angles and two negative angles that are coterminal with the given angle. Answers may vary.

13. 74°

14. −81°

15. 115.3°

16. 275°10′

17. −180°

18. −310°

Find the complement and the supplement.

19. 17.11°

20. 47°38′

21. 12°3′14″

22. 9.038°

23. 45.2°

24. 67.31°

Find the six trigonometric function values for the angle shown.

25.
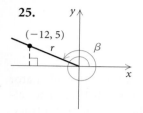

26.
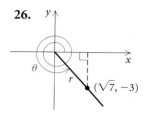

27.
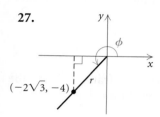

28.
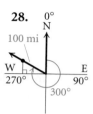

The terminal side of angle θ in standard position lies on the given line in the given quadrant. Find sin θ, cos θ, and tan θ.

29. $2x + 3y = 0$; quadrant IV

30. $4x + y = 0$; quadrant II

31. $5x - 4y = 0$; quadrant I

32. $y = 0.8x$; quadrant III

A function value and a quadrant are given. Find the other five function values. Give exact answers.

33. $\sin \theta = -\dfrac{1}{3}$, quadrant III

34. $\tan \beta = 5$, quadrant I

35. $\cot \theta = -2$, quadrant IV

36. $\cos \alpha = -\dfrac{4}{5}$, quadrant II

37. $\cos \phi = \dfrac{3}{5}$, quadrant IV

38. $\sin \theta = -\dfrac{5}{13}$, quadrant III

Find the reference angle and the exact function value if it exists.

39. $\cos 150°$

40. $\sec (-225°)$

41. $\tan (-135°)$

42. $\sin (-45°)$

43. $\sin 7560°$

44. $\tan 270°$

45. $\cos 495°$

46. $\tan 675°$

47. $\csc (-210°)$

48. $\sin 300°$

49. $\cot 570°$

50. $\cos (-120°)$

51. $\tan 330°$

52. $\cot 855°$

53. $\sec (-90°)$

54. $\sin 90°$

55. $\cos (-180°)$

56. $\csc 90°$

57. $\tan 240°$

58. $\cot (-180°)$

59. $\sin 495°$

60. $\sin 1050°$

61. $\csc 225°$

62. $\sin (-450°)$

63. $\cos 0°$

64. $\tan 480°$

65. $\cot (-90°)$

66. $\sec 315°$

67. $\cos 90°$

68. $\sin (-135°)$

69. $\cos 270°$

70. $\tan 0°$

Find the signs of the six trigonometric function values for the given angles.

71. $319°$

72. $-57°$

73. $194°$

74. $-620°$

75. $-215°$

76. $290°$

77. $-272°$

78. $91°$

Use a calculator in Exercises 79–82, but do not use the trigonometric function keys.

79. Given that

$$\sin 41° = 0.6561,$$
$$\cos 41° = 0.7547,$$
$$\tan 41° = 0.8693,$$

find the trigonometric function values for $319°$.

80. Given that

$$\sin 27° = 0.4540,$$
$$\cos 27° = 0.8910,$$
$$\tan 27° = 0.5095,$$

find the trigonometric function values for $333°$.

81. Given that

$$\sin 65° = 0.9063,$$
$$\cos 65° = 0.4226,$$
$$\tan 65° = 2.1445,$$

find the trigonometric function values for 115°.

82. Given that

$$\sin 35° = 0.5736,$$
$$\cos 35° = 0.8192,$$
$$\tan 35° = 0.7002,$$

find the trigonometric function values for 215°.

Aerial Navigation. *In aerial navigation, directions are given in degrees clockwise from north. Thus, east is 90°, south is 180°, and west is 270°. Several aerial directions or* **bearings** *are given below.*

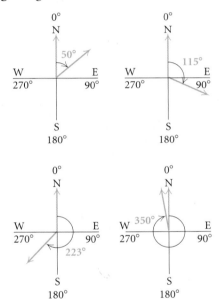

83. An airplane flies 150 km from an airport in a direction of 120°. How far east of the airport is the plane then? How far south?

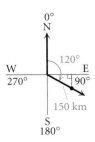

84. An airplane leaves an airport and travels for 100 mi in a direction of 300°. How far north of the airport is the plane then? How far west?

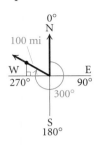

85. An airplane travels at 150 km/h for 2 hr in a direction of 138° from Omaha. At the end of this time, how far south of Omaha is the plane?

86. An airplane travels at 120 km/h for 2 hr in a direction of 319° from Chicago. At the end of this time, how far north of Chicago is the plane?

Find the function value. Round to four decimal places.

87. tan 310.8°

88. cos 205.5°

89. cot 146.15°

90. sin (−16.4°)

91. sin 118°42′

92. cos 273°45′

93. cos (−295.8°)

94. tan 1086.2°

95. cos 5417°

96. sec 240°55′

97. csc 520°

98. sin 3824°

Given the function value and the quadrant restriction, find θ.

FUNCTION VALUE	INTERVAL	θ
99. sin θ = −0.9956	(270°, 360°)	
100. tan θ = 0.2460	(180°, 270°)	
101. cos θ = −0.9388	(180°, 270°)	
102. sec θ = −1.0485	(90°, 180°)	
103. tan θ = −3.0545	(270°, 360°)	
104. sin θ = −0.4313	(180°, 270°)	
105. csc θ = 1.0480	(0°, 90°)	
106. cos θ = −0.0990	(90°, 180°)	

Collaborative Discussion and Writing

107. Why do the function values of θ depend only on the angle and not on the choice of a point on the terminal side?

108. Why is the domain of the tangent function different from the domains of the sine and the cosine functions?

Skill Maintenance

Graph the function. Sketch and label any vertical asymptotes.

109. $f(x) = \dfrac{1}{x^2 - 25}$

110. $g(x) = x^3 - 2x + 1$

Determine the domain and the range of the function.

111. $f(x) = \dfrac{x - 4}{x + 2}$

112. $g(x) = \dfrac{x^2 - 9}{2x^2 - 7x - 15}$

Find the zeros of the function.

113. $f(x) = 12 - x$

114. $g(x) = x^2 - x - 6$

Find the x-intercepts of the graph of the function.

115. $f(x) = 12 - x$

116. $g(x) = x^2 - x - 6$

Synthesis

117. *Valve Cap on a Bicycle.* The valve cap on a bicycle wheel is 12.5 in. from the center of the wheel. From the position shown, the wheel starts to roll. After the wheel has turned 390°, how far above the ground is the valve cap? Assume that the outer radius of the tire is 13.375 in.

118. *Seats of a Ferris Wheel.* The seats of a ferris wheel are 35 ft from the center of the wheel. When you board the wheel, you are 5 ft above the ground. After you have rotated through an angle of 765°, how far above the ground are you?

1.4 Radians, Arc Length, and Angular Speed

✦ Find points on the unit circle determined by real numbers.
✦ Convert between radian measure and degree measure; find coterminal, complementary, and supplementary angles.
✦ Find the length of an arc of a circle; find the measure of a central angle of a circle.
✦ Convert between linear speed and angular speed.

Another useful unit of angle measure is called a *radian*. To introduce radian measure, we use a circle centered at the origin with a radius of length 1. Such a circle is called a **unit circle.** Its equation is $x^2 + y^2 = 1$.

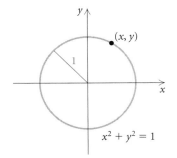

✦ Distances on the Unit Circle

The circumference of a circle of radius r is $2\pi r$. Thus for the unit circle, where $r = 1$, the circumference is 2π. If a point starts at A and travels around the circle (Fig. 1), it will travel a distance of 2π. If it travels halfway around the circle (Fig. 2), it will travel a distance of $\frac{1}{2} \cdot 2\pi$, or π.

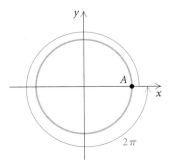

Figure 1

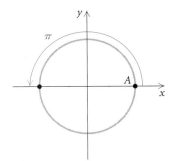

Figure 2

If a point C travels $\frac{1}{8}$ of the way around the circle (Fig. 3), it will travel a distance of $\frac{1}{8} \cdot 2\pi$, or $\pi/4$. Note that C is $\frac{1}{4}$ of the way from A to B. If a point D travels $\frac{1}{6}$ of the way around the circle (Fig. 4), it will travel a distance of $\frac{1}{6} \cdot 2\pi$, or $\pi/3$. Note that D is $\frac{1}{3}$ of the way from A to B.

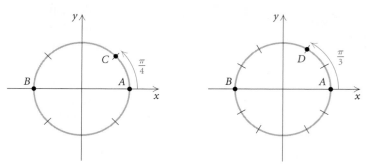

Figure 3   Figure 4

EXAMPLE 1   How far will a point travel if it goes (a) $\frac{1}{4}$, (b) $\frac{1}{12}$, (c) $\frac{3}{8}$, and (d) $\frac{5}{6}$ of the way around the unit circle?

Solution

a) $\frac{1}{4}$ of the total distance around the circle is $\frac{1}{4} \cdot 2\pi$, which is $\frac{1}{2} \cdot \pi$, or $\pi/2$.

b) The distance will be $\frac{1}{12} \cdot 2\pi$, which is $\frac{1}{6}\pi$, or $\pi/6$.

c) The distance will be $\frac{3}{8} \cdot 2\pi$, which is $\frac{3}{4}\pi$, or $3\pi/4$.

d) The distance will be $\frac{5}{6} \cdot 2\pi$, which is $\frac{5}{3}\pi$, or $5\pi/3$. Think of $5\pi/3$ as $\pi + \frac{2}{3}\pi$.

These distances are illustrated in the following figures.

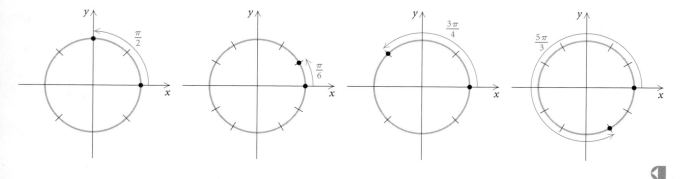

A point may travel completely around the circle and then continue. For example, if it goes around once and then continues $\frac{1}{4}$ of the way around, it will have traveled a distance of $2\pi + \frac{1}{4} \cdot 2\pi$, or $5\pi/2$ (Fig. 5). *Every* real number determines a point on the unit circle. For the positive number 10, for example, we start at A and travel counterclockwise a distance of 10. The point at which we stop is the point "determined" by the number 10. Note that $2\pi \approx 6.28$ and that $10 \approx 1.6(2\pi)$. Thus the point for 10 travels around the unit circle about $1\frac{3}{5}$ times (Fig. 6).

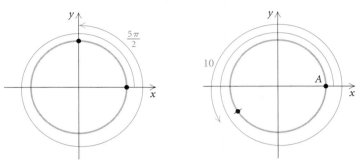

Figure 5   Figure 6

For a negative number, we move clockwise around the circle. Points for $-\pi/4$ and $-3\pi/2$ are shown in the figure below. The number 0 determines the point A.

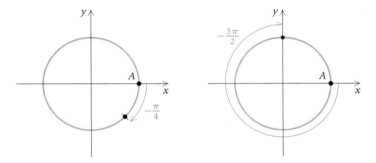

EXAMPLE 2   On the unit circle, mark the point determined by each of the following real numbers.

a) $\dfrac{9\pi}{4}$ **b)** $-\dfrac{7\pi}{6}$

Solution

a) Think of $9\pi/4$ as $2\pi + \frac{1}{4}\pi$. (See the figure below.) Since $9\pi/4 > 0$, the point moves counterclockwise. The point goes completely around once and then continues $\frac{1}{4}$ of the way from A to B.

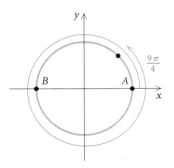

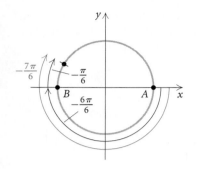

b) The number $-7\pi/6$ is negative, so the point moves clockwise. From A to B, the distance is π, or $\frac{6}{6}\pi$, so we need to go beyond B another distance of $\pi/6$, clockwise. (See the figure at left.)

▶ **Now Try Exercise 1.**

◆ Radian Measure

Degree measure is a common unit of angle measure in many everyday applications. But in many scientific fields and in mathematics (calculus, in particular), there is another commonly used unit of measure called the *radian*.

Consider the unit circle. Recall that this circle has radius 1. Suppose we measure, moving counterclockwise, an arc of length 1, and mark a point T on the circle.

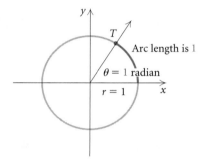

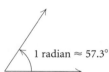

If we draw a ray from the origin through T, we have formed an angle. The measure of that angle is 1 **radian.** The word radian comes from the word *radius*. Thus measuring 1 "radius" along the circumference of the circle determines an angle whose measure is 1 *radian*. One radian is about 57.3°. Angles that measure 2 radians, 3 radians, and 6 radians are shown below.

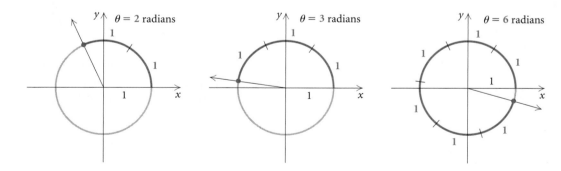

When we make a complete (counterclockwise) revolution, the terminal side coincides with the initial side on the positive x-axis. We then have an angle whose measure is 2π radians, or about 6.28 radians, which is the circumference of the circle:

$$2\pi r = 2\pi(1) = 2\pi.$$

Thus a rotation of 360° (1 revolution) has a measure of 2π radians. A half revolution is a rotation of 180°, or π radians. A quarter revolution is a rotation of 90°, or $\pi/2$ radians, and so on.

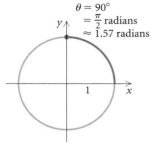
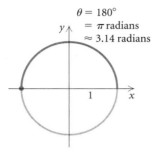
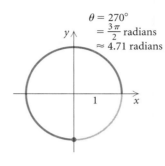
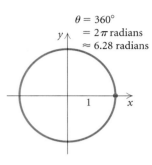

To convert between degrees and radians, we first note that

$$360° = 2\pi \text{ radians}.$$

It follows that

$$180° = \pi \text{ radians}.$$

To make conversions, we multiply by 1, noting that:

Converting between Degree Measure and Radian Measure

$$\frac{\pi \text{ radians}}{180°} = \frac{180°}{\pi \text{ radians}} = 1.$$

To convert from degree to radian measure, multiply by $\dfrac{\pi \text{ radians}}{180°}$.

To convert from radian to degree measure, multiply by $\dfrac{180°}{\pi \text{ radians}}$.

EXAMPLE 3   Convert each of the following to radians.

a) 120°   **b)** $-297.25°$

Solution

a) $120° = 120° \cdot \dfrac{\pi \text{ radians}}{180°}$ Multiplying by 1

$ = \dfrac{120°}{180°} \pi \text{ radians}$

$ = \dfrac{2\pi}{3} \text{ radians, or about 2.09 radians}$

TECHNOLOGY ·················
CONNECTION

To convert degrees to radians, we set the calculator in RADIAN mode. Then we enter the angle measure followed by ° (degrees) from the ANGLE menu. Example 3 is shown here.

```
120°
           2.094395102
−297.25°
          −5.187991202
```

To convert radians to degrees, we set the calculator in DEGREE mode. Then we enter the angle measure followed by ʳ (radians) from the ANGLE menu. Example 4 is shown here.

```
(3π/4)ʳ
                 135
8.5ʳ
          487.0141259
```

b) $-297.25° = -297.25° \cdot \dfrac{\pi \text{ radians}}{180°}$

$$= -\dfrac{297.25°}{180°}\,\pi \text{ radians}$$

$$= -\dfrac{297.25\,\pi}{180} \text{ radians}$$

$$\approx -5.19 \text{ radians}$$

▷ Now Try Exercises 23 and 35.

EXAMPLE 4   Convert each of the following to degrees.

a) $\dfrac{3\pi}{4}$ radians   **b)** 8.5 radians

Solution

a) $\dfrac{3\pi}{4}$ radians $= \dfrac{3\pi}{4}$ radians $\cdot \dfrac{180°}{\pi \text{ radians}}$ Multiplying by 1

$$= \dfrac{3\pi}{4\pi} \cdot 180° = \dfrac{3}{4} \cdot 180° = 135°$$

b) 8.5 radians $= 8.5$ radians $\cdot \dfrac{180°}{\pi \text{ radians}}$

$$= \dfrac{8.5(180°)}{\pi} \approx 487.01°$$

▷ Now Try Exercises 47 and 55.

The radian–degree equivalents of the most commonly used angle measures are illustrated in the following figures.

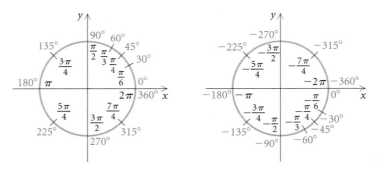

When a rotation is given in radians, the word "radians" is optional and is most often omitted. **Thus if no unit is given for a rotation, the rotation is understood to be in radians.**

We can also find coterminal, complementary, and supplementary angles in radian measure just as we did for degree measure in Section 1.3.

EXAMPLE 5   Find a positive angle and a negative angle that are coterminal with $2\pi/3$. Many answers are possible.

Solution To find angles coterminal with a given angle, we add or subtract multiples of 2π:

$$\frac{2\pi}{3} + 2\pi = \frac{2\pi}{3} + \frac{6\pi}{3} = \frac{8\pi}{3},$$

$$\frac{2\pi}{3} - 3(2\pi) = \frac{2\pi}{3} - \frac{18\pi}{3} = -\frac{16\pi}{3}.$$

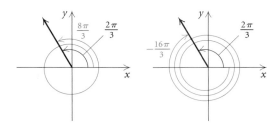

Thus, $8\pi/3$ and $-16\pi/3$ are two of the many angles coterminal with $2\pi/3$.

▶ **Now Try Exercise 11.**

EXAMPLE 6 Find the complement and the supplement of $\pi/6$.

Solution Since $90°$ equals $\pi/2$ radians, the complement of $\pi/6$ is

$$\frac{\pi}{2} - \frac{\pi}{6} = \frac{3\pi}{6} - \frac{\pi}{6} = \frac{2\pi}{6}, \quad \text{or} \quad \frac{\pi}{3}.$$

Since $180°$ equals π radians, the supplement of $\pi/6$ is

$$\pi - \frac{\pi}{6} = \frac{6\pi}{6} - \frac{\pi}{6} = \frac{5\pi}{6}.$$

Thus the complement of $\pi/6$ is $\pi/3$ and the supplement is $5\pi/6$.

▶ **Now Try Exercise 15.**

◆ Arc Length and Central Angles

Radian measure can be determined using a circle other than a unit circle. In the figure at left, a unit circle (with radius 1) is shown along with another circle (with radius r, $r \neq 1$). The angle shown is a **central angle** of both circles.

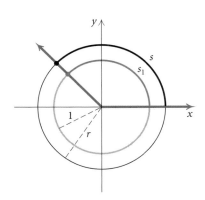

From geometry, we know that the arcs that the angle subtends have their lengths in the same ratio as the radii of the circles. The radii of the circles are r and 1. The corresponding arc lengths are s and s_1. Thus we have the proportion

$$\frac{s}{s_1} = \frac{r}{1},$$

which also can be written as

$$\frac{s_1}{1} = \frac{s}{r}.$$

Now s_1 is the *radian measure* of the rotation in question. It is common to use a Greek letter, such as θ, for the measure of an angle or rotation and the letter s for arc length. Adopting this convention, we rewrite the proportion above as

$$\theta = \frac{s}{r}.$$

In any circle, the measure (in radians) of a central angle, the arc length the angle subtends, and the length of the radius are related in this fashion. Or, in general, the following is true.

Radian Measure

The **radian measure** θ of a rotation is the ratio of the distance s traveled by a point at a radius r from the center of rotation, to the length of the radius r:

$$\theta = \frac{s}{r}.$$

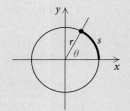

When using the formula $\theta = s/r$, be sure that θ is given in radians and s and r are expressed in the same unit.

EXAMPLE 7   Find the measure of a rotation in radians when a point 2 m from the center of rotation travels 4 m.

Solution   We have

$$\theta = \frac{s}{r}$$

$$= \frac{4 \text{ m}}{2 \text{ m}} = 2. \qquad \text{The unit is understood to be radians.}$$

▶ Now Try Exercise 65.

EXAMPLE 8   Find the length of an arc of a circle of radius 5 cm associated with an angle of $\pi/3$ radians.

Solution   We have

$$\theta = \frac{s}{r}, \quad \text{or} \quad s = r\theta.$$

Thus $s = 5 \text{ cm} \cdot \pi/3$, or about 5.24 cm.

▶ Now Try Exercise 63.

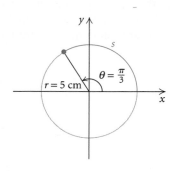

◆ Linear Speed and Angular Speed

Linear speed is defined as distance traveled per unit of time. If we use v for linear speed, s for distance, and t for time, then

$$v = \frac{s}{t}.$$

Similarly, **angular speed** is defined as amount of rotation per unit of time. For example, we might speak of the angular speed of a bicycle wheel as 150 revolutions per minute or the angular speed of the earth as 2π radians per day. The Greek letter ω (omega) is generally used for angular speed. Thus for a rotation θ and time t, angular speed is defined as

$$\omega = \frac{\theta}{t}.$$

As an example of how these definitions can be applied, let's consider the refurbished carousel at the Children's Museum in Indianapolis, Indiana. It consists of three circular rows of animals. All animals, regardless of the row, travel at the same angular speed. But the animals in the outer row travel at a greater linear speed than those in the inner rows. What is the relationship between the linear speed v and the angular speed ω?

To develop the relationship we seek, recall that, for rotations measured in radians, $\theta = s/r$. This is equivalent to

$$s = r\theta.$$

We divide by time, t, to obtain

$$\frac{s}{t} = \frac{r\theta}{t} \qquad \text{Dividing by } t$$

$$\frac{s}{t} = r \cdot \frac{\theta}{t}$$

$$\underset{v}{\downarrow} \qquad \underset{\omega}{\downarrow}$$

Now s/t is linear speed v and θ/t is angular speed ω. Thus we have the relationship we seek,

$$v = r\omega.$$

© 2006 The Children's Museum of Indianapolis

> ### Linear Speed in Terms of Angular Speed
>
> The **linear speed** v of a point a distance r from the center of rotation is given by
>
> $$v = r\omega,$$
>
> where ω is the **angular speed** in radians per unit of time.

For the formula $v = r\omega$, the units of distance for v and r must be the same, ω must be in radians per unit of time, and the units of time for v and ω must be the same.

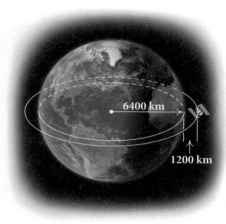

EXAMPLE 9 *Linear Speed of an Earth Satellite.* An Earth satellite in circular orbit 1200 km high makes one complete revolution every 90 min. What is its linear speed? Use 6400 km for the length of a radius of the Earth.

Solution To use the formula $v = r\omega$, we need to know r and ω:

$$r = 6400 \text{ km} + 1200 \text{ km} \qquad \text{Radius of Earth plus height of satellite}$$

$$= 7600 \text{ km},$$

$$\omega = \frac{\theta}{t} = \frac{2\pi}{90 \text{ min}} = \frac{\pi}{45 \text{ min}}. \qquad \text{We have, as usual, omitted the word radians.}$$

Now, using $v = r\omega$, we have

$$v = 7600 \text{ km} \cdot \frac{\pi}{45 \text{ min}} = \frac{7600\pi}{45} \cdot \frac{\text{km}}{\text{min}} \approx 531 \frac{\text{km}}{\text{min}}.$$

Thus the linear speed of the satellite is approximately 531 km/min.

▶ **Now Try Exercise 71.**

EXAMPLE 10 *Angular Speed of a Capstan.* An anchor is hoisted at a rate of 2 ft/sec as the chain is wound around a capstan with a 1.8-yd diameter. What is the angular speed of the capstan?

Solution We will use the formula $v = r\omega$ in the form $\omega = v/r$, taking care to use the proper units. Since v is given in feet per second, we need r in feet:

$$r = \frac{d}{2} = \frac{1.8}{2} \text{ yd} \cdot \frac{3 \text{ ft}}{1 \text{ yd}} = 2.7 \text{ ft}.$$

Then ω will be in radians per second:

$$\omega = \frac{v}{r} = \frac{2 \text{ ft/sec}}{2.7 \text{ ft}} = \frac{2 \text{ ft}}{\text{sec}} \cdot \frac{1}{2.7 \text{ ft}} \approx 0.741/\text{sec}.$$

Thus the angular speed is approximately 0.741 radian/sec.

▶ **Now Try Exercise 73.**

The formulas $\theta = \omega t$ and $v = r\omega$ can be used in combination to find distances and angles in various situations involving rotational motion.

EXAMPLE 11 *Angle of Revolution.* A 2006 Acura MDX is traveling at a speed of 70 mph. Its tires have an outside diameter of 28.56 in. Find the angle through which a tire turns in 10 sec.

28.56 in.

Solution Recall that $\omega = \theta/t$, or $\theta = \omega t$. Thus we can find θ if we know ω and t. To find ω, we use the formula $v = r\omega$. The linear speed v of a point on the outside of the tire is the speed of the Acura, 70 mph. For convenience, we first convert 70 mph to feet per second:

$$v = 70 \frac{\text{mi}}{\text{hr}} \cdot \frac{1 \text{ hr}}{60 \text{ min}} \cdot \frac{1 \text{ min}}{60 \text{ sec}} \cdot \frac{5280 \text{ ft}}{1 \text{ mi}}$$

$$\approx 102.667 \frac{\text{ft}}{\text{sec}}.$$

The radius of the tire is half the diameter. Now $r = d/2 = 28.56/2 = 14.28$ in. We will convert to feet, since v is in feet per second:

$$r = 14.28 \text{ in.} \cdot \frac{1 \text{ ft}}{12 \text{ in.}}$$

$$= \frac{14.28}{12} \text{ ft}$$

$$\approx 1.19 \text{ ft.}$$

Using $v = r\omega$, we have

$$102.667 \, \frac{\text{ft}}{\text{sec}} = 1.19 \, \text{ft} \cdot \omega,$$

so

$$\omega = \frac{102.667 \, \text{ft/sec}}{1.19 \, \text{ft}} \approx \frac{86.27}{\text{sec}}.$$

Then in 10 sec,

$$\theta = \omega t = \frac{86.27}{\text{sec}} \cdot 10 \, \text{sec} \approx 863.$$

Thus the angle, in radians, through which a tire turns in 10 sec is 863.

▶ Now Try Exercise 77.

1.4 EXERCISE SET

For each of Exercises 1–4, sketch a unit circle and mark the points determined by the given real numbers.

1. a) $\dfrac{\pi}{4}$ b) $\dfrac{3\pi}{2}$ c) $\dfrac{3\pi}{4}$

   d) π e) $\dfrac{11\pi}{4}$ f) $\dfrac{17\pi}{4}$

2. a) $\dfrac{\pi}{2}$ b) $\dfrac{5\pi}{4}$ c) 2π

   d) $\dfrac{9\pi}{4}$ e) $\dfrac{13\pi}{4}$ f) $\dfrac{23\pi}{4}$

3. a) $\dfrac{\pi}{6}$ b) $\dfrac{2\pi}{3}$ c) $\dfrac{7\pi}{6}$

   d) $\dfrac{10\pi}{6}$ e) $\dfrac{14\pi}{6}$ f) $\dfrac{23\pi}{4}$

4. a) $-\dfrac{\pi}{2}$ b) $-\dfrac{3\pi}{4}$ c) $-\dfrac{5\pi}{6}$

   d) $-\dfrac{5\pi}{2}$ e) $-\dfrac{17\pi}{6}$ f) $-\dfrac{9\pi}{4}$

Find two real numbers between -2π and 2π that determine each of the points on the unit circle.

5.

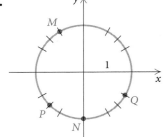

6.

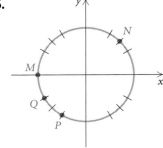

For Exercises 7 and 8, sketch a unit circle and mark the approximate location of the point determined by the given real number.

7. a) 2.4   **b)** 7.5
   c) 32   **d)** 320

8. a) 0.25   **b)** 1.8
   c) 47   **d)** 500

Find a positive angle and a negative angle that are coterminal with the given angle. Answers may vary.

9. $\dfrac{\pi}{4}$ **10.** $\dfrac{5\pi}{3}$

11. $\dfrac{7\pi}{6}$ **12.** π

13. $-\dfrac{2\pi}{3}$ **14.** $-\dfrac{3\pi}{4}$

Find the complement and the supplement.

15. $\dfrac{\pi}{3}$ **16.** $\dfrac{5\pi}{12}$

17. $\dfrac{3\pi}{8}$ **18.** $\dfrac{\pi}{4}$

19. $\dfrac{\pi}{12}$ **20.** $\dfrac{\pi}{6}$

Convert to radian measure. Leave the answer in terms of π.

21. 75°   **22.** 30°

23. 200°   **24.** −135°

25. −214.6°   **26.** 37.71°

27. −180°   **28.** 90°

29. 12.5°   **30.** 6.3°

31. −340°   **32.** −60°

Convert to radian measure. Round the answer to two decimal places.

33. 240°   **34.** 15°

35. −60°   **36.** 145°

37. 117.8°   **38.** −231.2°

39. 1.354°   **40.** 584°

41. 345°   **42.** −75°

43. 95°   **44.** 24.8°

Convert to degree measure. Round the answer to two decimal places.

45. $-\dfrac{3\pi}{4}$ **46.** $\dfrac{7\pi}{6}$

47. 8π **48.** $-\dfrac{\pi}{3}$

49. 1   **50.** −17.6

51. 2.347   **52.** 25

53. $\dfrac{5\pi}{4}$ **54.** -6π

55. −90   **56.** 37.12

57. $\dfrac{2\pi}{7}$ **58.** $\dfrac{\pi}{9}$

59. Certain positive angles are marked here in degrees. Find the corresponding radian measures.

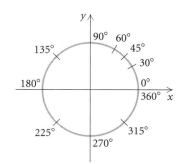

60. Certain negative angles are marked here in degrees. Find the corresponding radian measures.

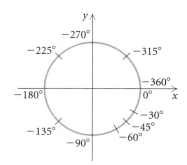

Arc Length and Central Angles.   *Complete the table.*
Round the answers to two decimal places.

	Distance, *s* (arc length)	Radius, *r*	Angle, θ
61.	8 ft	$3\frac{1}{2}$ ft	
62.	200 cm		45°
63.		4.2 in.	$\frac{5\pi}{12}$
64.	16 yd		5

65. In a circle with a 120-cm radius, an arc 132 cm long subtends an angle of how many radians? how many degrees, to the nearest degree?

66. In a circle with a 10-ft diameter, an arc 20 ft long subtends an angle of how many radians? how many degrees, to the nearest degree?

67. In a circle with a 2-yd radius, how long is an arc associated with an angle of 1.6 radians?

68. In a circle with a 5-m radius, how long is an arc associated with an angle of 2.1 radians?

69. *Angle of Revolution.*   Through how many radians does the minute hand of a clock rotate from 12:40 P.M. to 1:30 P.M.?

70. *Angle of Revolution.*   A tire on a 2006 Dodge Ram truck has an outside diameter of 31.125 in. Through what angle (in radians) does the tire turn while traveling 1 mi?

31.125 in.

71. *Linear Speed.*   A flywheel with a 15-cm diameter is rotating at a rate of 7 radians/sec. What is the linear speed of a point on its rim, in centimeters per minute?

72. *Linear Speed.*   A wheel with a 30-cm radius is rotating at a rate of 3 radians/sec. What is the linear speed of a point on its rim, in meters per minute?

73. *Angular Speed of a Printing Press.*   This text was printed on a four-color web heatset offset press. A cylinder on this press has a 21-in. diameter. The linear speed of a point on the cylinder's surface is 18.33 feet per second. What is the angular speed of the cylinder, in revolutions per hour? Printers often refer to the angular speed as impressions per hour (IPH). (*Source*: Von Hoffmann Press, St. Louis, Missouri)

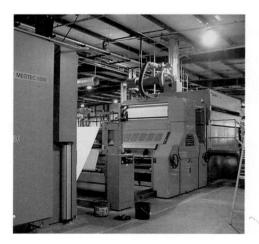

74. *Linear Speeds on a Carousel.*   When Alicia and Zoe ride the carousel described earlier in this section, Alicia always selects a horse on the outside row, whereas Zoe prefers the row closest to the center. These rows are 19 ft 3 in. and 13 ft 11 in. from the center, respectively (*Source*: The Children's Museum, Indianapolis, IN). The angular speed of the carousel is 2.4 revolutions per minute. What is the difference, in miles per hour, in the linear speeds of Alicia and Zoe?

75. *Linear Speed at the Equator.*   The Earth has a 4000-mi radius and rotates one revolution every 24 hr. What is the linear speed of a point on the equator, in miles per hour?

76. *Linear Speed of the Earth.* The Earth is about 93,000,000 mi from the Sun and traverses its orbit, which is nearly circular, every 365.25 days. What is the linear velocity of the earth in its orbit, in miles per hour?

77. *Tour of Flanders.* Tom Boonen of Belgium won the 2005 Tour of Flanders bicycle race. The wheel of his bicycle had a 67-cm diameter. His overall average linear speed during the race was 40.423 km/h. (*Source:* Toby Holsman, Bicycle Garage Indy, Indianapolis, Indiana; www.velonews.com) What was the angular speed of the wheel, in revolutions per hour?

78. *Determining the Speed of a River.* A water wheel has a 10-ft radius. To get a good approximation of the speed of the river, you count the revolutions of the wheel and find that it makes 14 revolutions per minute (rpm). What is the speed of the river, in miles per hour?

←10 ft→

79. *John Deere Tractor.* A rear wheel on a John Deere 8300 farm tractor has a 23-in. radius. Find the angle (in radians) through which a wheel rotates in 12 sec if the tractor is traveling at a speed of 22 mph.

23 in.

Technology Connection

80. In each of Exercises 33–44, convert to radian measure using a graphing calculator.

81. In each of Exercises 45–58, convert to degree measure using a graphing calculator.

Collaborative Discussion and Writing

82. Explain in your own words why it is preferable to omit the word, or unit, *radians* in radian measures.

83. In circular motion with a fixed angular speed, the length of the radius is directly proportional to the linear speed. Explain why with an example.

84. Two new cars are each driven at an average speed of 60 mph for an extended highway test drive of 2000 mi. The diameter of the wheels of the two cars are 15 in. and 16 in., respectively. If the cars use tires of equal durability and profile, differing only by the diameter, which car will probably need new tires first? Explain your answer.

Skill Maintenance

In each of Exercises 85–92, fill in the blanks with the correct terms. Some of the given choices will not be used.

inverse
a horizontal line
a vertical line
exponential function
logarithmic function
natural
common
logarithm
one-to-one
a relation
vertical asymptote
horizontal asymptote
even function
odd function
sine of θ
cosine of θ
tangent of θ

85. The domain of a(n) _____ function f is the range of the inverse f^{-1}.

86. The _____ is the length of the side adjacent to θ divided by the length of the hypotenuse.

87. The function $f(x) = a^x$, where x is a real number, $a > 0$ and $a \neq 1$, is called the _____, base a.

88. The graph of a rational function may or may not cross a(n) _____ .

89. If the graph of a function f is symmetric with respect to the origin, we say that it is a(n) _____ .

90. Logarithms, base e, are called _____ logarithms.

91. If it is possible for a(n) _____ to intersect the graph of a function more than once, then the function is not one-to-one and its _____ is not a function.

92. A(n) _____ is an exponent.

Synthesis

93. On the earth, one degree of latitude is how many kilometers? how many miles? (Assume that the radius of the earth is 6400 km, or 4000 mi, approximately.)

94. A point on the unit circle has y-coordinate $-\sqrt{21}/5$. What is its x-coordinate? Check using a calculator.

95. A **mil** is a unit of angle measure. A right angle has a measure of 1600 mils. Convert each of the following to degrees, minutes, and seconds.

    a) 100 mils   **b)** 350 mils

96. A **grad** is a unit of angle measure similar to a degree. A right angle has a measure of 100 grads. Convert each of the following to grads.

    a) 48°   **b)** $\dfrac{5\pi}{7}$

97. *Angular Speed of a Gear Wheel.* One gear wheel turns another, the teeth being on the rims. The wheels have 40-cm and 50-cm radii, and the smaller wheel rotates at 20 rpm. Find the angular speed of the larger wheel, in radians per second.

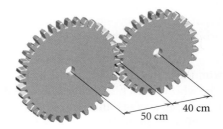

98. *Angular Speed of a Pulley.* Two pulleys, 50 cm and 30 cm in diameter, respectively, are connected by a belt. The larger pulley makes 12 revolutions per minute. Find the angular speed of the smaller pulley, in radians per second.

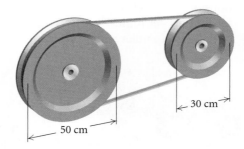

99. *Distance between Points on the Earth.* To find the distance between two points on the Earth when their latitude and longitude are known, we can use a right triangle for an excellent approximation if the points are not too far apart. Point A is at latitude 38°27′30″ N, longitude 82°57′15″ W; and point B is at latitude 38°28′45″ N, longitude 82°56′30″ W. Find the distance from A to B in nautical miles. (One minute of latitude is one nautical mile.)

100. *Hands of a Clock.* At what time between noon and 1:00 P.M. are the hands of a clock perpendicular?

Circular Functions: Graphs and Properties

◆ Given the coordinates of a point on the unit circle, find its reflections across the *x*-axis, the *y*-axis, and the origin.

◆ Determine the six trigonometric function values for a real number when the coordinates of the point on the unit circle determined by that real number are given.

◆ Find function values for any real number using a calculator.

◆ Graph the six circular functions and state their properties.

The domains of the trigonometric functions, defined in Sections 1.1 and 1.3, have been sets of angles or rotations measured in a real number of degree units. We can also consider the domains to be sets of real numbers, or radians, introduced in Section 1.4. Many applications in calculus that use the trigonometric functions refer only to radians.

Let's again consider radian measure and the unit circle. We defined radian measure for θ as

$$\theta = \frac{s}{r}.$$

When $r = 1$,

$$\theta = \frac{s}{1}, \quad \text{or} \quad \theta = s.$$

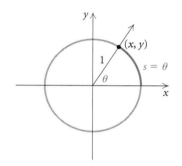

The arc length *s* on the unit circle is the same as the radian measure of the angle *θ*.

In the figure above, the point (x, y) is the point where the terminal side of the angle with radian measure *s* intersects the unit circle. We can now extend our definitions of the trigonometric functions using domains composed of real numbers, or radians.

In the definitions, *s can be considered the radian measure of an angle or the measure of an arc length on the unit circle. Either way, s is a real number.* To each real number *s*, there corresponds an arc length *s* on the unit circle. Trigonometric functions with domains composed of real numbers are called **circular functions.**

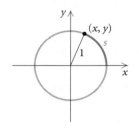

Basic Circular Functions

For a real number s that determines a point (x, y) on the unit circle:

$$\sin s = \text{second coordinate} = y,$$

$$\cos s = \text{first coordinate} = x,$$

$$\tan s = \frac{\text{second coordinate}}{\text{first coordinate}} = \frac{y}{x} \ (x \neq 0),$$

$$\csc s = \frac{1}{\text{second coordinate}} = \frac{1}{y} \ (y \neq 0),$$

$$\sec s = \frac{1}{\text{first coordinate}} = \frac{1}{x} \ (x \neq 0),$$

$$\cot s = \frac{\text{first coordinate}}{\text{second coordinate}} = \frac{x}{y} \ (y \neq 0).$$

We can consider the domains of trigonometric functions to be real numbers rather than angles. We can determine these values for a specific real number if we know the coordinates of the point on the unit circle determined by that number. As with degree measure, we can also find these function values directly using a calculator.

◆ Reflections on the Unit Circle

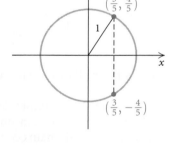

Let's consider the unit circle and a few of its points. For any point (x, y) on the unit circle, $x^2 + y^2 = 1$, we know that $-1 \leq x \leq 1$ and $-1 \leq y \leq 1$. If we know the x- or y-coordinate of a point on the unit circle, we can find the other coordinate. If $x = \frac{3}{5}$, then

$$\left(\tfrac{3}{5}\right)^2 + y^2 = 1$$

$$y^2 = 1 - \tfrac{9}{25} = \tfrac{16}{25}$$

$$y = \pm\tfrac{4}{5}.$$

Thus, $\left(\frac{3}{5}, \frac{4}{5}\right)$ and $\left(\frac{3}{5}, -\frac{4}{5}\right)$ are points on the unit circle. There are two points with an x-coordinate of $\frac{3}{5}$.

Now let's consider the radian measure $\pi/3$ and determine the coordinates of the point on the unit circle determined by $\pi/3$. We construct a right triangle by dropping a perpendicular segment from the point to the x-axis.

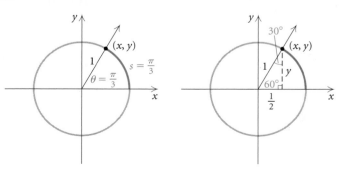

Since $\pi/3 = 60°$, we have a 30°–60° right triangle in which the side opposite the 30° angle is one half of the hypotenuse. The hypotenuse, or radius, is 1, so the side opposite the 30° angle is $\frac{1}{2} \cdot 1$, or $\frac{1}{2}$. Using the Pythagorean theorem, we can find the other side:

$$\left(\frac{1}{2}\right)^2 + y^2 = 1$$

$$y^2 = 1 - \frac{1}{4} = \frac{3}{4}$$

$$y = \sqrt{\frac{3}{4}} = \frac{\sqrt{3}}{2}.$$

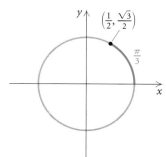

We know that y is positive since the point is in the first quadrant. Thus the coordinates of the point determined by $\pi/3$ are $x = 1/2$ and $y = \sqrt{3}/2$, or $(1/2, \sqrt{3}/2)$. We can always check to see if a point is on the unit circle by substituting into the equation $x^2 + y^2 = 1$:

$$\left(\frac{1}{2}\right)^2 + \left(\frac{\sqrt{3}}{2}\right)^2 = \frac{1}{4} + \frac{3}{4} = 1.$$

Because a unit circle is symmetric with respect to the x-axis, the y-axis, and the origin, we can use the coordinates of one point on the unit circle to find coordinates of its reflections.

EXAMPLE 1   Each of the following points lies on the unit circle. Find their reflections across the x-axis, the y-axis, and the origin.

a) $\left(\dfrac{3}{5}, \dfrac{4}{5}\right)$ b) $\left(\dfrac{\sqrt{2}}{2}, \dfrac{\sqrt{2}}{2}\right)$ c) $\left(\dfrac{1}{2}, \dfrac{\sqrt{3}}{2}\right)$

Solution

a)
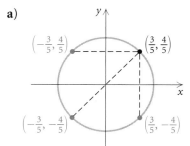

b)
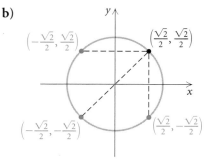

c)
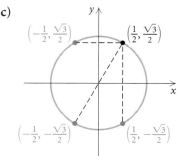

▶ Now Try Exercise 1.

✦ Finding Function Values

Knowing the coordinates of only a few points on the unit circle along with their reflections allows us to find trigonometric function values of the most frequently used real numbers, or radians.

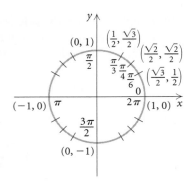

EXAMPLE 2   Find each of the following function values.

a) $\tan \dfrac{\pi}{3}$

b) $\cos \dfrac{3\pi}{4}$

c) $\sin\left(-\dfrac{\pi}{6}\right)$

d) $\cos \dfrac{4\pi}{3}$

e) $\cot \pi$

f) $\csc\left(-\dfrac{7\pi}{2}\right)$

Solution   We locate the point on the unit circle determined by the rotation, and then find its coordinates using reflection if necessary.

a) The coordinates of the point determined by $\pi/3$ are $(1/2, \sqrt{3}/2)$.

b) The reflection of $(\sqrt{2}/2, \sqrt{2}/2)$ across the y-axis is $(-\sqrt{2}/2, \sqrt{2}/2)$.

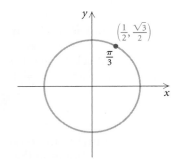

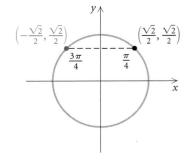

Thus, $\tan \dfrac{\pi}{3} = \dfrac{y}{x} = \dfrac{\sqrt{3}/2}{1/2} = \sqrt{3}.$

Thus, $\cos \dfrac{3\pi}{4} = x = -\dfrac{\sqrt{2}}{2}.$

c) The reflection of $(\sqrt{3}/2, 1/2)$ across the x-axis is $(\sqrt{3}/2, -1/2)$.

d) The reflection of $(1/2, \sqrt{3}/2)$ across the origin is $(-1/2, -\sqrt{3}/2)$.

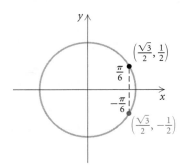

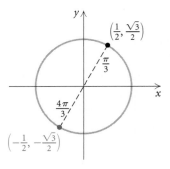

Thus, $\sin\left(-\dfrac{\pi}{6}\right) = y = -\dfrac{1}{2}.$

Thus, $\cos \dfrac{4\pi}{3} = x = -\dfrac{1}{2}.$

e) The coordinates of the point determined by π are $(-1, 0)$.

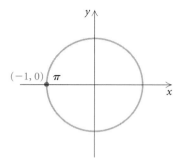

Thus, $\cot \pi = \dfrac{x}{y} = \dfrac{-1}{0}$, which is not defined.

We can also think of $\cot \pi$ as the reciprocal of $\tan \pi$. Since $\tan \pi = y/x = 0/-1 = 0$ and the reciprocal of 0 is not defined, we know that $\cot \pi$ is not defined.

f) The coordinates of the point determined by $-7\pi/2$ are $(0, 1)$.

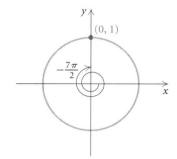

Thus, $\csc\left(-\dfrac{7\pi}{2}\right) = \dfrac{1}{y} = \dfrac{1}{1} = 1$.

▷ Now Try Exercises 9 and 11.

TECHNOLOGY ··············
    CONNECTION

To find trigonometric function values of angles measured in radians, we set the calculator in RADIAN mode.

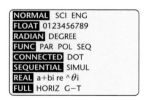

Parts (a)–(c) of Example 3 are shown in the window below.

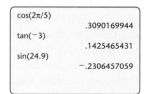

Using a calculator, we can find trigonometric function values of any real number without knowing the coordinates of the point that it determines on the unit circle. Most calculators have both degree and radian modes. When finding function values of radian measures, or real numbers, we *must* set the calculator in RADIAN mode.

EXAMPLE 3 Find each of the following function values of radian measures using a calculator. Round the answers to four decimal places.

a) $\cos \dfrac{2\pi}{5}$

b) $\tan(-3)$

c) $\sin 24.9$

d) $\sec \dfrac{\pi}{7}$

Solution   Using a calculator set in RADIAN mode, we find the values.

a) $\cos \dfrac{2\pi}{5} \approx 0.3090$

b) $\tan(-3) \approx 0.1425$

c) $\sin 24.9 \approx -0.2306$

d) $\sec \dfrac{\pi}{7} = \dfrac{1}{\cos \dfrac{\pi}{7}} \approx 1.1099$

Note in part (d) that the secant function value can be found by taking the reciprocal of the cosine value. Thus we can enter $\cos \pi/7$ and use the reciprocal key.

▷ Now Try Exercises 25 and 33.

TECHNOLOGY ··
       CONNECTION

Exploration

We can graph the unit circle using a graphing calculator. We use PARAMETRIC mode with the following window and let $X1T = \cos T$ and $Y1T = \sin T$. Here we use DEGREE mode.

WINDOW

    Tmin $= 0$
    Tmax $= 360$
    Tstep $= 15$
    Xmin $= -1.5$
    Xmax $= 1.5$
    Xscl $= 1$
    Ymin $= -1$
    Ymax $= 1$
    Yscl $= 1$

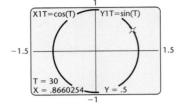

Using the trace key and an arrow key to move the cursor around the unit circle, we see the T, X, and Y values appear on the screen. What do they represent? Repeat this exercise in RADIAN mode. What do the T, X, and Y values represent?

From the definitions on p. 60, we can relabel any point (x, y) on the unit circle as $(\cos s, \sin s)$, where s is any real number.

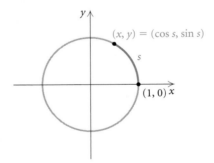

◆ Graphs of the Sine and Cosine Functions

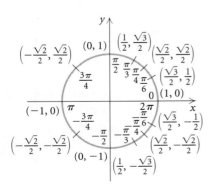

Properties of functions can be observed from their graphs. We begin by graphing the sine and cosine functions. We make a table of values, plot the points, and then connect those points with a smooth curve. It is helpful to first draw a unit circle and label a few points with coordinates. We can either use the coordinates as the function values or find approximate sine and cosine values directly with a calculator.

s	sin s	cos s
0	0	1
$\pi/6$	0.5	0.8660
$\pi/4$	0.7071	0.7071
$\pi/3$	0.8660	0.5
$\pi/2$	1	0
$3\pi/4$	0.7071	-0.7071
π	0	-1
$5\pi/4$	-0.7071	-0.7071
$3\pi/2$	-1	0
$7\pi/4$	-0.7071	0.7071
2π	0	1

s	sin s	cos s
0	0	1
$-\pi/6$	-0.5	0.8660
$-\pi/4$	-0.7071	0.7071
$-\pi/3$	-0.8660	0.5
$-\pi/2$	-1	0
$-3\pi/4$	-0.7071	-0.7071
$-\pi$	0	-1
$-5\pi/4$	0.7071	-0.7071
$-3\pi/2$	1	0
$-7\pi/4$	0.7071	0.7071
-2π	0	1

The graphs are as follows.

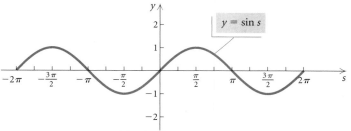

The sine function

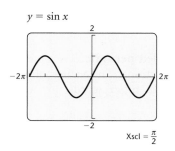
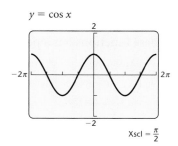
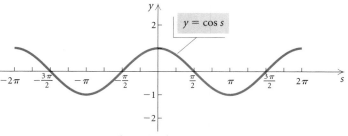

The cosine function

The sine and cosine functions are continuous functions. Note in the graph of the sine function that function values increase from 0 at $s = 0$ to 1 at $s = \pi/2$, then decrease to 0 at $s = \pi$, decrease further to -1 at $s = 3\pi/2$, and increase to 0 at 2π. The reverse pattern follows when s decreases from 0 to -2π. Note in the graph of the cosine function that function values start at 1 when $s = 0$, and decrease to 0 at $s = \pi/2$. They decrease further to -1 at $s = \pi$, then increase to 0 at $s = 3\pi/2$, and increase further to 1 at $s = 2\pi$. An identical pattern follows when s decreases from 0 to -2π.

From the unit circle and the graphs of the functions, we know that the domain of both the sine and cosine functions is the entire set of real numbers, $(-\infty, \infty)$. The range of each function is the set of all real numbers from -1 to 1, $[-1, 1]$.

> ### Domain and Range of Sine and Cosine Functions
> The *domain* of the sine and cosine functions is $(-\infty, \infty)$.
> The *range* of the sine and cosine functions is $[-1, 1]$.

TECHNOLOGY CONNECTION

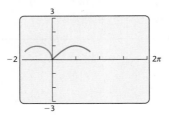

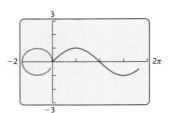

Another way to construct the sine and cosine graphs is by considering the unit circle and transferring vertical distances for the sine function and horizontal distances for the cosine function. Using a graphing calculator, we can visualize the transfer of these distances. We use the calculator set in PARAMETRIC and RADIAN modes and let $X_{1T} = \cos T - 1$ and $Y_{1T} = \sin T$ for the unit circle centered at $(-1, 0)$ and $X_{2T} = T$ and $Y_{2T} = \sin T$ for the sine curve. Use the following window settings.

Tmin = 0	Xmin = −2	Ymin = −3
Tmax = 2π	Xmax = 2π	Ymax = 3
Tstep = .1	Xscl = $\pi/2$	Yscl = 1

With the calculator set in SIMULTANEOUS mode, we can actually watch the sine function (in red) "unwind" from the unit circle (in blue). In the two screens at left, we partially illustrate this animated procedure.

Consult your calculator's instruction manual for specific keystrokes and graph both the sine curve and the cosine curve in this manner.

A function with a repeating pattern is called **periodic.** The sine and cosine functions are examples of periodic functions. The values of these functions repeat themselves every 2π units. In other words, for any s, we have

$$\sin (s + 2\pi) = \sin s \quad \text{and} \quad \cos (s + 2\pi) = \cos s.$$

To see this another way, think of the part of the graph between 0 and 2π and note that the rest of the graph consists of copies of it. If we translate the graph of $y = \sin x$ or $y = \cos x$ to the left or right 2π units, we will obtain the original graph. We say that each of these functions has a period of 2π.

> **Periodic Function**
>
> A function f is said to be **periodic** if there exists a positive constant p such that
>
> $$f(s + p) = f(s)$$
>
> for all s in the domain of f. The smallest such positive number p is called the period of the function.

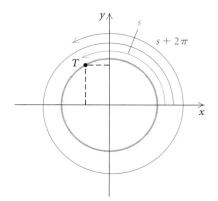

The period p can be thought of as the length of the shortest recurring interval.

We can also use the unit circle to verify that the period of the sine and cosine functions is 2π. Consider any real number s and the point T that it determines on a unit circle, as shown at left. If we increase s by 2π, the point determined by $s + 2\pi$ is again the point T. Hence for any real number s,

$$\sin(s + 2\pi) = \sin s \quad \text{and} \quad \cos(s + 2\pi) = \cos s.$$

It is also true that $\sin(s + 4\pi) = \sin s$, $\sin(s + 6\pi) = \sin s$, and so on. In fact, for *any* integer k, the following equations are identities:

$$\sin[s + k(2\pi)] = \sin s \quad \text{and} \quad \cos[s + k(2\pi)] = \cos s,$$

or

$$\sin s = \sin(s + 2k\pi) \quad \text{and} \quad \cos s = \cos(s + 2k\pi).$$

The **amplitude** of a periodic function is defined as one half of the distance between its maximum and minimum function values. It is always positive. Both the graphs and the unit circle verify that the maximum value of the sine and cosine functions is 1, whereas the minimum value of each is -1. Thus,

the amplitude of the sine function $= \frac{1}{2}\left|1 - (-1)\right| = 1$

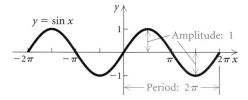

and

the amplitude of the cosine function is $\frac{1}{2}\left|1 - (-1)\right| = 1$.

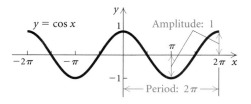

TECHNOLOGY
CONNECTION

Exploration
Using the TABLE feature on a graphing calculator, compare the y-values for $y_1 = \sin x$ and $y_2 = \sin(-x)$ and for $y_3 = \cos x$ and $y_4 = \cos(-x)$. We set TblMin $= 0$ and $\triangle$Tbl $= \pi/12$.

X	Y1	Y2
0	0	0
.2618	.25882	−.2588
.5236	.5	−.5
.7854	.70711	−.7071
1.0472	.86603	−.866
1.309	.96593	−.9659
1.5708	1	−1

X = 0

X	Y3	Y4
0	1	1
.2618	.96593	.96593
.5236	.86603	.86603
.7854	.70711	.70711
1.0472	.5	.5
1.309	.25882	.25882
1.5708	0	0

X = 0

What appears to be the relationship between $\sin x$ and $\sin(-x)$ and between $\cos x$ and $\cos(-x)$?

Consider any real number s and its opposite, $-s$. These numbers determine points T and T_1 on a unit circle that are symmetric with respect to the x-axis.

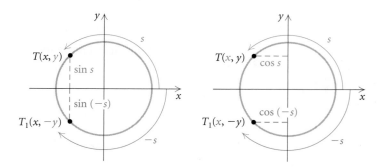

Because their second coordinates are opposites of each other, we know that for any number s,

$$\sin(-s) = -\sin s.$$

Because their first coordinates are the same, we know that for any number s,

$$\cos(-s) = \cos s.$$

Thus we have shown the following.

Even and Odd Functions

If the graph of a function f is symmetric with respect to the y-axis, we say that it is an **even function.** That is, for each x in the domain of f, $f(x) = f(-x)$.

If the graph of a function f is symmetric with respect to the origin, we say that it is an **odd function.** That is, for each x in the domain of f, $f(-x) = -f(x)$.

The sine function is *odd.*

The cosine function is *even.*

The following is a summary of the properties of the sine and cosine functions.

CONNECTING THE CONCEPTS

Comparing the Sine and Cosine Functions

SINE FUNCTION

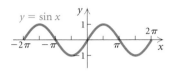

1. Continuous
2. Period: 2π
3. Domain: All real numbers
4. Range: $[-1, 1]$
5. Amplitude: 1
6. Odd: $\sin(-s) = -\sin s$

COSINE FUNCTION

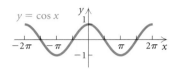

1. Continuous
2. Period: 2π
3. Domain: All real numbers
4. Range: $[-1, 1]$
5. Amplitude: 1
6. Even: $\cos(-s) = \cos s$

● ● ●

✦ Graphs of the Tangent, Cotangent, Cosecant, and Secant Functions

To graph the tangent function, we could make a table of values using a calculator, but in this case it is easier to begin with the definition of tangent and the coordinates of a few points on the unit circle. We recall that

$$\tan s = \frac{y}{x} = \frac{\sin s}{\cos s}.$$

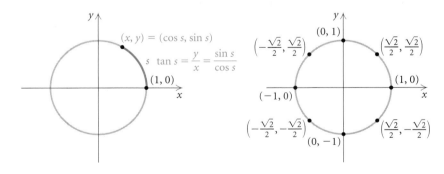

The tangent function is not defined when x, the first coordinate, is 0. That is, it is not defined for any number s whose cosine is 0:

$$s = \pm\frac{\pi}{2}, \pm\frac{3\pi}{2}, \pm\frac{5\pi}{2}, \ldots .$$

We draw vertical asymptotes at these locations (see Fig. 1 below).

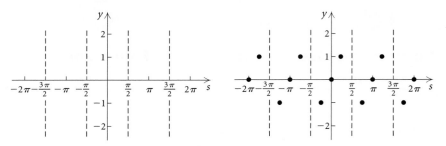

Figure 1   **Figure 2**

We also note that

$$\tan s = 0 \text{ at } s = 0, \pm\pi, \pm2\pi, \pm3\pi, \ldots,$$

$$\tan s = 1 \text{ at } s = \ldots -\frac{7\pi}{4}, -\frac{3\pi}{4}, \frac{\pi}{4}, \frac{5\pi}{4}, \frac{9\pi}{4}, \ldots,$$

$$\tan s = -1 \text{ at } s = \ldots -\frac{9\pi}{4}, -\frac{5\pi}{4}, -\frac{\pi}{4}, \frac{3\pi}{4}, \frac{7\pi}{4}, \ldots.$$

We can add these ordered pairs to the graph (see Fig. 2 above) and investigate the values in $(-\pi/2, \pi/2)$ using a calculator. Note that the function value is 0 when $s = 0$, and the values increase without bound as s increases toward $\pi/2$. The graph gets closer and closer to an asymptote as s gets closer to $\pi/2$, but it never touches the line. As s decreases from 0 to $-\pi/2$, the values decrease without bound. Again the graph gets closer and closer to an asymptote, but it never touches it. We now complete the graph.

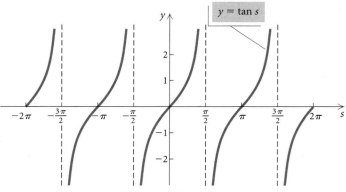

$y = \tan s$

The tangent function

From the graph, we see that the tangent function is continuous except where it is not defined. The period of the tangent function is π. Note that although there is a period, there is no amplitude because there are no maximum and minimum values. When $\cos s = 0$, $\tan s$ is not defined

($\tan s = \sin s / \cos s$). Thus the domain of the tangent function is the set of all real numbers except $(\pi/2) + k\pi$, where k is an integer. The range of the function is the set of all real numbers.

The cotangent function ($\cot s = \cos s / \sin s$) is not defined when y, the second coordinate, is 0—that is, it is not defined for any number s whose sine is 0. Thus the cotangent is not defined for $s = 0, \pm\pi, \pm 2\pi, \pm 3\pi, \ldots$. The graph of the function is shown below.

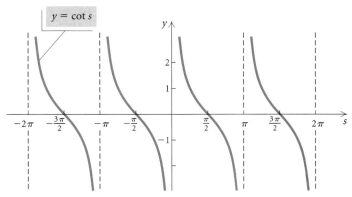

The cotangent function

The cosecant and sine functions are reciprocal functions, as are the secant and cosine functions. The graphs of the cosecant and secant functions can be constructed by finding the reciprocals of the values of the sine and cosine functions, respectively. Thus the functions will be positive together and negative together. The cosecant function is not defined for those numbers s whose sine is 0. The secant function is not defined for those numbers s whose cosine is 0. In the graphs below, the sine and cosine functions are shown by the gray curves for reference.

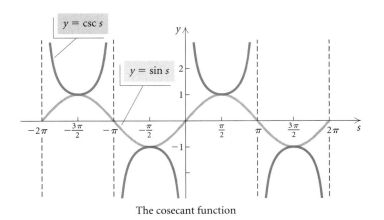

The cosecant function

TECHNOLOGY CONNECTION

When graphing trigonometric functions that are not defined for all real numbers, it is best to use DOT mode for the graph. Here we illustrate $y = \tan x$ and $y = \csc x$.

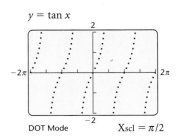

DOT Mode $X_{scl} = \pi/2$

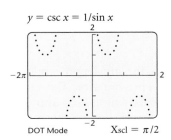

DOT Mode $X_{scl} = \pi/2$

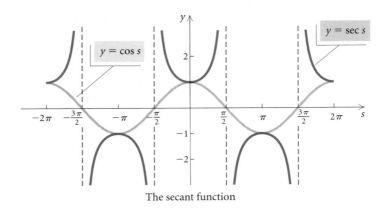

The secant function

The following is a summary of the basic properties of the tangent, cotangent, cosecant, and secant functions. These functions are continuous except where they are not defined.

CONNECTING THE CONCEPTS

Comparing the Tangent, Cotangent, Cosecant, and Secant Functions

TANGENT FUNCTION

1. Period: π
2. Domain: All real numbers except $(\pi/2) + k\pi$, where k is an integer
3. Range: All real numbers

COTANGENT FUNCTION

1. Period: π
2. Domain: All real numbers except $k\pi$, where k is an integer
3. Range: All real numbers

COSECANT FUNCTION

1. Period: 2π
2. Domain: All real numbers except $k\pi$, where k is an integer
3. Range: $(-\infty, -1] \cup [1, \infty)$

SECANT FUNCTION

1. Period: 2π
2. Domain: All real numbers except $(\pi/2) + k\pi$, where k is an integer
3. Range: $(-\infty, -1] \cup [1, \infty)$ ● ● ●

In this chapter, we have used the letter s for arc length and have avoided the letters x and y, which generally represent first and second coordinates. Nevertheless, we can represent the arc length on a unit circle by any variable, such as s, t, x, or θ. Each arc length determines a point that can be labeled with an ordered pair. The first coordinate of that ordered pair is the cosine of the arc length, and the second coordinate is the sine of the arc length. The identities we have developed hold no matter what symbols are used for variables—for example, $\cos(-s) = \cos s$, $\cos(-x) = \cos x$, $\cos(-\theta) = \cos \theta$, and $\cos(-t) = \cos t$.

1.5 ◆ EXERCISE SET

*The following points are on the unit circle. Find the coordinates of their reflections across (**a**) the x-axis, (**b**) the y-axis, and (**c**) the origin.*

1. $\left(-\dfrac{3}{4}, \dfrac{\sqrt{7}}{4}\right)$

2. $\left(\dfrac{2}{3}, \dfrac{\sqrt{5}}{3}\right)$

3. $\left(\dfrac{2}{5}, -\dfrac{\sqrt{21}}{5}\right)$

4. $\left(-\dfrac{\sqrt{3}}{2}, -\dfrac{1}{2}\right)$

5. The number $\pi/4$ determines a point on the unit circle with coordinates $\left(\sqrt{2}/2, \sqrt{2}/2\right)$. What are the coordinates of the point determined by $-\pi/4$?

6. A number β determines a point on the unit circle with coordinates $\left(-2/3, \sqrt{5}/3\right)$. What are the coordinates of the point determined by $-\beta$?

Find the function value using coordinates of points on the unit circle. Give exact answers.

7. $\sin \pi$

8. $\cos\left(-\dfrac{\pi}{3}\right)$

9. $\cot \dfrac{7\pi}{6}$

10. $\tan \dfrac{11\pi}{4}$

11. $\sin(-3\pi)$

12. $\csc \dfrac{3\pi}{4}$

13. $\cos \dfrac{5\pi}{6}$

14. $\tan\left(-\dfrac{\pi}{4}\right)$

15. $\sec \dfrac{\pi}{2}$

16. $\cos 10\pi$

17. $\cos \dfrac{\pi}{6}$

18. $\sin \dfrac{2\pi}{3}$

19. $\sin \dfrac{5\pi}{4}$

20. $\cos \dfrac{11\pi}{6}$

21. $\sin(-5\pi)$

22. $\tan \dfrac{3\pi}{2}$

23. $\cot \dfrac{5\pi}{2}$

24. $\tan \dfrac{5\pi}{3}$

Find the function value using a calculator set in RADIAN mode. Round the answer to four decimal places, where appropriate.

25. $\tan \dfrac{\pi}{7}$

26. $\cos\left(-\dfrac{2\pi}{5}\right)$

27. $\sec 37$

28. $\sin 11.7$

29. $\cot 342$

30. $\tan 1.3$

31. $\cos 6\pi$

32. $\sin \dfrac{\pi}{10}$

33. $\csc 4.16$

34. $\sec \dfrac{10\pi}{7}$

35. $\tan \dfrac{7\pi}{4}$

36. $\cos 2000$

37. $\sin\left(-\dfrac{\pi}{4}\right)$

38. $\cot 7\pi$

39. $\sin 0$

40. $\cos(-29)$

41. $\tan \dfrac{2\pi}{9}$

42. $\sin \dfrac{8\pi}{3}$

43. a) Sketch a graph of $y = \sin x$.
   b) By reflecting the graph in part (a), sketch a graph of $y = \sin(-x)$.
   c) By reflecting the graph in part (a), sketch a graph of $y = -\sin x$.
   d) How do the graphs in parts (b) and (c) compare?

44. a) Sketch a graph of $y = \cos x$.
   b) By reflecting the graph in part (a), sketch a graph of $y = \cos(-x)$.
   c) By reflecting the graph in part (a), sketch a graph of $y = -\cos x$.
   d) How do the graphs in parts (a) and (b) compare?

45. **a)** Sketch a graph of $y = \sin x$.
 b) By translating, sketch a graph of
 $y = \sin (x + \pi)$.
 c) By reflecting the graph of part (a), sketch a
 graph of $y = -\sin x$.
 d) How do the graphs of parts (b) and (c)
 compare?

46. **a)** Sketch a graph of $y = \sin x$.
 b) By translating, sketch a graph of
 $y = \sin (x - \pi)$.
 c) By reflecting the graph of part (a), sketch a
 graph of $y = -\sin x$.
 d) How do the graphs of parts (b) and (c)
 compare?

47. **a)** Sketch a graph of $y = \cos x$.
 b) By translating, sketch a graph of
 $y = \cos (x + \pi)$.
 c) By reflecting the graph of part (a), sketch a
 graph of $y = -\cos x$.
 d) How do the graphs of parts (b) and (c)
 compare?

48. **a)** Sketch a graph of $y = \cos x$.
 b) By translating, sketch a graph of
 $y = \cos (x - \pi)$.
 c) By reflecting the graph of part (a), sketch a
 graph of $y = -\cos x$.
 d) How do the graphs of parts (b) and (c)
 compare?

49. **a)** Sketch a graph of $y = \tan x$.
 b) By reflecting the graph of part (a), sketch a
 graph of $y = \tan (-x)$.
 c) By reflecting the graph of part (a), sketch a
 graph of $y = -\tan x$.
 d) How do the graphs in parts (b) and (c)
 compare?

50. **a)** Sketch a graph of $y = \sec x$.
 b) By reflecting the graph of part (a), sketch a
 graph of $y = \sec (-x)$.
 c) By reflecting the graph of part (a), sketch a
 graph of $y = -\sec x$.
 d) How do the graphs in parts (a) and (b)
 compare?

51. Of the six circular functions, which are even?
 Which are odd?

52. Of the six circular functions, which have
 period π? Which have period 2π?

Consider the coordinates on the unit circle for
Exercises 53–56.

53. In which quadrants is the tangent function
 positive? negative?

54. In which quadrants is the sine function positive?
 negative?

55. In which quadrants is the cosine function
 positive? negative?

56. In which quadrants is the cosecant function
 positive? negative?

Technology Connection

Use a graphing calculator to determine the domain, the
range, the period, and the amplitude of the function.

57. $y = (\sin x)^2$ **58.** $y = |\cos x| + 1$

59. Using a calculator, consider $(\sin x)/x$, where x is
 between 0 and $\pi/2$. As x approaches 0, this function approaches a limiting value. What is it?

60. Using graphs, determine all numbers x that
 satisfy $\sin x < \cos x$.

Collaborative Discussion and Writing

61. Describe how the graphs of the sine and cosine
 functions are related.

62. Explain why both the sine and cosine functions
 are continuous, but the tangent function, defined
 as sine/cosine, is not continuous.

Skill Maintenance

Graph both functions on the same set of axes, and
describe how g is a transformation of f.

63. $f(x) = x^2$, $g(x) = 2x^2 - 3$

64. $f(x) = x^2$, $g(x) = (x - 2)^2$

65. $f(x) = |x|$, $g(x) = \frac{1}{2}|x - 4| + 1$

66. $f(x) = x^3$, $g(x) = -x^3$

Write an equation for a function that has a graph with
the given characteristics.

67. The shape of $y = x^3$, but reflected across the
 x-axis, shifted right 2 units, and shifted down
 1 unit

68. The shape of $y = 1/x$, but shrunk vertically by a
 factor of $\frac{1}{4}$ and shifted up 3 units

Synthesis

Complete. (For example, $\sin(x + 2\pi) = \sin x$.)

69. $\cos(-x) = $ _____

70. $\sin(-x) = $ _____

71. $\sin(x + 2k\pi), k \in \mathbb{Z} = $ _____

72. $\cos(x + 2k\pi), k \in \mathbb{Z} = $ _____

73. $\sin(\pi - x) = $ _____

74. $\cos(\pi - x) = $ _____

75. $\cos(x - \pi) = $ _____

76. $\cos(x + \pi) = $ _____

77. $\sin(x + \pi) = $ _____

78. $\sin(x - \pi) = $ _____

79. Find all numbers x that satisfy the following.

 a) $\sin x = 1$
 b) $\cos x = -1$
 c) $\sin x = 0$

80. Find $f \circ g$ and $g \circ f$, where $f(x) = x^2 + 2x$ and $g(x) = \cos x$.

Determine the domain of the function.

81. $f(x) = \sqrt{\cos x}$

82. $g(x) = \dfrac{1}{\sin x}$

83. $f(x) = \dfrac{\sin x}{\cos x}$

84. $g(x) = \log(\sin x)$

Graph.

85. $y = 3 \sin x$

86. $y = \sin |x|$

87. $y = \sin x + \cos x$

88. $y = |\cos x|$

89. One of the motivations for developing trigonometry with a unit circle is that you can actually "see" $\sin \theta$ and $\cos \theta$ on the circle. Note in the figure below that $AP = \sin \theta$ and $OA = \cos \theta$. It turns out that you can also "see" the other four trigonometric functions. Prove each of the following.

 a) $BD = \tan \theta$ b) $OD = \sec \theta$
 c) $OE = \csc \theta$ d) $CE = \cot \theta$

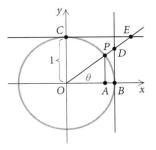

1.6

Graphs of Transformed Sine and Cosine Functions

 ◆ Graph transformations of $y = \sin x$ and $y = \cos x$ in the form

$$y = A \sin(Bx - C) + D$$

 and

$$y = A \cos(Bx - C) + D$$

 and determine the amplitude, the period, and the phase shift.
 ◆ Graph sums of functions.
 ◆ Graph functions (damped oscillations) found by multiplying trigonometric functions by other functions.

◆ Variations of Basic Graphs

In Section 1.5, we graphed all six trigonometric functions. In this section, we will consider variations of the graphs of the sine and cosine functions.

For example, we will graph equations like the following:

$$y = 5 \sin \tfrac{1}{2}x, \qquad y = \cos(2x - \pi), \quad \text{and} \quad y = \tfrac{1}{2}\sin x - 3.$$

In particular, we are interested in graphs of functions in the form

$$y = A \sin(Bx - C) + D$$

and

$$y = A \cos(Bx - C) + D,$$

where A, B, C, and D are constants. These constants have the effect of translating, reflecting, stretching, and shrinking the basic graphs. Let's first examine the effect of each constant individually. Then we will consider the combined effects of more than one constant.

The Constant D

Let's observe the effect of the constant D in the graphs below.

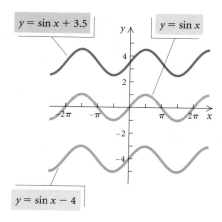

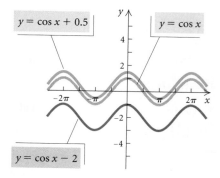

The constant D in

$$y = A \sin(Bx - C) + D \quad \text{and} \quad y = A \cos(Bx - C) + D$$

translates the graphs up D units if $D > 0$ or down $|D|$ units if $D < 0$.

EXAMPLE 1 Sketch a graph of $y = \sin x + 3$.

Solution The graph of $y = \sin x + 3$ is a *vertical* translation of the graph of $y = \sin x$ up 3 units. One way to sketch the graph is to first consider $y = \sin x$ on an interval of length 2π, say, $[0, 2\pi]$. The zeros of the function and the maximum and minimum values can be considered key points. These are

$$(0, 0), \quad \left(\frac{\pi}{2}, 1\right), \quad (\pi, 0), \quad \left(\frac{3\pi}{2}, -1\right), \quad (2\pi, 0).$$

These key points are transformed up 3 units to obtain the key points of the graph of $y = \sin x + 3$. These are

$$(0, 3), \quad \left(\frac{\pi}{2}, 4\right), \quad (\pi, 3), \quad \left(\frac{3\pi}{2}, 2\right), \quad (2\pi, 3).$$

The graph of $y = \sin x + 3$ can be sketched on the interval $[0, 2\pi]$ and extended to obtain the rest of the graph by repeating the graph on intervals of length 2π.

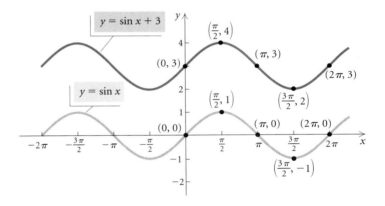

The Constant A

Next, we consider the effect of the constant A. What can we observe in the following graphs? What is the effect of the constant A on the graph of the basic function when **(a)** $0 < A < 1$? **(b)** $A > 1$? **(c)** $-1 < A < 0$? **(d)** $A < -1$?

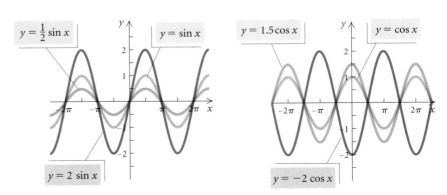

If $|A| > 1$, then there will be a vertical stretching. If $|A| < 1$, then there will be a vertical shrinking. If $A < 0$, the graph is also reflected across the x-axis.

> **Amplitude**
>
> The **amplitude** of the graphs of $y = A \sin(Bx - C) + D$ and $y = A \cos(Bx - C) + D$ is $|A|$.

EXAMPLE 2   Sketch a graph of $y = 2 \cos x$. What is the amplitude?

Solution   The constant 2 in $y = 2 \cos x$ has the effect of stretching the graph of $y = \cos x$ vertically by a factor of 2 units. Since the function values of $y = \cos x$ are such that $-1 \leq \cos x \leq 1$, the function values of $y = 2 \cos x$ are such that $-2 \leq 2 \cos x \leq 2$. The maximum value of $y = 2 \cos x$ is 2, and the minimum value is -2. Thus the *amplitude*, A, is $\frac{1}{2}|2 - (-2)|$, or 2.

We draw the graph of $y = \cos x$ and consider its key points,

$$(0, 1), \quad \left(\frac{\pi}{2}, 0\right), \quad (\pi, -1), \quad \left(\frac{3\pi}{2}, 0\right), \quad (2\pi, 1),$$

on the interval $[0, 2\pi]$.

We then multiply the second coordinates by 2 to obtain the key points of $y = 2 \cos x$. These are

$$(0, 2), \quad \left(\frac{\pi}{2}, 0\right), \quad (\pi, -2), \quad \left(\frac{3\pi}{2}, 0\right), \quad (2\pi, 2).$$

We plot these points and sketch the graph on the interval $[0, 2\pi]$. Then we repeat this part of the graph on adjacent intervals of length 2π.

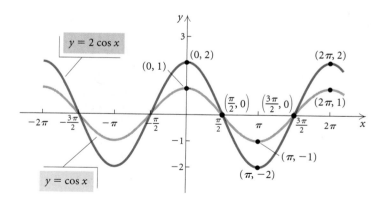

EXAMPLE 3   Sketch a graph of $y = -\frac{1}{2} \sin x$. What is the amplitude?

Solution   The amplitude of the graph is $\left|-\frac{1}{2}\right|$, or $\frac{1}{2}$. The graph of $y = -\frac{1}{2} \sin x$ is a vertical shrinking and a reflection of the graph of $y = \sin x$ across the x-axis. In graphing, the key points of $y = \sin x$,

$$(0, 0), \quad \left(\frac{\pi}{2}, 1\right), \quad (\pi, 0), \quad \left(\frac{3\pi}{2}, -1\right), \quad (2\pi, 0),$$

are transformed to

$$(0, 0), \quad \left(\frac{\pi}{2}, -\frac{1}{2}\right), \quad (\pi, 0), \quad \left(\frac{3\pi}{2}, \frac{1}{2}\right), \quad (2\pi, 0).$$

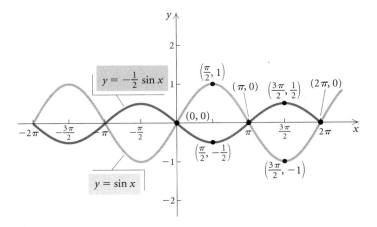

The Constant B

Now, we consider the effect of the constant B. Changes in the constants A and D *do not* change the period. But what effect, if any, does a change in B have on the period of the function? Let's observe the period of each of the following graphs.

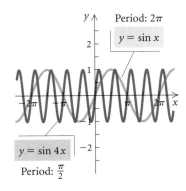

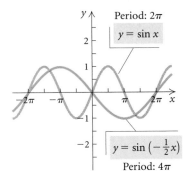

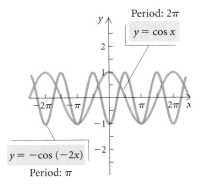

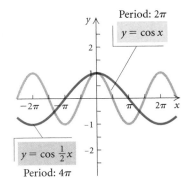

If $|B| < 1$, then there will be a horizontal stretching. If $|B| > 1$, then there will be a horizontal shrinking. If $B < 0$, the graph is also reflected across the y-axis.

> **Period**
>
> The **period** of the graphs of $y = A \sin(Bx - C) + D$ and $y = A \cos(Bx - C) + D$ is $\left| \dfrac{2\pi}{B} \right|$.*

EXAMPLE 4   Sketch a graph of $y = \sin 4x$. What is the period?

Solution   The constant B has the effect of changing the period. The graph of $y = f(4x)$ is obtained from the graph of $y = f(x)$ by shrinking the graph horizontally. The period of $y = \sin 4x$ is $|2\pi/4|$, or $\pi/2$. The new graph is obtained by dividing the first coordinate of each ordered-pair solution of $y = f(x)$ by 4. The key points of $y = \sin x$ are

$$(0, 0), \quad \left(\frac{\pi}{2}, 1 \right), \quad (\pi, 0), \quad \left(\frac{3\pi}{2}, -1 \right), \quad (2\pi, 0).$$

These are transformed to the key points of $y = \sin 4x$, which are

$$(0, 0), \quad \left(\frac{\pi}{8}, 1 \right), \quad \left(\frac{\pi}{4}, 0 \right), \quad \left(\frac{3\pi}{8}, -1 \right), \quad \left(\frac{\pi}{2}, 0 \right).$$

We plot these key points and sketch in the graph on the shortened interval $[0, \pi/2]$. Then we repeat the graph on other intervals of length $\pi/2$.

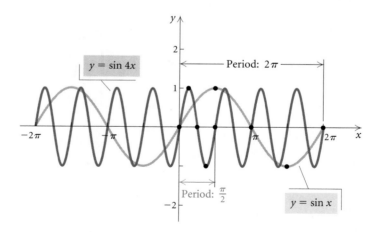

---

*The period of the graphs of $y = A \tan(Bx - C) + D$ and $y = A \cot(Bx - C) + D$ is $|\pi/B|$.
The period of the graphs of $y = A \sec(Bx - C) + D$ and $y = A \csc(Bx - C) + D$ is $|2\pi/B|$.

The Constant C

Next, we examine the effect of the constant C. The curve in each of the following graphs has an amplitude of 1 and a period of 2π, but there are six distinct graphs. What is the effect of the constant C?

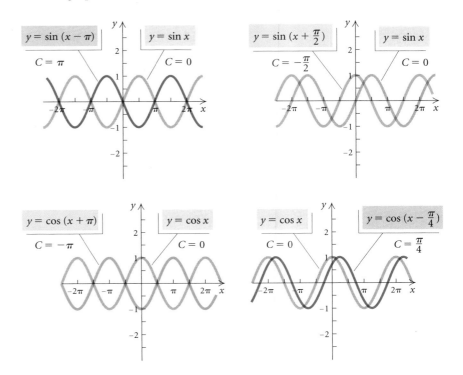

For each of the functions of the form

$$y = A \sin (Bx - C) + D \quad \text{and} \quad y = A \cos (Bx - C) + D$$

that are graphed above, the coefficient of x, which is B, is 1. In this case, the effect of the constant C on the graph of the basic function is a horizontal translation of $|C|$ units. In Example 5, which follows, $B = 1$. We will consider functions where $B \neq 1$ in Examples 6 and 7. When $B \neq 1$, the horizontal translation will be $|C/B|$.

EXAMPLE 5 Sketch a graph of $y = \sin\left(x - \dfrac{\pi}{2}\right)$.

Solution The amplitude is 1, and the period is 2π. The graph of $y = f(x - c)$ is obtained from the graph of $y = f(x)$ by translating the graph horizontally—to the right c units if $c > 0$ and to the left $|c|$ units if $c < 0$. The graph of $y = \sin (x - \pi/2)$ is a translation of the graph of

$y = \sin x$ to the right $\pi/2$ units. The value $\pi/2$ is called the **phase shift.** The key points of $y = \sin x$,

$$(0, 0), \quad \left(\frac{\pi}{2}, 1\right), \quad (\pi, 0), \quad \left(\frac{3\pi}{2}, -1\right), \quad (2\pi, 0),$$

are transformed by adding $\pi/2$ to each of the first coordinates to obtain the following key points of $y = \sin(x - \pi/2)$:

$$\left(\frac{\pi}{2}, 0\right), \quad (\pi, 1), \quad \left(\frac{3\pi}{2}, 0\right), \quad (2\pi, -1), \quad \left(\frac{5\pi}{2}, 0\right).$$

We plot these key points and sketch the curve on the interval $[\pi/2, 5\pi/2]$. Then we repeat the graph on other intervals of length 2π.

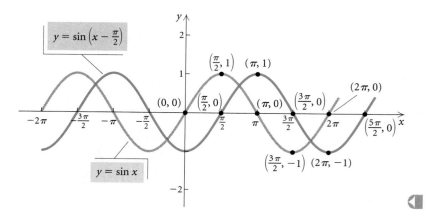

Combined Transformations

Now we consider combined transformations of graphs. It is helpful to rewrite

$$y = A \sin(Bx - C) + D \qquad \text{and} \quad y = A \cos(Bx - C) + D$$

as

$$y = A \sin\left[B\left(x - \frac{C}{B}\right)\right] + D \quad \text{and} \quad y = A \cos\left[B\left(x - \frac{C}{B}\right)\right] + D.$$

EXAMPLE 6   Sketch a graph of $y = \cos(2x - \pi)$.

Solution   The graph of

$$y = \cos(2x - \pi)$$

is the same as the graph of

$$y = 1 \cdot \cos\left[2\left(x - \frac{\pi}{2}\right)\right] + 0.$$

The amplitude is 1. The factor 2 shrinks the period by half, making the period $|2\pi/2|$, or π. The phase shift $\pi/2$ translates the graph of $y = \cos 2x$ to the right $\pi/2$ units. Thus, to form the graph, we first graph $y = \cos x$, followed by $y = \cos 2x$ and then $y = \cos[2(x - \pi/2)]$.

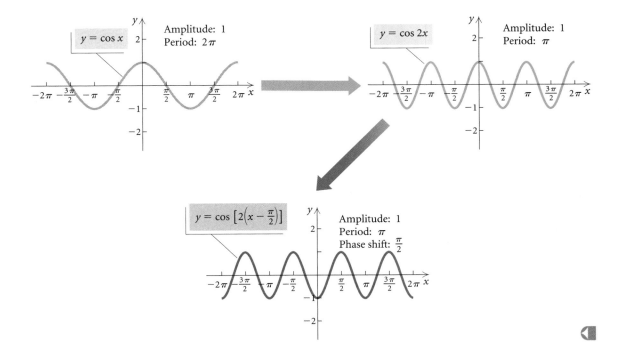

Phase Shift
The **phase shift** of the graphs

$$y = A \sin(Bx - C) + D = A \sin\left[B\left(x - \frac{C}{B}\right)\right] + D$$

and

$$y = A \cos(Bx - C) + D = A \cos\left[B\left(x - \frac{C}{B}\right)\right] + D$$

is the quantity $\dfrac{C}{B}$.

If $C/B > 0$, the graph is translated to the right C/B units. If $C/B < 0$, the graph is translated to the left $|C/B|$ units. Be sure that the horizontal stretching or shrinking based on the constant B is done before the translation based on the phase shift C/B.

Let's now summarize the effect of the constants. When graphing, we carry out the procedures in the order listed.

When graphing transformations of the tangent and cotangent functions, note that the period is $|\pi/B|$. When graphing transformations of the secant and cosecant functions, note that the period is $|2\pi/B|$.

Transformations of Sine and Cosine Functions

To graph

$$y = A \sin (Bx - C) + D = A \sin \left[B\left(x - \frac{C}{B} \right) \right] + D$$

and

$$y = A \cos (Bx - C) + D = A \cos \left[B\left(x - \frac{C}{B} \right) \right] + D,$$

follow the steps listed below in the order in which they are listed.

1. Stretch or shrink the graph horizontally according to B.

   $|B| < 1$ Stretch horizontally
   $|B| > 1$ Shrink horizontally
   $B < 0$ Reflect across the y-axis

   The *period* is $\left| \dfrac{2\pi}{B} \right|$.

2. Stretch or shrink the graph vertically according to A.

   $|A| < 1$ Shrink vertically
   $|A| > 1$ Stretch vertically
   $A < 0$ Reflect across the x-axis

   The *amplitude* is $|A|$.

3. Translate the graph horizontally according to C/B.

   $\dfrac{C}{B} < 0$ $\left| \dfrac{C}{B} \right|$ units to the left

   $\dfrac{C}{B} > 0$ $\dfrac{C}{B}$ units to the right

   The *phase shift* is $\dfrac{C}{B}$.

4. Translate the graph vertically according to D.

   $D < 0$ $|D|$ units down
   $D > 0$ D units up

EXAMPLE 7   Sketch a graph of $y = 3 \sin (2x + \pi/2) + 1$. Find the amplitude, the period, and the phase shift.

Solution   We first note that

$$y = 3 \sin \left(2x + \frac{\pi}{2} \right) + 1 = 3 \sin \left[2\left(x - \left(-\frac{\pi}{4} \right) \right) \right] + 1.$$

Then we have the following:

$$\text{Amplitude} = |A| = |3| = 3,$$

$$\text{Period} = \left|\frac{2\pi}{B}\right| = \left|\frac{2\pi}{2}\right| = \pi,$$

$$\text{Phase shift} = \frac{C}{B} = \frac{-\pi/2}{2} = -\frac{\pi}{4}.$$

To create the final graph, we begin with the basic sine curve, $y = \sin x$. Then we sketch graphs of each of the following equations in sequence.

1. $y = \sin 2x$ **2.** $y = 3 \sin 2x$

3. $y = 3 \sin\left[2\left(x - \left(-\frac{\pi}{4}\right)\right)\right]$ **4.** $y = 3 \sin\left[2\left(x - \left(-\frac{\pi}{4}\right)\right)\right] + 1$

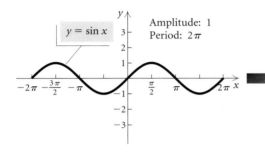

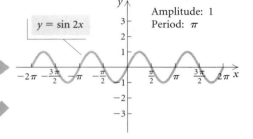

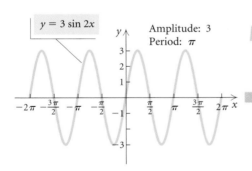

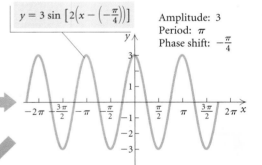

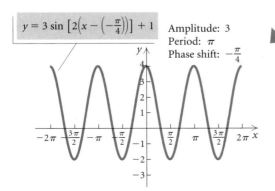

▶ **Now Try Exercise 27.**

All the graphs in Examples 1–7 can be checked using a graphing calculator. Even though it is faster and more accurate to graph using a calculator, graphing by hand gives us a greater understanding of the effect of changing the constants A, B, C, and D.

Graphing calculators are especially convenient when a period or a phase shift is not a multiple of $\pi/4$.

EXAMPLE 8   Graph $y = 3 \cos 2\pi x - 1$. Find the amplitude, the period, and the phase shift.

Solution   First we note the following:

$$\text{Amplitude} = |A| = |3| = 3,$$

$$\text{Period} = \left|\frac{2\pi}{B}\right| = \left|\frac{2\pi}{2\pi}\right| = |1| = 1,$$

$$\text{Phase shift} = \frac{C}{B} = \frac{0}{2\pi} = 0.$$

There is no phase shift in this case because the constant $C = 0$. The graph has a vertical translation of the graph of the cosine function down 1 unit, an amplitude of 3, and a period of 1, so we can use $[-4, 4, -5, 5]$ as the viewing window.

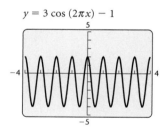

$$y = 3 \cos (2\pi x) - 1$$

▶ **Now Try Exercise 29.**

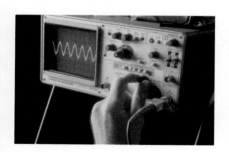

The transformation techniques that we learned in this section for graphing the sine and cosine functions can also be applied in the same manner to the other trigonometric functions. Transformations of this type appear in the synthesis exercises in Exercise Set 1.6.

An **oscilloscope** is an electronic device that converts electrical signals into graphs like those in the preceding examples. These graphs are often called sine waves. By manipulating the controls, we can change the amplitude, the period, and the phase of sine waves. The oscilloscope has many applications, and the trigonometric functions play a major role in many of them.

◆ Graphs of Sums: Addition of Ordinates

The output of an electronic synthesizer used in the recording and playing of music can be converted into sine waves by an oscilloscope. The following graphs illustrate simple tones of different frequencies. The frequency of a simple tone is the number of vibrations in the signal of the tone per second. The loudness or intensity of the tone is reflected in the height of the graph (its amplitude). The three tones in the diagrams below all have the same intensity but different frequencies.

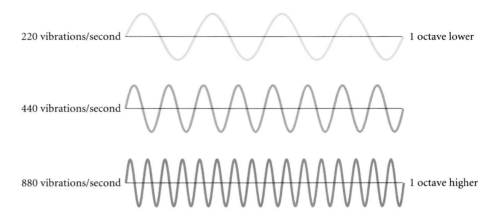

220 vibrations/second — 1 octave lower

440 vibrations/second

880 vibrations/second — 1 octave higher

Musical instruments can generate extremely complex sine waves. On a single instrument, overtones can become superimposed on a simple tone. When multiple notes are played simultaneously, graphs become very complicated. This can happen when multiple notes are played on a single instrument or a group of instruments, or even when the same simple note is played on different instruments.

Combinations of simple tones produce interesting curves. Consider two tones whose graphs are $y_1 = 2 \sin x$ and $y_2 = \sin 2x$. The combination of the two tones produces a new sound whose graph is $y = 2 \sin x + \sin 2x$, as shown in the following example.

EXAMPLE 9   Graph: $y = 2 \sin x + \sin 2x$.

Solution   We graph $y = 2 \sin x$ and $y = \sin 2x$ using the same set of axes.

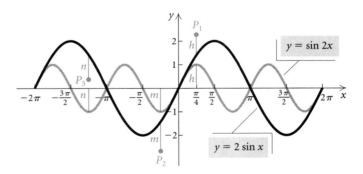

Now we graphically add some y-coordinates, or ordinates, to obtain points on the graph that we seek. At $x = \pi/4$, we transfer the distance h, which is the value of $\sin 2x$, up to add it to the value of $2 \sin x$. Point P_1 is on the graph that we seek. At $x = -\pi/4$, we use a similar procedure, but this time both ordinates are negative. Point P_2 is on the graph. At $x = -5\pi/4$, we add the negative ordinate of $\sin 2x$ to the positive ordinate of $2 \sin x$. Point P_3 is also on the graph. We continue to plot points in this fashion and then connect them to get the desired graph, shown below. This method is called **addition of ordinates,** because we add the y-values (ordinates) of $y = \sin 2x$ to the y-values (ordinates) of $y = 2 \sin x$. Note that the period of $2 \sin x$ is 2π and the period of $\sin 2x$ is π. The period of the sum $2 \sin x + \sin 2x$ is 2π, the least common multiple of 2π and π.

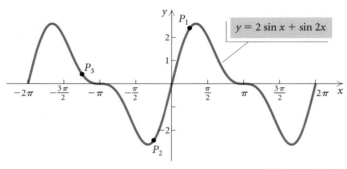

▶ **Now Try Exercise 45.**

✦ Damped Oscillation: Multiplication of Ordinates

Suppose that a weight is attached to a spring and the spring is stretched and put into motion. The weight oscillates up and down. If we could assume falsely that the weight will bob up and down forever, then its height h after time t, in seconds, might be approximated by a function like

$$h(t) = 5 + 2 \sin (6\pi t).$$

Over a short time period, this might be a valid model, but experience tells us that eventually the spring will come to rest. A more appropriate model is provided by the following example, which illustrates **damped oscillation**.

EXAMPLE 10   Sketch a graph of $f(x) = e^{-x/2} \sin x$.

Solution   The function f is the product of two functions g and h, where

$$g(x) = e^{-x/2} \quad \text{and} \quad h(x) = \sin x.$$

Thus, to find function values, we can **multiply ordinates.** Let's do more analysis before graphing. Note that for any real number x,

$$-1 \leq \sin x \leq 1.$$

Because all values of the exponential function are positive, we can multiply by $e^{-x/2}$ and obtain the inequality

$$-e^{-x/2} \leq e^{-x/2} \sin x \leq e^{-x/2}.$$

The direction of the inequality symbols does not change since $e^{-x/2} > 0$. This also tells us that the original function crosses the x-axis only at values for which $\sin x = 0$. These are the numbers $k\pi$, for any integer k.

The inequality tells us that the function f is constrained between the graphs of $y = -e^{-x/2}$ and $y = e^{-x/2}$. We start by graphing these functions using dashed lines. Since we also know that $f(x) = 0$ when $x = k\pi$, k an integer, we mark these points on the graph. Then we use a calculator and compute other function values. The graph is as follows.

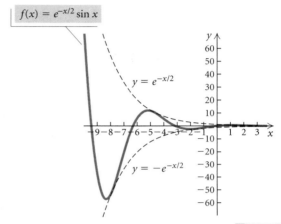

$f(x) = e^{-x/2} \sin x$

$y = e^{-x/2}$

$y = -e^{-x/2}$

▶ Now Try Exercise 53.

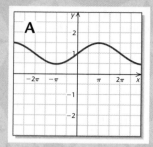

A

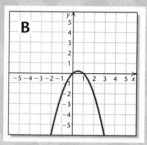

B

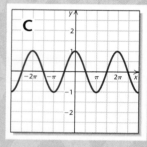

C

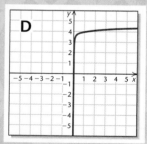

D

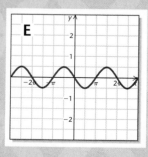

E

Visualizing the Graph

Match the function with its graph.

1. $f(x) = -\sin x$

2. $f(x) = 2x^3 - x + 1$

3. $y = \dfrac{1}{2}\cos\left(x + \dfrac{\pi}{2}\right)$

4. $f(x) = \cos\left(\dfrac{1}{2}x\right)$

5. $y = -x^2 + x$

6. $y = \dfrac{1}{2}\log x + 4$

7. $f(x) = 2^{x-1}$

8. $f(x) = \dfrac{1}{2}\sin\left(\dfrac{1}{2}x\right) + 1$

9. $f(x) = -\cos(x - \pi)$

10. $f(x) = -\dfrac{1}{2}x^4$

Answers on page A-5

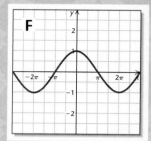

F

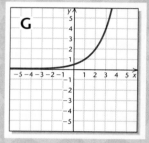

G

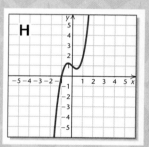

H

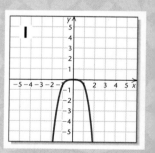

I

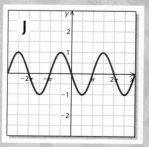

J

1.6 EXERCISE SET

Determine the amplitude, the period, and the phase shift of the function and sketch the graph of the function.

1. $y = \sin x + 1$

2. $y = \dfrac{1}{4} \cos x$

3. $y = -3 \cos x$

4. $y = \sin (-2x)$

5. $y = \dfrac{1}{2} \cos x$

6. $y = \sin \left(\dfrac{1}{2} x \right)$

7. $y = \sin (2x)$

8. $y = \cos x - 1$

9. $y = 2 \sin \left(\dfrac{1}{2} x \right)$

10. $y = \cos \left(x - \dfrac{\pi}{2} \right)$

11. $y = \dfrac{1}{2} \sin \left(x + \dfrac{\pi}{2} \right)$

12. $y = \cos x - \dfrac{1}{2}$

13. $y = 3 \cos (x - \pi)$

14. $y = -\sin \left(\dfrac{1}{4} x \right) + 1$

15. $y = \dfrac{1}{3} \sin x - 4$

16. $y = \cos \left(\dfrac{1}{2} x + \dfrac{\pi}{2} \right)$

17. $y = -\cos (-x) + 2$

18. $y = \dfrac{1}{2} \sin \left(2x - \dfrac{\pi}{4} \right)$

Determine the amplitude, the period, and the phase shift of the function.

19. $y = 2 \cos \left(\dfrac{1}{2} x - \dfrac{\pi}{2} \right)$

20. $y = 4 \sin \left(\dfrac{1}{4} x + \dfrac{\pi}{8} \right)$

21. $y = -\dfrac{1}{2} \sin \left(2x + \dfrac{\pi}{2} \right)$

22. $y = -3 \cos (4x - \pi) + 2$

23. $y = 2 + 3 \cos (\pi x - 3)$

24. $y = 5 - 2 \cos \left(\dfrac{\pi}{2} x + \dfrac{\pi}{2} \right)$

25. $y = -\dfrac{1}{2} \cos (2 \pi x) + 2$

26. $y = -2 \sin (-2x + \pi) - 2$

27. $y = -\sin \left(\dfrac{1}{2} x - \dfrac{\pi}{2} \right) + \dfrac{1}{2}$

28. $y = \dfrac{1}{3} \cos (-3x) + 1$

29. $y = \cos (-2 \pi x) + 2$

30. $y = \dfrac{1}{2} \sin (2 \pi x + \pi)$

31. $y = -\dfrac{1}{4} \cos (\pi x - 4)$

32. $y = 2 \sin (2 \pi x + 1)$

In Exercises 33–40, match the function with one of the graphs (a)–(h), which follow.

a)

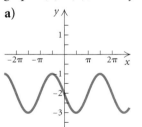

b)

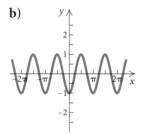

c)

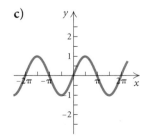

d)

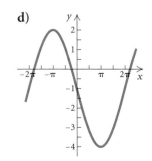

e)

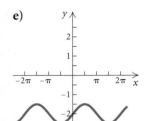

f)

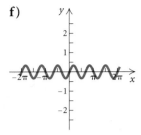

g)

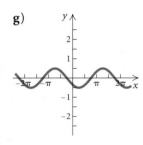

h)

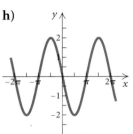

33. $y = -\cos 2x$

34. $y = \dfrac{1}{2} \sin x - 2$

35. $y = 2 \cos \left(x + \dfrac{\pi}{2} \right)$

36. $y = -3 \sin \dfrac{1}{2} x - 1$

37. $y = \sin (x - \pi) - 2$

38. $y = -\dfrac{1}{2} \cos \left(x - \dfrac{\pi}{4} \right)$

39. $y = \dfrac{1}{3} \sin 3x$

40. $y = \cos \left(x - \dfrac{\pi}{2} \right)$

In Exercises 41–44, determine the equation of the function that is graphed.

41.

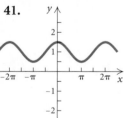

42.

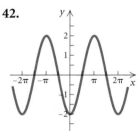

43.

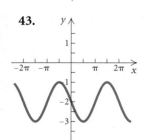

44.

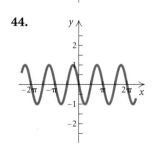

Graph using addition of ordinates.

45. $y = 2 \cos x + \cos 2x$ **46.** $y = 3 \cos x + \cos 3x$

47. $y = \sin x + \cos 2x$ **48.** $y = 2 \sin x + \cos 2x$

49. $y = \sin x - \cos x$ **50.** $y = 3 \cos x - \sin x$

51. $y = 3 \cos x + \sin 2x$ **52.** $y = 3 \sin x - \cos 2x$

Graph each of the following.

53. $f(x) = e^{-x/2} \cos x$ **54.** $f(x) = e^{-0.4x} \sin x$

55. $f(x) = 0.6x^2 \cos x$ **56.** $f(x) = e^{-x/4} \sin x$

57. $f(x) = x \sin x$ **58.** $f(x) = |x| \cos x$

59. $f(x) = 2^{-x} \sin x$ **60.** $f(x) = 2^{-x} \cos x$

Technology Connection

Use a graphing calculator to graph the function.

61. $y = x + \sin x$ **62.** $y = -x - \sin x$

63. $y = \cos x - x$ **64.** $y = -(\cos x - x)$

65. $y = \cos 2x + 2x$ **66.** $y = \cos 3x + \sin 3x$

67. $y = 4 \cos 2x - 2 \sin x$

68. $y = 7.5 \cos x + \sin 2x$

Use a graphing calculator to graph each of the following on the given interval and approximate the zeros.

69. $f(x) = \dfrac{\sin x}{x}; \ [-12, 12]$

70. $f(x) = \dfrac{\cos x - 1}{x}; \ [-12, 12]$

71. $f(x) = x^3 \sin x; \ [-5, 5]$

72. $f(x) = \dfrac{(\sin x)^2}{x}; \ [-4, 4]$

73. *Temperature during an Illness.*   The temperature T of a patient during a 12-day illness is given by

$$T(t) = 101.6° + 3° \sin \left(\dfrac{\pi}{8} t \right).$$

a) Graph the function on the interval $[0, 12]$.
b) What are the maximum and the minimum temperatures during the illness?

74. *Periodic Sales.* A company in a northern climate has sales of skis as given by

$$S(t) = 10\left(1 - \cos\frac{\pi}{6}t\right),$$

where t is the time, in months ($t = 0$ corresponds to July 1), and $S(t)$ is in thousands of dollars.

a) Graph the function on a 12-month interval $[0, 12]$.
b) What is the period of the function?
c) What is the minimum amount of sales and when does it occur?
d) What is the maximum amount of sales and when does it occur?

Collaborative Discussion and Writing

75. In the equations $y = A \sin(Bx - C) + D$ and $y = A \cos(Bx - C) + D$, which constants translate the graphs and which constants stretch and shrink the graphs? Describe in your own words the effect of each constant.

76. In the transformation steps listed in this section, why must step (1) precede step (3)? Give an example that illustrates this.

Skill Maintenance

Classify the function as linear, quadratic, cubic, quartic, rational, exponential, logarithmic, or trigonometric.

77. $f(x) = \dfrac{x + 4}{x}$

78. $y = \dfrac{1}{2}\log x - 4$

79. $y = x^4 - x - 2$

80. $\dfrac{3}{4}x + \dfrac{1}{2}y = -5$

81. $f(x) = \sin x - 3$

82. $f(x) = 0.5e^{x-2}$

83. $y = \dfrac{2}{5}$

84. $y = \sin x + \cos x$

85. $y = x^2 - x^3$

86. $f(x) = \left(\dfrac{1}{2}\right)^x$

Synthesis

Find the maximum and minimum values of the function.

87. $y = 2\cos\left[3\left(x - \dfrac{\pi}{2}\right)\right] + 6$

88. $y = \dfrac{1}{2}\sin(2x - 6\pi) - 4$

The transformation techniques that we learned in this section for graphing the sine and cosine functions can also be applied to the other trigonometric functions. Sketch a graph of each of the following.

89. $y = -\tan x$

90. $y = \tan(-x)$

91. $y = -2 + \cot x$

92. $y = -\dfrac{3}{2}\csc x$

93. $y = 2\tan\dfrac{1}{2}x$

94. $y = \cot 2x$

95. $y = 2\sec(x - \pi)$

96. $y = 4\tan\left(\dfrac{1}{4}x + \dfrac{\pi}{8}\right)$

97. $y = 2\csc\left(\dfrac{1}{2}x - \dfrac{3\pi}{4}\right)$

98. $y = 4\sec(2x - \pi)$

99. *Satellite Location.* A satellite circles the earth in such a way that it is y miles from the equator (north or south, height not considered) t minutes after its launch, where

$$y(t) = 3000\left[\cos\frac{\pi}{45}(t-10)\right].$$

$$y = 3000\left[\cos\frac{\pi}{45}(x-10)\right]$$

What are the amplitude, the period, and the phase shift?

100. *Water Wave.*   The cross-section of a water wave is given by

$$y = 3\sin\left(\frac{\pi}{4}x + \frac{\pi}{4}\right),$$

where y is the vertical height of the water wave and x is the distance from the origin to the wave.

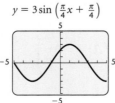

$$y = 3\sin\left(\frac{\pi}{4}x + \frac{\pi}{4}\right)$$

What are the amplitude, the period, and the phase shift?

101. *Damped Oscillations.*   Suppose that the motion of a spring is given by

$$d(t) = 6e^{-0.8t}\cos(6\pi t) + 4,$$

where d is the distance, in inches, of a weight from the point at which the spring is attached to a ceiling, after t seconds. How far do you think the spring is from the ceiling when the spring stops bobbing?

102. *Rotating Beacon.*   A police car is parked 10 ft from a wall. On top of the car is a beacon rotating in such a way that the light is at a distance $d(t)$ from point Q after t seconds, where

$$d(t) = 10\tan(2\pi t).$$

When d is positive, as shown in the figure, the light is pointing north of Q, and when d is negative, the light is pointing south of Q.

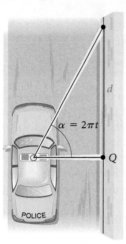

Explain the meaning of the values of t for which the function is not defined.

CHAPTER 1 SUMMARY AND REVIEW

Important Properties and Formulas

Trigonometric Function Values of an Acute Angle θ

Let θ be an acute angle of a right triangle. The six
trigonometric functions of θ are as follows:

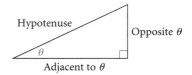

$$\sin \theta = \frac{\text{opp}}{\text{hyp}}, \qquad \cos \theta = \frac{\text{adj}}{\text{hyp}}, \qquad \tan \theta = \frac{\text{opp}}{\text{adj}},$$

$$\csc \theta = \frac{\text{hyp}}{\text{opp}}, \qquad \sec \theta = \frac{\text{hyp}}{\text{adj}}, \qquad \cot \theta = \frac{\text{adj}}{\text{opp}}.$$

Reciprocal Functions

$$\csc \theta = \frac{1}{\sin \theta}, \qquad \sec \theta = \frac{1}{\cos \theta}, \qquad \cot \theta = \frac{1}{\tan \theta}$$

Function Values of Special Angles

	0°	30°	45°	60°	90°
sin	0	1/2	$\sqrt{2}/2$	$\sqrt{3}/2$	1
cos	1	$\sqrt{3}/2$	$\sqrt{2}/2$	1/2	0
tan	0	$\sqrt{3}/3$	1	$\sqrt{3}$	Not defined

Cofunction Identities

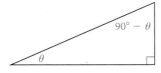

$$\sin \theta = \cos (90° - \theta), \qquad \cos \theta = \sin (90° - \theta),$$
$$\tan \theta = \cot (90° - \theta), \qquad \cot \theta = \tan (90° - \theta),$$
$$\sec \theta = \csc (90° - \theta), \qquad \csc \theta = \sec (90° - \theta)$$

Trigonometric Functions of Any Angle θ

If $P(x, y)$ is any point on the terminal side of any angle θ in standard position,
and r is the distance from the origin to $P(x, y)$, where $r = \sqrt{x^2 + y^2}$, then

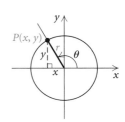

$$\sin \theta = \frac{y}{r}, \qquad \cos \theta = \frac{x}{r}, \qquad \tan \theta = \frac{y}{x},$$

$$\csc \theta = \frac{r}{y}, \qquad \sec \theta = \frac{r}{x}, \qquad \cot \theta = \frac{x}{y}.$$

Signs of Function Values

The signs of the function values depend only on the coordinates of the point P on the terminal side of an angle.

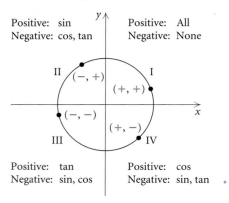

Positive: sin   Positive: All
Negative: cos, tan   Negative: None

Positive: tan   Positive: cos
Negative: sin, cos   Negative: sin, tan

Radian–Degree Equivalents

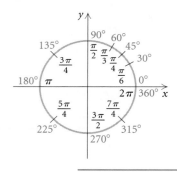

Linear Speed in Terms of Angular Speed

$$v = r\omega$$

Basic Circular Functions

For a real number s that determines a point (x, y) on the unit circle:

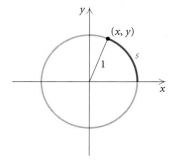

$$\sin s = y,$$
$$\cos s = x,$$
$$\tan s = \frac{y}{x}.$$

Sine is an odd function: $\sin(-s) = -\sin s$
Cosine is an even function: $\cos(-s) = \cos s$

Transformations of Sine and Cosine Functions

To graph $y = A\sin(Bx - C) + D$ and $y = A\cos(Bx - C) + D$:

1. Stretch or shrink the graph horizontally according to B. $\left(\text{Period} = \left|\dfrac{2\pi}{B}\right|\right)$

2. Stretch or shrink the graph vertically according to A. (Amplitude = $|A|$)

3. Translate the graph horizontally according to C/B. $\left(\text{Phase shift} = \dfrac{C}{B}\right)$

4. Translate the graph vertically according to D.

REVIEW EXERCISES

Determine whether the statement is true or false.

1. Given that $(-a, b)$ is a point on the unit circle and θ is in the second quadrant, then $\cos \theta$ is a.   [1.3]

2. The lengths of corresponding sides in similar triangles are in the same ratio.   [1.1]

3. The measure 300° is greater than the measure 5 radians.   [1.4]

4. If $\sec \theta > 0$ and $\cot \theta < 0$, then θ is in the fourth quadrant.   [1.3]

5. The amplitude of $y = \frac{1}{2} \sin x$ is twice as large as the amplitude of $y = \sin \frac{1}{2}x$.   [1.6]

6. The supplement of $\frac{9}{13}\pi$ is greater than the complement of $\frac{\pi}{6}$.   [1.3]

7. Find the six trigonometric function values of θ.   [1.1]

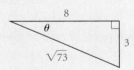

8. Given that β is acute and $\sin \beta = \frac{\sqrt{91}}{10}$, find the other five trigonometric function values.   [1.1]

Find the exact function value, if it exists.

9. $\cos 45°$ [1.1]   10. $\cot 60°$ [1.1]

11. $\cos 495°$ [1.3]   12. $\sin 150°$ [1.3]

13. $\sec (-270°)$ [1.3]   14. $\tan (-600°)$ [1.3]

15. $\csc 60°$ [1.1]   16. $\cot (-45°)$ [1.1]

17. Convert 22.27° to degrees, minutes, and seconds. Round to the nearest second.   [1.1]

18. Convert 47°33′27″ to decimal degree notation. Round to two decimal places.   [1.1]

Find the function value. Round to four decimal places.   [1.3]

19. $\tan 2184°$ 20. $\sec 27.9°$

21. $\cos 18°13′42″$ 22. $\sin 245°24′$

23. $\cot (-33.2°)$ 24. $\sin 556.13°$

Find θ in the interval indicated. Round the answer to the nearest tenth of a degree.   [1.3]

25. $\cos \theta = -0.9041$, $(180°, 270°)$

26. $\tan \theta = 1.0799$, $(0°, 90°)$

Find the exact acute angle θ, in degrees, given the function value.   [1.1]

27. $\sin \theta = \dfrac{\sqrt{3}}{2}$ 28. $\tan \theta = \sqrt{3}$

29. $\cos \theta = \dfrac{\sqrt{2}}{2}$ 30. $\sec \theta = \dfrac{2\sqrt{3}}{3}$

31. Given that $\sin 59.1° \approx 0.8581$, $\cos 59.1° \approx 0.5135$, and $\tan 59.1° \approx 1.6709$, find the six function values for 30.9°.   [1.1]

Solve each of the following right triangles. Standard lettering has been used.   [1.2]

32. $a = 7.3$, $c = 8.6$

33. $a = 30.5$, $B = 51.17°$

34. One leg of a right triangle bears east. The hypotenuse is 734 m long and bears N57°23′E. Find the perimeter of the triangle.

35. An observer's eye is 6 ft above the floor. A mural is being viewed. The bottom of the mural is at floor level. The observer looks down 13° to see the bottom and up 17° to see the top. How tall is the mural?

For angles of the following measures, state in which quadrant the terminal side lies.   [1.3]

36. $142°11′5″$ 37. $-635.2°$

38. $-392°$

Find a positive angle and a negative angle that are coterminal with the given angle. Answers may vary.

39. 65°   [1.3]

40. $\dfrac{7\pi}{3}$ [1.4]

Find the complement and the supplement.

41. 13.4°   [1.3]

42. $\dfrac{\pi}{6}$ [1.4]

43. Find the six trigonometric function values for the angle θ shown.   [1.3]

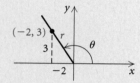

44. Given that $\tan \theta = 2/\sqrt{5}$ and that the terminal side is in quadrant III, find the other five function values.   [1.3]

45. An airplane travels at 530 mph for $3\frac{1}{2}$ hr in a direction of 160° from Minneapolis, Minnesota. At the end of that time, how far south of Minneapolis is the airplane?   [1.3]

46. On a unit circle, mark and label the points determined by $7\pi/6$, $-3\pi/4$, $-\pi/3$, and $9\pi/4$.   [1.4]

For angles of the following measures, convert to radian measure in terms of π, and convert to radian measure not in terms of π. Round the answer to two decimal places.   [1.4]

47. 145.2°

48. −30°

Convert to degree measure. Round the answer to two decimal places.   [1.4]

49. $\dfrac{3\pi}{2}$

50. 3

51. −4.5

52. 11π

53. Find the length of an arc of a circle, given a central angle of $\pi/4$ and a radius of 7 cm.   [1.4]

54. An arc 18 m long on a circle of radius 8 m subtends an angle of how many radians? how many degrees, to the nearest degree?   [1.4]

55. At one time, inside La Madeleine French Bakery and Cafe in Houston, Texas, there was one of the few remaining working watermills in the world. The 300-year-old French-built waterwheel had a radius of 7 ft and made one complete revolution in 70 sec. (*Source*: La Madeleine French Bakery and Cafe, Houston, TX) What was the linear speed, in feet per minute, of a point on the rim?   [1.4]

56. An automobile wheel has a diameter of 14 in. If the car travels at a speed of 55 mph, what is the angular velocity, in radians per hour, of a point on the edge of the wheel?   [1.4]

57. The point $\left(\frac{3}{5}, -\frac{4}{5}\right)$ is on a unit circle. Find the coordinates of its reflections across the *x*-axis, the *y*-axis, and the origin.   [1.5]

Find the exact function value, if it exists.   [1.5]

58. $\cos \pi$

59. $\tan \dfrac{5\pi}{4}$

60. $\sin \dfrac{5\pi}{3}$

61. $\sin \left(-\dfrac{7\pi}{6}\right)$

62. $\tan \dfrac{\pi}{6}$

63. $\cos \left(-13\pi\right)$

Find the function value. Round to four decimal places.   [1.5]

64. $\sin 24$

65. $\cos \left(-75\right)$

66. $\cot 16\pi$

67. $\tan \dfrac{3\pi}{7}$

68. $\sec 14.3$

69. $\cos \left(-\dfrac{\pi}{5}\right)$

70. Graph each of the six trigonometric functions from -2π to 2π.   [1.5]

71. What is the period of each of the six trigonometric functions?   [1.5]

72. Complete the following table.   [1.5]

Function	Domain	Range
sine		
cosine		
tangent		

73. Complete the following table with the sign of the specified trigonometric function value in each of the four quadrants.   [1.3]

Function	I	II	III	IV
sine				
cosine				
tangent				

Determine the amplitude, the period, and the phase shift of the function, and sketch the graph of the function.   [1.6]

74. $y = \sin\left(x + \dfrac{\pi}{2}\right)$

75. $y = 3 + \dfrac{1}{2}\cos\left(2x - \dfrac{\pi}{2}\right)$

In Exercises 76–79, match the function with one of the graphs (a)–(d), which follow.   [1.6]

a)
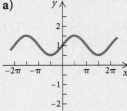

b)
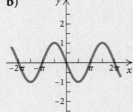

c)
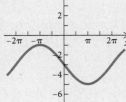

d)
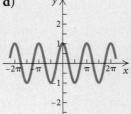

76. $y = \cos 2x$

77. $y = \dfrac{1}{2}\sin x + 1$

78. $y = -2\sin\dfrac{1}{2}x - 3$

79. $y = -\cos\left(x - \dfrac{\pi}{2}\right)$

80. Sketch a graph of $y = 3\cos x + \sin x$ for values of x between 0 and 2π.   [1.6]

81. Graph: $f(x) = e^{-0.7x}\cos x$.   [1.6]

82. Which of the following is the reflection of $\left(-\dfrac{1}{2}, \dfrac{\sqrt{3}}{2}\right)$ across the y-axis?   [1.5]

A. $\left(\dfrac{1}{2}, -\dfrac{\sqrt{3}}{2}\right)$ B. $\left(\dfrac{\sqrt{3}}{2}, \dfrac{1}{2}\right)$

C. $\left(\dfrac{1}{2}, \dfrac{\sqrt{3}}{2}\right)$ D. $\left(\dfrac{\sqrt{3}}{2}, -\dfrac{1}{2}\right)$

83. Which of the following is the domain of the cosine function?   [1.5]

A. $(-1, 1)$ B. $(-\infty, \infty)$

C. $[0, \infty)$ D. $[-1, 1]$

Collaborative Discussion and Writing

84. Compare the terms radian and degree. [1.1], [1.4]

85. Describe the shape of the graph of the cosine function. How many maximum values are there of the cosine function? Where do they occur?   [1.5]

86. Does $5\sin x = 7$ have a solution for x? Why or why not?   [1.5]

Synthesis

87. Graph $y = 3\sin(x/2)$, and determine the domain, the range, and the period.   [1.6]

88. In the graph below, $y_1 = \sin x$ is shown and y_2 is shown in red. Express y_2 as a transformation of the graph of y_1.   [1.6]

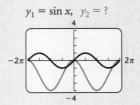

89. Find the domain of $y = \log(\cos x)$.   [1.6]

90. Given that $\sin x = 0.6144$ and that the terminal side is in quadrant II, find the other basic circular function values.   [1.3]

CHAPTER 1 TEST

1. Find the six trigonometric function values of θ.

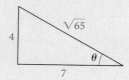

Find the exact function value, if it exists.

2. $\sin 120°$

3. $\tan (-45°)$

4. $\cos 3\pi$

5. $\sec \dfrac{5\pi}{4}$

6. Convert $38°27'56''$ to decimal degree notation. Round to two decimal places.

Find the function values. Round to four decimal places.

7. $\tan 526.4°$

8. $\sin (-12°)$

9. $\sec \dfrac{5\pi}{9}$

10. $\cos 76.07$

11. Find the exact acute angle θ, in degrees, for which $\sin \theta = \frac{1}{2}$.

12. Given that $\sin 28.4° \approx 0.4756$, $\cos 28.4° \approx 0.8796$, and $\tan 28.4° \approx 0.5407$, find the six trigonometric function values for $61.6°$.

13. Solve the right triangle with $b = 45.1$ and $A = 35.9°$. Standard lettering has been used.

14. Find a positive angle and a negative angle coterminal with a $112°$ angle.

15. Find the supplement of $\dfrac{5\pi}{6}$.

16. Given that $\sin \theta = -4/\sqrt{41}$ and that the terminal side is in quadrant IV, find the other five trigonometric function values.

17. Convert $210°$ to radian measure in terms of π.

18. Convert $\dfrac{3\pi}{4}$ to degree measure.

19. Find the length of an arc of a circle given a central angle of $\pi/3$ and a radius of 16 cm.

Consider the function $y = -\sin (x - \pi/2) + 1$ for Exercises 20–23.

20. Find the amplitude.

21. Find the period.

22. Find the phase shift.

23. Which is the graph of the function?

a)

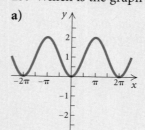

b)

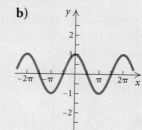

c)

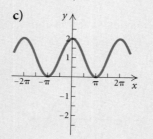

d)

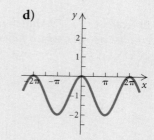

24. *Height of a Kite.* The angle of elevation of a kite is $65°$ with 490 ft of string out. Assuming the string is taut, how high is the kite?

25. *Location.* A pickup-truck camper travels at 50 mph for 6 hr in a direction of $115°$ from Flagstaff, Arizona. At the end of that time, how far east of Flagstaff is the camper?

26. *Linear Speed.* A ferris wheel has a radius of 6 m and revolves at 1.5 rpm. What is the linear speed, in meters per minute?

27. Graph: $f(x) = \frac{1}{2}x^2 \sin x$.

Synthesis

28. Determine the domain of $f(x) = \dfrac{-3}{\sqrt{\cos x}}$.

Trigonometric Identities, Inverse Functions, and Equations

2.1 Identities: Pythagorean and Sum and Difference

2.2 Identities: Cofunction, Double-Angle, and Half-Angle

2.3 Proving Trigonometric Identities

2.4 Inverses of the Trigonometric Functions

2.5 Solving Trigonometric Equations

APPLICATION

For a rope course and climbing wall, a guy wire is attached 47 ft high on a vertical pole. Another guy wire is attached 40 ft above the ground on the same pole. (*Source: Experiential Resources, Inc., Todd Domeck, Owner*) Find the angle between the wires if they are attached to the ground 50 ft from the pole.

This problem appears as Exercise 89 in Section 2.1.

2.1

Identities: Pythagorean and Sum and Difference

✦ State the Pythagorean identities.

✦ Simplify and manipulate expressions containing trigonometric expressions.

✦ Use the sum and difference identities to find function values.

An **identity** is an equation that is true for all *possible* replacements of the variables. The following is a list of the identities studied in Chapter 1.

Basic Identities

$$\sin x = \frac{1}{\csc x}, \qquad \csc x = \frac{1}{\sin x}, \qquad \sin(-x) = -\sin x,$$
$$\cos(-x) = \cos x,$$

$$\cos x = \frac{1}{\sec x}, \qquad \sec x = \frac{1}{\cos x}, \qquad \tan(-x) = -\tan x,$$

$$\tan x = \frac{1}{\cot x}, \qquad \cot x = \frac{1}{\tan x}, \qquad \tan x = \frac{\sin x}{\cos x},$$

$$\cot x = \frac{\cos x}{\sin x}$$

In this section, we will develop some other important identities.

✦ Pythagorean Identities

We now consider three other identities that are fundamental to a study of trigonometry. They are called the *Pythagorean identities*. Recall that the equation of a unit circle in the *xy*-plane is

$$x^2 + y^2 = 1.$$

For any point on the unit circle, the coordinates x and y satisfy this equation. Suppose that a real number s determines a point on the unit circle with coordinates (x, y), or $(\cos s, \sin s)$. Then $x = \cos s$ and $y = \sin s$. Substituting $\cos s$ for x and $\sin s$ for y in the equation of the unit circle gives us the identity

$$(\cos s)^2 + (\sin s)^2 = 1, \qquad \text{Substituting } \cos s \text{ for } x \text{ and } \sin s \text{ for } y$$

which can be expressed as

$$\sin^2 s + \cos^2 s = 1.$$

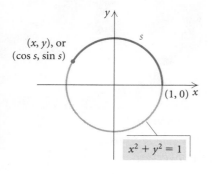

(x, y), or $(\cos s, \sin s)$

$x^2 + y^2 = 1$

It is conventional in trigonometry to use the notation $\sin^2 s$ rather than $(\sin s)^2$. Note that $\sin^2 s \neq \sin s^2$.

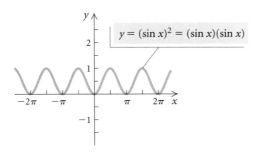

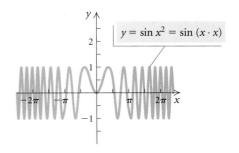

The identity $\sin^2 s + \cos^2 s = 1$ gives a relationship between the sine and the cosine of any real number s. It is an important **Pythagorean identity.**

We can divide by $\sin^2 s$ on both sides of the preceding identity:

$$\frac{\sin^2 s}{\sin^2 s} + \frac{\cos^2 s}{\sin^2 s} = \frac{1}{\sin^2 s}. \qquad \text{Dividing by } \sin^2 s$$

Simplifying gives us a second identity:

$$1 + \cot^2 s = \csc^2 s.$$

This equation is true for any replacement of s with a real number for which $\sin^2 s \neq 0$, since we divided by $\sin^2 s$. But the numbers for which $\sin^2 s = 0$ (or $\sin s = 0$) are exactly the ones for which the cotangent and cosecant functions are not defined. Hence our new equation holds for all real numbers s for which $\cot s$ and $\csc s$ are defined and is thus an identity.

The third Pythagorean identity can be obtained by dividing by $\cos^2 s$ on both sides of the first Pythagorean identity:

$$\frac{\sin^2 s}{\cos^2 s} + \frac{\cos^2 s}{\cos^2 s} = \frac{1}{\cos^2 s} \qquad \text{Dividing by } \cos^2 s$$

$$\tan^2 s + 1 = \sec^2 s. \qquad \text{Simplifying}$$

The identities we have developed hold no matter what symbols are used for the variables. For example, we could write $\sin^2 s + \cos^2 s = 1$, $\sin^2 \theta + \cos^2 \theta = 1$, or $\sin^2 x + \cos^2 x = 1$.

Pythagorean Identities

$$\sin^2 x + \cos^2 x = 1,$$
$$1 + \cot^2 x = \csc^2 x,$$
$$1 + \tan^2 x = \sec^2 x$$

It is often helpful to express the Pythagorean identities in equivalent forms.

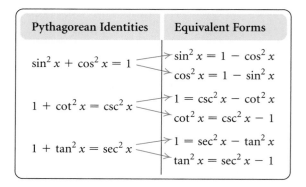

Pythagorean Identities	Equivalent Forms
$\sin^2 x + \cos^2 x = 1$	$\sin^2 x = 1 - \cos^2 x$ $\cos^2 x = 1 - \sin^2 x$
$1 + \cot^2 x = \csc^2 x$	$1 = \csc^2 x - \cot^2 x$ $\cot^2 x = \csc^2 x - 1$
$1 + \tan^2 x = \sec^2 x$	$1 = \sec^2 x - \tan^2 x$ $\tan^2 x = \sec^2 x - 1$

✦ Simplifying Trigonometric Expressions

We can factor, simplify, and manipulate trigonometric expressions in the same way that we manipulate strictly algebraic expressions.

EXAMPLE 1 Multiply and simplify: $\cos x (\tan x - \sec x)$.

Solution

$\cos x (\tan x - \sec x)$

$\qquad = \cos x \tan x - \cos x \sec x$ Multiplying

$\qquad = \cos x \dfrac{\sin x}{\cos x} - \cos x \dfrac{1}{\cos x}$ Recalling the identities $\tan x = \dfrac{\sin x}{\cos x}$ and $\sec x = \dfrac{1}{\cos x}$ and substituting

$\qquad = \sin x - 1$ Simplifying   ▶ Now Try Exercise 3.

There is no general procedure for manipulating trigonometric expressions, but it is often helpful to write everything in terms of sines and cosines, as we did in Example 1. We also look for the Pythagorean identity, $\sin^2 x + \cos^2 x = 1$, within a trigonometric expression.

EXAMPLE 2 Factor and simplify: $\sin^2 x \cos^2 x + \cos^4 x$.

Solution

$\sin^2 x \cos^2 x + \cos^4 x$

$\qquad = \cos^2 x (\sin^2 x + \cos^2 x)$ Removing a common factor

$\qquad = \cos^2 x \cdot (1)$ Using $\sin^2 x + \cos^2 x = 1$

$\qquad = \cos^2 x$ ▶ Now Try Exercise 5.

A graphing calculator can be used to perform a partial check of an identity. First, we graph the expression on the left side of the equals sign. Then we graph the expression on the right side using the same screen. If the two graphs are indistinguishable, then we have a partial verification that the equation is an identity. Of course, we can never see the entire graph, so there can always be some doubt. Also, the graphs may not overlap precisely, but you may not be able to tell because the difference between the graphs may be less than the width of a pixel. However, if the graphs are obviously different, we know that a mistake has been made.

Consider the identity in Example 1:

$$\cos x \,(\tan x - \sec x) = \sin x - 1.$$

Recalling that $\sec x = 1/\cos x$, we enter

$$y_1 = \cos x \,[\tan x - (1/\cos x)] \quad \text{and} \quad y_2 = \sin x - 1.$$

To graph, we first select SEQUENTIAL mode. Then we select the "line"-graph style for y_1 and the "path"-graph style, denoted by ─◯, for y_2. The calculator will graph y_1 first. Then it will graph y_2 as the circular cursor traces the leading edge of the graph, allowing us to determine whether the graphs coincide. As you can see in the first screen on the left, the graphs appear to be identical. Thus, $\cos x \,(\tan x - \sec x) = \sin x - 1$ is most likely an identity.

The TABLE feature can also be used to check identities. Note in the table at left that the function values are the same except for those values of x for which $\cos x = 0$. The domain of y_1 excludes these values. The domain of y_2 is the set of all real numbers. Thus all real numbers except $\pm\pi/2, \pm3\pi/2, \pm5\pi/2, \ldots$ are possible replacements for x in the identity. Recall that an identity is an equation that is true for all *possible* replacements.

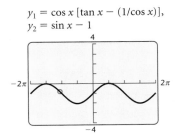

$y_1 = \cos x \,[\tan x - (1/\cos x)],$
$y_2 = \sin x - 1$

X	Y1	Y2
−6.283	−1	−1
−5.498	−.2929	−.2929
−4.712	ERROR	0
−3.927	−.2929	−.2929
−3.142	−1	−1
−2.356	−1.707	−1.707
−1.571	ERROR	−2

X = −6.28318530718

TblStart = -2π
ΔTbl = $\pi/4$

EXAMPLE 3   Simplify each of the following trigonometric expressions.

a) $\dfrac{\cot (-\theta)}{\csc (-\theta)}$

b) $\dfrac{2 \sin^2 t + \sin t - 3}{1 - \cos^2 t - \sin t}$

Solution

a) $\dfrac{\cot (-\theta)}{\csc (-\theta)} = \dfrac{\dfrac{\cos (-\theta)}{\sin (-\theta)}}{\dfrac{1}{\sin (-\theta)}}$ **Rewriting in terms of sines and cosines**

$\qquad\qquad = \dfrac{\cos (-\theta)}{\sin (-\theta)} \cdot \sin (-\theta)$ **Multiplying by the reciprocal**

$\qquad\qquad = \cos (-\theta) = \cos \theta$ **The cosine function is even.**

b) $\dfrac{2\sin^2 t + \sin t - 3}{1 - \cos^2 t - \sin t}$

$\quad = \dfrac{2\sin^2 t + \sin t - 3}{\sin^2 t - \sin t}$ Substituting $\sin^2 t$ for $1 - \cos^2 t$

$\quad = \dfrac{(2\sin t + 3)(\sin t - 1)}{\sin t\,(\sin t - 1)}$ Factoring in both numerator and denominator

$\quad = \dfrac{2\sin t + 3}{\sin t}$ Simplifying

$\quad = \dfrac{2\sin t}{\sin t} + \dfrac{3}{\sin t}$

$\quad = 2 + \dfrac{3}{\sin t}, \quad \text{or} \quad 2 + 3\csc t$

▶ Now Try Exercises 17 and 19.

We can add and subtract trigonometric rational expressions in the same way that we do algebraic expressions.

EXAMPLE 4 Add and simplify: $\dfrac{\cos x}{1 + \sin x} + \tan x$.

Solution

$\dfrac{\cos x}{1 + \sin x} + \tan x = \dfrac{\cos x}{1 + \sin x} + \dfrac{\sin x}{\cos x}$ Using $\tan x = \dfrac{\sin x}{\cos x}$

$\quad = \dfrac{\cos x}{1 + \sin x} \cdot \dfrac{\cos x}{\cos x} + \dfrac{\sin x}{\cos x} \cdot \dfrac{1 + \sin x}{1 + \sin x}$

  Multiplying by forms of 1

$\quad = \dfrac{\cos^2 x + \sin x + \sin^2 x}{\cos x\,(1 + \sin x)}$ Adding

$\quad = \dfrac{1 + \sin x}{\cos x\,(1 + \sin x)}$ Using $\sin^2 x + \cos^2 x = 1$

$\quad = \dfrac{1}{\cos x}, \quad \text{or} \quad \sec x$ Simplifying

▶ Now Try Exercise 27.

When radicals occur, the use of absolute value is sometimes necessary, but it can be difficult to determine when to use it. In Examples 5 and 6, we will assume that all radicands are nonnegative. This means that the identities are meant to be confined to certain quadrants.

EXAMPLE 5 Multiply and simplify: $\sqrt{\sin^3 x \cos x} \cdot \sqrt{\cos x}$.

Solution

$\sqrt{\sin^3 x \cos x} \cdot \sqrt{\cos x} = \sqrt{\sin^3 x \cos^2 x}$

$\qquad\qquad = \sqrt{\sin^2 x \cos^2 x \sin x}$

$\qquad\qquad = \sin x \cos x \sqrt{\sin x}$

▶ Now Try Exercise 31.

EXAMPLE 6 Rationalize the denominator: $\sqrt{\dfrac{2}{\tan x}}$.

Solution

$$\sqrt{\frac{2}{\tan x}} = \sqrt{\frac{2}{\tan x} \cdot \frac{\tan x}{\tan x}}$$

$$= \sqrt{\frac{2\tan x}{\tan^2 x}}$$

$$= \frac{\sqrt{2\tan x}}{\tan x}$$

▶ Now Try Exercise 37.

Often in calculus, a substitution is a useful manipulation, as we show in the following example.

EXAMPLE 7 Express $\sqrt{9 + x^2}$ as a trigonometric function of θ without using radicals by letting $x = 3\tan\theta$. Assume that $0 < \theta < \pi/2$. Then find $\sin\theta$ and $\cos\theta$.

Solution We have

$$\sqrt{9 + x^2} = \sqrt{9 + (3\tan\theta)^2} \qquad \text{Substituting } 3\tan\theta \text{ for } x$$

$$= \sqrt{9 + 9\tan^2\theta}$$

$$= \sqrt{9(1 + \tan^2\theta)} \qquad \text{Factoring}$$

$$= \sqrt{9\sec^2\theta} \qquad \text{Using } 1 + \tan^2 x = \sec^2 x$$

$$= 3|\sec\theta| = 3\sec\theta. \qquad \text{For } 0 < \theta < \pi/2, \sec\theta > 0, \text{ so } |\sec\theta| = \sec\theta.$$

We can express $\sqrt{9 + x^2} = 3\sec\theta$ as

$$\sec\theta = \frac{\sqrt{9 + x^2}}{3}.$$

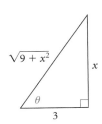

In a right triangle, we know that $\sec\theta$ is hypotenuse/adjacent, when θ is one of the acute angles. Using the Pythagorean theorem, we can determine that the side opposite θ is x. Then from the right triangle, we see that

$$\sin\theta = \frac{x}{\sqrt{9 + x^2}} \quad \text{and} \quad \cos\theta = \frac{3}{\sqrt{9 + x^2}}.$$

▶ Now Try Exercise 45.

◆ Sum and Difference Identities

We now develop some important identities involving sums or differences of two numbers (or angles), beginning with an identity for the cosine of the difference of two numbers. We use the letters u and v for these numbers.

Let's consider a real number u in the interval $[\pi/2, \pi]$ and a real number v in the interval $[0, \pi/2]$. These determine points A and B on the unit circle, as shown below. The arc length s is $u - v$, and we know that $0 \le s \le \pi$. Recall that the coordinates of A are $(\cos u, \sin u)$, and the coordinates of B are $(\cos v, \sin v)$.

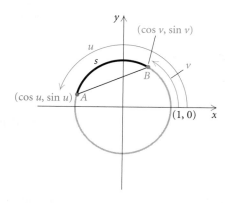

Using the distance formula $d = \sqrt{(x_2 - x_1)^2 + (y_2 - y_1)^2}$, we can write an expression for the distance AB:

$$AB = \sqrt{(\cos u - \cos v)^2 + (\sin u - \sin v)^2}$$

This can be simplified as follows:

$$AB = \sqrt{\cos^2 u - 2\cos u \cos v + \cos^2 v + \sin^2 u - 2\sin u \sin v + \sin^2 v}$$
$$= \sqrt{(\sin^2 u + \cos^2 u) + (\sin^2 v + \cos^2 v) - 2(\cos u \cos v + \sin u \sin v)}$$
$$= \sqrt{2 - 2(\cos u \cos v + \sin u \sin v)}.$$

Now let's imagine rotating the circle above so that point B is at $(1, 0)$. Although the coordinates of point A are now $(\cos s, \sin s)$, the distance AB has not changed.

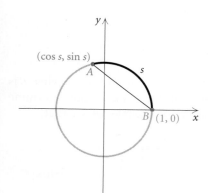

Again we use the distance formula to write an expression for the distance AB:

$$AB = \sqrt{(\cos s - 1)^2 + (\sin s - 0)^2}.$$

This can be simplified as follows:

$$AB = \sqrt{\cos^2 s - 2\cos s + 1 + \sin^2 s}$$
$$= \sqrt{(\sin^2 s + \cos^2 s) + 1 - 2\cos s}$$
$$= \sqrt{2 - 2\cos s}.$$

Equating our two expressions for AB, we obtain

$$\sqrt{2 - 2(\cos u \cos v + \sin u \sin v)} = \sqrt{2 - 2\cos s}.$$

Solving this equation for $\cos s$ gives

$$\cos s = \cos u \cos v + \sin u \sin v. \tag{1}$$

But $s = u - v$, so we have the equation

$$\cos(u - v) = \cos u \cos v + \sin u \sin v. \tag{2}$$

Formula (1) above holds when s is the length of the shortest arc from A to B. Given any real numbers u and v, the length of the shortest arc from A to B is not always $u - v$. In fact, it could be $v - u$. However, since $\cos(-x) = \cos x$, we know that $\cos(v - u) = \cos(u - v)$. Thus, $\cos s$ is always equal to $\cos(u - v)$. Formula (2) holds for all real numbers u and v. That formula is thus the identity we sought:

$$\cos(u - v) = \cos u \cos v + \sin u \sin v.$$

The cosine sum formula follows easily from the one we have just derived. Let's consider $\cos(u + v)$. This is equal to $\cos[u - (-v)]$, and by the identity above, we have

$$\cos(u + v) = \cos[u - (-v)]$$
$$= \cos u \cos(-v) + \sin u \sin(-v).$$

But $\cos(-v) = \cos v$ and $\sin(-v) = -\sin v$, so the identity we seek is the following:

$$\cos(u + v) = \cos u \cos v - \sin u \sin v.$$

EXAMPLE 8 Find $\cos(5\pi/12)$ exactly.

Solution We can express $5\pi/12$ as a difference of two numbers whose sine and cosine values are known:

$$\frac{5\pi}{12} = \frac{9\pi}{12} - \frac{4\pi}{12}, \quad \text{or} \quad \frac{3\pi}{4} - \frac{\pi}{3}.$$

Then, using $\cos(u - v) = \cos u \cos v + \sin u \sin v$, we have

$$\cos\frac{5\pi}{12} = \cos\left(\frac{3\pi}{4} - \frac{\pi}{3}\right) = \cos\frac{3\pi}{4}\cos\frac{\pi}{3} + \sin\frac{3\pi}{4}\sin\frac{\pi}{3}$$
$$= -\frac{\sqrt{2}}{2} \cdot \frac{1}{2} + \frac{\sqrt{2}}{2} \cdot \frac{\sqrt{3}}{2}$$
$$= -\frac{\sqrt{2}}{4} + \frac{\sqrt{6}}{4}$$
$$= \frac{\sqrt{6} - \sqrt{2}}{4}.$$

▶ Now Try Exercise 55.

TECHNOLOGY ⋯⋯⋯⋯⋯
CONNECTION

We can check the result of Example 8 using a graphing calculator set in RADIAN mode.

```
cos(5π/12)
            .2588190451
(√(6)−√(2))/4
            .2588190451
```

Consider $\cos(\pi/2 - \theta)$. We can use the identity for the cosine of a difference to simplify as follows:

$$\cos\left(\frac{\pi}{2} - \theta\right) = \cos\frac{\pi}{2}\cos\theta + \sin\frac{\pi}{2}\sin\theta$$
$$= 0 \cdot \cos\theta + 1 \cdot \sin\theta$$
$$= \sin\theta.$$

Thus we have developed the identity

$$\sin\theta = \cos\left(\frac{\pi}{2} - \theta\right). \qquad \begin{array}{l}\text{This cofunction identity first}\\\text{appeared in Section 1.1.}\end{array} \qquad (3)$$

This identity holds for any real number θ. From it we can obtain an identity for the cosine function. We first let α be any real number. Then we replace θ in $\sin \theta = \cos (\pi/2 - \theta)$ with $\pi/2 - \alpha$. This gives us

$$\sin \left(\frac{\pi}{2} - \alpha \right) = \cos \left[\frac{\pi}{2} - \left(\frac{\pi}{2} - \alpha \right) \right] = \cos \alpha,$$

which yields the identity

$$\cos \alpha = \sin \left(\frac{\pi}{2} - \alpha \right). \tag{4}$$

Using identities (3) and (4) and the identity for the cosine of a difference, we can obtain an identity for the sine of a sum. We start with identity (3) and substitute $u + v$ for θ:

$$\sin \theta = \cos \left(\frac{\pi}{2} - \theta \right) \qquad \text{Identity (3)}$$

$$\sin (u + v) = \cos \left[\frac{\pi}{2} - (u + v) \right] \qquad \text{Substituting } u + v \text{ for } \theta$$

$$= \cos \left[\left(\frac{\pi}{2} - u \right) - v \right]$$

$$= \cos \left(\frac{\pi}{2} - u \right) \cos v + \sin \left(\frac{\pi}{2} - u \right) \sin v$$

Using the identity for the cosine of a difference

$$= \sin u \cos v + \cos u \sin v. \qquad \text{Using identities (3) and (4)}$$

Thus the identity we seek is

$$\sin (u + v) = \sin u \cos v + \cos u \sin v.$$

To find a formula for the sine of a difference, we can use the identity just derived, substituting $-v$ for v:

$$\sin (u + (-v)) = \sin u \cos (-v) + \cos u \sin (-v).$$

Simplifying gives us

$$\sin (u - v) = \sin u \cos v - \cos u \sin v.$$

EXAMPLE 9   Find $\sin 105°$ exactly.

Solution   We express $105°$ as the sum of two measures:

$$105° = 45° + 60°.$$

Then

$$\sin 105° = \sin (45° + 60°)$$

$$= \sin 45° \cos 60° + \cos 45° \sin 60°$$

Using $\sin (u + v) = \sin u \cos v + \cos u \sin v$

$$= \frac{\sqrt{2}}{2} \cdot \frac{1}{2} + \frac{\sqrt{2}}{2} \cdot \frac{\sqrt{3}}{2} = \frac{\sqrt{2} + \sqrt{6}}{4}.$$

▶ **Now Try Exercise 51.**

Formulas for the tangent of a sum or a difference can be derived using identities already established. A summary of the sum and difference identities follows.

Sum and Difference Identities

$$\sin(u \pm v) = \sin u \cos v \pm \cos u \sin v,$$

$$\cos(u \pm v) = \cos u \cos v \mp \sin u \sin v,$$

$$\tan(u \pm v) = \frac{\tan u \pm \tan v}{1 \mp \tan u \tan v}$$

There are six identities here, half of them obtained by using the signs shown in color.

EXAMPLE 10 Find $\tan 15°$ exactly.

Solution We rewrite $15°$ as $45° - 30°$ and use the identity for the tangent of a difference:

$$\tan 15° = \tan(45° - 30°) = \frac{\tan 45° - \tan 30°}{1 + \tan 45° \tan 30°}$$

$$= \frac{1 - \sqrt{3}/3}{1 + 1 \cdot \sqrt{3}/3} = \frac{3 - \sqrt{3}}{3 + \sqrt{3}}.$$

▶ Now Try Exercise 53.

EXAMPLE 11 Assume that $\sin \alpha = \frac{2}{3}$ and $\sin \beta = \frac{1}{3}$ and that α and β are between 0 and $\pi/2$. Then evaluate $\sin(\alpha + \beta)$.

Solution Using the identity for the sine of a sum, we have

$$\sin(\alpha + \beta) = \sin \alpha \cos \beta + \cos \alpha \sin \beta$$

$$= \tfrac{2}{3} \cos \beta + \tfrac{1}{3} \cos \alpha.$$

To finish, we need to know the values of $\cos \beta$ and $\cos \alpha$. Using reference triangles and the Pythagorean theorem, we can determine these values from the diagrams:

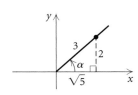

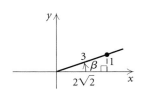

$$\cos \alpha = \frac{\sqrt{5}}{3} \quad \text{and} \quad \cos \beta = \frac{2\sqrt{2}}{3}.$$

Cosine values are positive in the first quadrant.

Substituting these values gives us

$$\sin(\alpha + \beta) = \frac{2}{3} \cdot \frac{2\sqrt{2}}{3} + \frac{1}{3} \cdot \frac{\sqrt{5}}{3}$$

$$= \frac{4}{9}\sqrt{2} + \frac{1}{9}\sqrt{5}, \quad \text{or} \quad \frac{4\sqrt{2} + \sqrt{5}}{9}.$$

▶ Now Try Exercise 65.

2.1 EXERCISE SET

Multiply and simplify.

1. $(\sin x - \cos x)(\sin x + \cos x)$

2. $\tan x (\cos x - \csc x)$

3. $\cos y \sin y (\sec y + \csc y)$

4. $(\sin x + \cos x)(\sec x + \csc x)$

5. $(\sin \phi - \cos \phi)^2$

6. $(1 + \tan x)^2$

7. $(\sin x + \csc x)(\sin^2 x + \csc^2 x - 1)$

8. $(1 - \sin t)(1 + \sin t)$

Factor and simplify.

9. $\sin x \cos x + \cos^2 x$

10. $\tan^2 \theta - \cot^2 \theta$

11. $\sin^4 x - \cos^4 x$

12. $4 \sin^2 y + 8 \sin y + 4$

13. $2 \cos^2 x + \cos x - 3$

14. $3 \cot^2 \beta + 6 \cot \beta + 3$

15. $\sin^3 x + 27$

16. $1 - 125 \tan^3 s$

Simplify.

17. $\dfrac{\sin^2 x \cos x}{\cos^2 x \sin x}$

18. $\dfrac{30 \sin^3 x \cos x}{6 \cos^2 x \sin x}$

19. $\dfrac{\sin^2 x + 2 \sin x + 1}{\sin x + 1}$

20. $\dfrac{\cos^2 \alpha - 1}{\cos \alpha + 1}$

21. $\dfrac{4 \tan t \sec t + 2 \sec t}{6 \tan t \sec t + 2 \sec t}$

22. $\dfrac{\csc (-x)}{\cot (-x)}$

23. $\dfrac{\sin^4 x - \cos^4 x}{\sin^2 x - \cos^2 x}$

24. $\dfrac{4 \cos^3 x}{\sin^2 x} \cdot \left(\dfrac{\sin x}{4 \cos x} \right)^2$

25. $\dfrac{5 \cos \phi}{\sin^2 \phi} \cdot \dfrac{\sin^2 \phi - \sin \phi \cos \phi}{\sin^2 \phi - \cos^2 \phi}$

26. $\dfrac{\tan^2 y}{\sec y} \div \dfrac{3 \tan^3 y}{\sec y}$

27. $\dfrac{1}{\sin^2 s - \cos^2 s} - \dfrac{2}{\cos s - \sin s}$

28. $\left(\dfrac{\sin x}{\cos x} \right)^2 - \dfrac{1}{\cos^2 x}$

29. $\dfrac{\sin^2 \theta - 9}{2 \cos \theta + 1} \cdot \dfrac{10 \cos \theta + 5}{3 \sin \theta + 9}$

30. $\dfrac{9 \cos^2 \alpha - 25}{2 \cos \alpha - 2} \cdot \dfrac{\cos^2 \alpha - 1}{6 \cos \alpha - 10}$

Simplify. Assume that all radicands are nonnegative.

31. $\sqrt{\sin^2 x \cos x} \cdot \sqrt{\cos x}$

32. $\sqrt{\cos^2 x \sin x} \cdot \sqrt{\sin x}$

33. $\sqrt{\cos \alpha \sin^2 \alpha} - \sqrt{\cos^3 \alpha}$

34. $\sqrt{\tan^2 x - 2 \tan x \sin x + \sin^2 x}$

35. $(1 - \sqrt{\sin y})(\sqrt{\sin y} + 1)$

36. $\sqrt{\cos \theta}(\sqrt{2 \cos \theta} + \sqrt{\sin \theta \cos \theta})$

Rationalize the denominator.

37. $\sqrt{\dfrac{\sin x}{\cos x}}$

38. $\sqrt{\dfrac{\cos x}{\tan x}}$

39. $\sqrt{\dfrac{\cos^2 y}{2 \sin^2 y}}$

40. $\sqrt{\dfrac{1 - \cos \beta}{1 + \cos \beta}}$

Rationalize the numerator.

41. $\sqrt{\dfrac{\cos x}{\sin x}}$

42. $\sqrt{\dfrac{\sin x}{\cot x}}$

43. $\sqrt{\dfrac{1 + \sin y}{1 - \sin y}}$

44. $\sqrt{\dfrac{\cos^2 x}{2 \sin^2 x}}$

Use the given substitution to express the given radical expression as a trigonometric function without radicals. Assume that a > 0 and 0 < θ < π/2. Then find expressions for the indicated trigonometric functions.

45. Let $x = a \sin \theta$ in $\sqrt{a^2 - x^2}$. Then find $\cos \theta$ and $\tan \theta$.

46. Let $x = 2 \tan \theta$ in $\sqrt{4 + x^2}$. Then find $\sin \theta$ and $\cos \theta$.

47. Let $x = 3 \sec \theta$ in $\sqrt{x^2 - 9}$. Then find $\sin \theta$ and $\cos \theta$.

48. Let $x = a \sec \theta$ in $\sqrt{x^2 - a^2}$. Then find $\sin \theta$ and $\cos \theta$.

Use the given substitution to express the given radical expression as a trigonometric function without radicals. Assume that 0 < θ < π/2.

49. Let $x = \sin \theta$ in $\dfrac{x^2}{\sqrt{1 - x^2}}$.

50. Let $x = 4 \sec \theta$ in $\dfrac{\sqrt{x^2 - 16}}{x^2}$.

Use the sum and difference identities to evaluate exactly.

51. $\sin \dfrac{\pi}{12}$

52. $\cos 75°$

53. $\tan 105°$

54. $\tan \dfrac{5\pi}{12}$

55. $\cos 15°$

56. $\sin \dfrac{7\pi}{12}$

First write each of the following as a trigonometric function of a single angle; then evaluate.

57. $\sin 37° \cos 22° + \cos 37° \sin 22°$

58. $\cos 83° \cos 53° + \sin 83° \sin 53°$

59. $\cos 19° \cos 5° - \sin 19° \sin 5°$

60. $\sin 40° \cos 15° - \cos 40° \sin 15°$

61. $\dfrac{\tan 20° + \tan 32°}{1 - \tan 20° \tan 32°}$

62. $\dfrac{\tan 35° - \tan 12°}{1 + \tan 35° \tan 12°}$

63. Derive the formula for the tangent of a sum.

64. Derive the formula for the tangent of a difference.

Assuming that $\sin u = \frac{3}{5}$ and $\sin v = \frac{4}{5}$ and that u and v are between 0 and π/2, evaluate each of the following exactly.

65. $\cos (u + v)$

66. $\tan (u - v)$

67. $\sin (u - v)$

68. $\cos (u - v)$

Assuming that $\sin \theta = 0.6249$ and $\cos \phi = 0.1102$ and that both θ and φ are first-quadrant angles, evaluate each of the following.

69. $\tan (\theta + \phi)$

70. $\sin (\theta - \phi)$

71. $\cos (\theta - \phi)$

72. $\cos (\theta + \phi)$

Simplify.

73. $\sin (\alpha + \beta) + \sin (\alpha - \beta)$

74. $\cos (\alpha + \beta) - \cos (\alpha - \beta)$

75. $\cos (u + v) \cos v + \sin (u + v) \sin v$

76. $\sin (u - v) \cos v + \cos (u - v) \sin v$

Technology Connection

77. Check your answers to each of Exercises 17–30 by graphing the original expression and the simplified result in the same window.

78. Check your solutions to each of Exercises 51–56 with a calculator.

Collaborative Discussion and Writing

79. What is the difference between a trigonometric equation that is an identity and a trigonometric equation that is not an identity? Give an example of each.

80. Why is it possible to use a graph to *disprove* that an equation is an identity but not to *prove* that one is?

Skill Maintenance

Solve.

81. $2x - 3 = 2\left(x - \frac{3}{2}\right)$

82. $x - 7 = x + 3.4$

Given that $\sin 31° = 0.5150$ and $\cos 31° = 0.8572$, find the specified function value.

83. $\sec 59°$

84. $\tan 59°$

Synthesis

Angles between Lines. *One of the identities gives an easy way to find an angle formed by two lines. Consider two lines with equations l_1: $y = m_1x + b_1$ and l_2: $y = m_2x + b_2$.*

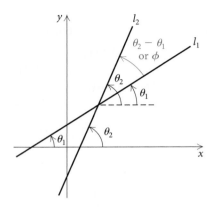

The slopes m_1 and m_2 are the tangents of the angles θ_1 and θ_2 that the lines form with the positive direction of the x-axis. Thus we have $m_1 = \tan\theta_1$ and $m_2 = \tan\theta_2$. To find the measure of $\theta_2 - \theta_1$, or ϕ, we proceed as follows:

$$\tan\phi = \tan(\theta_2 - \theta_1)$$
$$= \frac{\tan\theta_2 - \tan\theta_1}{1 + \tan\theta_2\tan\theta_1}$$
$$= \frac{m_2 - m_1}{1 + m_2m_1}.$$

This formula also holds when the lines are taken in the reverse order. When ϕ is acute, $\tan\phi$ will be positive. When ϕ is obtuse, $\tan\phi$ will be negative.

Find the measure of the angle from l_1 to l_2.

85. l_1: $2x = 3 - 2y$,
$\quad l_2$: $x + y = 5$

86. l_1: $3y = \sqrt{3}x + 3$,
$\quad l_2$: $y = \sqrt{3}x + 2$

87. l_1: $y = 3$,
$\quad l_2$: $x + y = 5$

88. l_1: $2x + y - 4 = 0$,
$\quad l_2$: $y - 2x + 5 = 0$

89. *Rope Course and Climbing Wall.* For a rope course and climbing wall, a guy wire R is attached 47 ft high on a vertical pole. Another guy wire S is attached 40 ft above the ground on the same pole. (*Source*: Experiential Resources, Inc., Todd Domeck, Owner) Find the angle α between the wires if they are attached to the ground 50 ft from the pole.

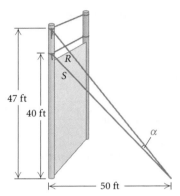

90. *Circus Guy Wire.* In a circus, a guy wire A is attached to the top of a 30-ft pole. Wire B is used for performers to walk up to the tight wire, 10 ft above the ground. Find the angle ϕ between the wires if they are attached to the ground 40 ft from the pole.

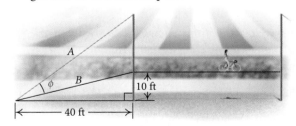

91. Given that $f(x) = \cos x$, show that
$$\frac{f(x + h) - f(x)}{h} = \cos x\left(\frac{\cos h - 1}{h}\right) - \sin x\left(\frac{\sin h}{h}\right).$$

92. Given that $f(x) = \sin x$, show that
$$\frac{f(x + h) - f(x)}{h} = \sin x\left(\frac{\cos h - 1}{h}\right) + \cos x\left(\frac{\sin h}{h}\right).$$

Show that each of the following is not an identity by finding a replacement or replacements for which the sides of the equation do not name the same number.

93. $\dfrac{\sin 5x}{x} = \sin 5$

94. $\sqrt{\sin^2 \theta} = \sin \theta$

95. $\cos (2\alpha) = 2 \cos \alpha$

96. $\sin (-x) = \sin x$

97. $\dfrac{\cos 6x}{\cos x} = 6$

98. $\tan^2 \theta + \cot^2 \theta = 1$

Find the slope of line l_1, where m_2 is the slope of line l_2 and ϕ is the smallest positive angle from l_1 to l_2.

99. $m_2 = \frac{2}{3}, \quad \phi = 30°$

100. $m_2 = \frac{4}{3}, \quad \phi = 45°$

101. Line l_1 contains the points $(-3, 7)$ and $(-3, -2)$. Line l_2 contains $(0, -4)$ and $(2, 6)$. Find the smallest positive angle from l_1 to l_2.

102. Line l_1 contains the points $(-2, 4)$ and $(5, -1)$. Find the slope of line l_2 such that the angle from l_1 to l_2 is $45°$.

103. Find an identity for $\cos 2\theta$. (*Hint:* $2\theta = \theta + \theta$.)

104. Find an identity for $\sin 2\theta$. (*Hint:* $2\theta = \theta + \theta$.)

Derive the identity.

105. $\tan \left(x + \dfrac{\pi}{4} \right) = \dfrac{1 + \tan x}{1 - \tan x}$

106. $\sin \left(x - \dfrac{3\pi}{2} \right) = \cos x$

107. $\sin (\alpha + \beta) + \sin (\alpha - \beta) = 2 \sin \alpha \cos \beta$

108. $\dfrac{\sin (\alpha + \beta)}{\cos (\alpha - \beta)} = \dfrac{\tan \alpha + \tan \beta}{1 + \tan \alpha \tan \beta}$

2.2

Identities: Cofunction, Double-Angle, and Half-Angle

◆ Use cofunction identities to derive other identities.

◆ Use the double-angle identities to find function values of twice an angle when one function value is known for that angle.

◆ Use the half-angle identities to find function values of half an angle when one function value is known for that angle.

◆ Simplify trigonometric expressions using the double-angle and half-angle identities.

✦ Cofunction Identities

Each of the identities listed below yields a conversion to a *cofunction*. For this reason, we call them cofunction identities.

Cofunction Identities

$$\sin \left(\frac{\pi}{2} - x \right) = \cos x, \qquad \cos \left(\frac{\pi}{2} - x \right) = \sin x,$$

$$\tan \left(\frac{\pi}{2} - x \right) = \cot x, \qquad \cot \left(\frac{\pi}{2} - x \right) = \tan x,$$

$$\sec \left(\frac{\pi}{2} - x \right) = \csc x, \qquad \csc \left(\frac{\pi}{2} - x \right) = \sec x$$

We verified the first two of these identities in Section 2.1. The other four can be proved using the first two and the definitions of the trigonometric functions. These identities hold for all real numbers, and thus, for all angle measures, but if we restrict θ to values such that $0° < \theta < 90°$, or $0 < \theta < \pi/2$, then we have a special application to the acute angles of a right triangle.

Comparing graphs can lead to possible identities. On the left below, we see that the graph of $y = \sin(x + \pi/2)$ is a translation of the graph of $y = \sin x$ to the left $\pi/2$ units. On the right, we see the graph of $y = \cos x$.

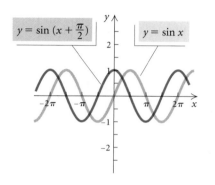
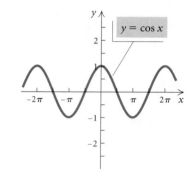

Comparing the graphs, we note a possible identity:

$$\sin\left(x + \frac{\pi}{2}\right) = \cos x.$$

The identity can be proved using the identity for the sine of a sum developed in Section 6.1.

EXAMPLE 1 Prove the identity $\sin(x + \pi/2) = \cos x$.

Solution

$$\sin\left(x + \frac{\pi}{2}\right) = \sin x \cos \frac{\pi}{2} + \cos x \sin \frac{\pi}{2} \qquad \text{Using } \sin(u + v) = \\ \sin u \cos v + \cos u \sin v$$
$$= \sin x \cdot 0 + \cos x \cdot 1$$
$$= \cos x \qquad \blacktriangleleft$$

We now state four more cofunction identities. These new identities that involve the sine and cosine functions can be verified using previously established identities as seen in Example 1.

Cofunction Identities for the Sine and Cosine

$$\sin\left(x \pm \frac{\pi}{2}\right) = \pm \cos x, \qquad \cos\left(x \pm \frac{\pi}{2}\right) = \mp \sin x$$

EXAMPLE 2 Find an identity for each of the following.

a) $\tan\left(x + \dfrac{\pi}{2}\right)$ **b)** $\sec(x - 90°)$

Solution

a) We have

$$\tan\left(x + \frac{\pi}{2}\right) = \frac{\sin\left(x + \dfrac{\pi}{2}\right)}{\cos\left(x + \dfrac{\pi}{2}\right)} \qquad \text{Using } \tan x = \frac{\sin x}{\cos x}$$

$$= \frac{\cos x}{-\sin x} \qquad \text{Using cofunction identities}$$

$$= -\cot x.$$

Thus the identity we seek is

$$\tan\left(x + \frac{\pi}{2}\right) = -\cot x.$$

b) We have

$$\sec(x - 90°) = \frac{1}{\cos(x - 90°)} = \frac{1}{\sin x} = \csc x.$$

Thus, $\sec(x - 90°) = \csc x.$ ▶ Now Try Exercises 5 and 7.

◆ Double-Angle Identities

If we double an angle of measure x, the new angle will have measure $2x$. **Double-angle identities** give trigonometric function values of $2x$ in terms of function values of x. To develop these identities, we will use the sum formulas from the preceding section. We first develop a formula for $\sin 2x$. Recall that

$$\sin(u + v) = \sin u \cos v + \cos u \sin v.$$

We will consider a number x and substitute it for both u and v in this identity. Doing so gives us

$$\sin(x + x) = \sin 2x$$

$$= \sin x \cos x + \cos x \sin x$$

$$= 2 \sin x \cos x.$$

Our first double-angle identity is thus

$$\textbf{sin } 2x = 2 \textbf{ sin } x \textbf{ cos } x.$$

TECHNOLOGY ·············
CONNECTION

Graphing calculators provide visual partial checks of identities. We can graph

$$y_1 = \sin 2x, \quad \text{and}$$
$$y_2 = 2 \sin x \cos x$$

using the "line"-graph style for y_1 and the "path"-graph style for y_2 and see that they appear to have the same graph. We can also use the TABLE feature.

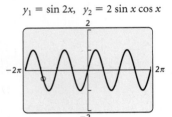

$y_1 = \sin 2x, \quad y_2 = 2 \sin x \cos x$

X	Y1	Y2
−6.283	2E-13	0
−5.498	1	1
−4.712	0	0
−3.927	−1	−1
−3.142	0	0
−2.356	1	1
−1.571	0	0

X = −1.57079632679

ΔTbl = π/4

Double-angle identities for the cosine and tangent functions can be derived in much the same way as the identity above:

$$\cos 2x = \cos^2 x - \sin^2 x, \qquad \tan 2x = \frac{2 \tan x}{1 - \tan^2 x}.$$

EXAMPLE 3 Given that $\tan \theta = -\frac{3}{4}$ and θ is in quadrant II, find each of the following.

a) $\sin 2\theta$ **b)** $\cos 2\theta$

c) $\tan 2\theta$ **d)** The quadrant in which 2θ lies

Solution By drawing a reference triangle as shown, we find that

$$\sin \theta = \frac{3}{5}$$

and

$$\cos \theta = -\frac{4}{5}.$$

Thus we have the following.

a) $\sin 2\theta = 2 \sin \theta \cos \theta = 2 \cdot \dfrac{3}{5} \cdot \left(-\dfrac{4}{5}\right) = -\dfrac{24}{25}$

b) $\cos 2\theta = \cos^2 \theta - \sin^2 \theta = \left(-\dfrac{4}{5}\right)^2 - \left(\dfrac{3}{5}\right)^2 = \dfrac{16}{25} - \dfrac{9}{25} = \dfrac{7}{25}$

c) $\tan 2\theta = \dfrac{2 \tan \theta}{1 - \tan^2 \theta} = \dfrac{2 \cdot \left(-\frac{3}{4}\right)}{1 - \left(-\frac{3}{4}\right)^2} = \dfrac{-\frac{3}{2}}{1 - \frac{9}{16}} = -\dfrac{3}{2} \cdot \dfrac{16}{7} = -\dfrac{24}{7}$

Note that $\tan 2\theta$ could have been found more easily in this case by simply dividing:

$$\tan 2\theta = \frac{\sin 2\theta}{\cos 2\theta} = \frac{-\frac{24}{25}}{\frac{7}{25}} = -\frac{24}{7}.$$

d) Since $\sin 2\theta$ is negative and $\cos 2\theta$ is positive, we know that 2θ is in quadrant IV.

▶ **Now Try Exercise 9.**

Two other useful identities for $\cos 2x$ can be derived easily, as follows.

$$\cos 2x = \cos^2 x - \sin^2 x \qquad\qquad \cos 2x = \cos^2 x - \sin^2 x$$
$$= (1 - \sin^2 x) - \sin^2 x \qquad\qquad = \cos^2 x - (1 - \cos^2 x)$$
$$= 1 - 2 \sin^2 x \qquad\qquad\qquad\qquad = 2 \cos^2 x - 1$$

> **Double-Angle Identities**
>
> $$\sin 2x = 2 \sin x \cos x, \qquad\qquad \cos 2x = \cos^2 x - \sin^2 x$$
> $$\tan 2x = \frac{2 \tan x}{1 - \tan^2 x} \qquad\qquad\qquad = 1 - 2 \sin^2 x$$
> $$\qquad\qquad\qquad\qquad\qquad\qquad = 2 \cos^2 x - 1$$

Solving the last two cosine double-angle identities for $\sin^2 x$ and $\cos^2 x$, respectively, we obtain two more identities:

$$\sin^2 x = \frac{1 - \cos 2x}{2} \quad \text{and} \quad \cos^2 x = \frac{1 + \cos 2x}{2}.$$

Using division and these two identities, we obtain the following useful identity:

$$\tan^2 x = \frac{1 - \cos 2x}{1 + \cos 2x}.$$

EXAMPLE 4   Find an equivalent expression for each of the following.

a) $\sin 3\theta$ in terms of function values of θ

b) $\cos^3 x$ in terms of function values of x or $2x$, raised only to the first power

Solution

a) $\sin 3\theta = \sin(2\theta + \theta)$
$$= \sin 2\theta \cos \theta + \cos 2\theta \sin \theta$$
$$= (2 \sin \theta \cos \theta) \cos \theta + (2 \cos^2 \theta - 1) \sin \theta$$
> Using $\sin 2\theta = 2 \sin \theta \cos \theta$ and $\cos 2\theta = 2 \cos^2 \theta - 1$
$$= 2 \sin \theta \cos^2 \theta + 2 \sin \theta \cos^2 \theta - \sin \theta$$
$$= 4 \sin \theta \cos^2 \theta - \sin \theta$$

We could also substitute $\cos^2 \theta - \sin^2 \theta$ or $1 - 2 \sin^2 \theta$ for $\cos 2\theta$. Each substitution leads to a different result, but all results are equivalent.

b) $\cos^3 x = \cos^2 x \cos x$
$$= \frac{1 + \cos 2x}{2} \cos x$$
$$= \frac{\cos x + \cos x \cos 2x}{2}$$

▶ Now Try Exercise 15.

✦ Half-Angle Identities

If we take half of an angle of measure x, the new angle will have measure $x/2$. **Half-angle identities** give trigonometric function values of $x/2$ in terms of function values of x. To develop these identities, we replace x with $x/2$ and take square roots. For example,

$$\sin^2 x = \frac{1 - \cos 2x}{2} \qquad \text{Solving the identity} \atop \cos 2x = 1 - 2\sin^2 x \text{ for } \sin^2 x$$

$$\sin^2 \frac{x}{2} = \frac{1 - \cos 2 \cdot \dfrac{x}{2}}{2} \qquad \text{Substituting } \frac{x}{2} \text{ for } x$$

$$\sin^2 \frac{x}{2} = \frac{1 - \cos x}{2}$$

$$\sin \frac{x}{2} = \pm \sqrt{\frac{1 - \cos x}{2}}. \qquad \text{Taking square roots}$$

The formula is called a *half-angle formula*. The use of $+$ and $-$ depends on the quadrant in which the angle $x/2$ lies. Half-angle identities for the cosine and tangent functions can be derived in a similar manner. Two additional formulas for the half-angle tangent identity are listed below.

Half-Angle Identities

$$\sin \frac{x}{2} = \pm \sqrt{\frac{1 - \cos x}{2}},$$

$$\cos \frac{x}{2} = \pm \sqrt{\frac{1 + \cos x}{2}},$$

$$\tan \frac{x}{2} = \pm \sqrt{\frac{1 - \cos x}{1 + \cos x}}$$

$$= \frac{\sin x}{1 + \cos x} = \frac{1 - \cos x}{\sin x}$$

EXAMPLE 5   Find $\tan(\pi/8)$ exactly.

Solution   We have

$$\tan \frac{\pi}{8} = \tan \frac{\dfrac{\pi}{4}}{2} = \frac{\sin \dfrac{\pi}{4}}{1 + \cos \dfrac{\pi}{4}} = \frac{\dfrac{\sqrt{2}}{2}}{1 + \dfrac{\sqrt{2}}{2}} = \frac{\dfrac{\sqrt{2}}{2}}{\dfrac{2 + \sqrt{2}}{2}}$$

$$= \frac{\sqrt{2}}{2 + \sqrt{2}} = \frac{\sqrt{2}}{2 + \sqrt{2}} \cdot \frac{2 - \sqrt{2}}{2 - \sqrt{2}}$$

$$= \sqrt{2} - 1. \qquad \qquad \text{▶ Now Try Exercise 21.}$$

The identities that we have developed are also useful for simplifying trigonometric expressions.

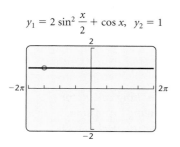

$y_1 = 2 \sin^2 \dfrac{x}{2} + \cos x, \quad y_2 = 1$

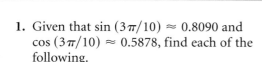

X	Y1	Y2
−6.283	1	1
−5.498	1	1
−4.712	1	1
−3.927	1	1
−3.142	1	1
−2.356	1	1
−1.571	1	1

X = −6.28318530718

ΔTbl = π/4

EXAMPLE 6 Simplify each of the following.

a) $\dfrac{\sin x \cos x}{\frac{1}{2} \cos 2x}$

b) $2 \sin^2 \dfrac{x}{2} + \cos x$

Solution

a) We can obtain $2 \sin x \cos x$ in the numerator by multiplying the expression by $\frac{2}{2}$:

$$\frac{\sin x \cos x}{\frac{1}{2} \cos 2x} = \frac{2}{2} \cdot \frac{\sin x \cos x}{\frac{1}{2} \cos 2x} = \frac{2 \sin x \cos x}{\cos 2x}$$

$$= \frac{\sin 2x}{\cos 2x} \qquad \text{Using } \sin 2x = 2 \sin x \cos x$$

$$= \tan 2x.$$

b) We have

$$2 \sin^2 \frac{x}{2} + \cos x = 2\left(\frac{1 - \cos x}{2} \right) + \cos x$$

$$\text{Using } \sin \frac{x}{2} = \pm \sqrt{\frac{1 - \cos x}{2}}, \text{ or } \sin^2 \frac{x}{2} = \frac{1 - \cos x}{2}$$

$$= 1 - \cos x + \cos x$$

$$= 1. \qquad \blacktriangleright \text{ Now Try Exercise 29.}$$

2.2   EXERCISE SET

1. Given that $\sin (3\pi/10) \approx 0.8090$ and $\cos (3\pi/10) \approx 0.5878$, find each of the following.

   a) The other four function values for $3\pi/10$
   b) The six function values for $\pi/5$

2. Given that

$$\sin \frac{\pi}{12} = \frac{\sqrt{2 - \sqrt{3}}}{2} \quad \text{and} \quad \cos \frac{\pi}{12} = \frac{\sqrt{2 + \sqrt{3}}}{2},$$

   find exact answers for each of the following.

   a) The other four function values for $\pi/12$
   b) The six function values for $5\pi/12$

3. Given that $\sin \theta = \frac{1}{3}$ and that the terminal side is in quadrant II, find exact answers for each of the following.

   a) The other function values for θ

   b) The six function values for $\pi/2 - \theta$
   c) The six function values for $\theta - \pi/2$

4. Given that $\cos \phi = \frac{4}{5}$ and that the terminal side is in quadrant IV, find exact answers for each of the following.

   a) The other function values for ϕ
   b) The six function values for $\pi/2 - \phi$
   c) The six function values for $\phi + \pi/2$

Find an equivalent expression for each of the following.

5. $\sec \left(x + \dfrac{\pi}{2} \right)$

6. $\cot \left(x - \dfrac{\pi}{2} \right)$

7. $\tan \left(x - \dfrac{\pi}{2} \right)$

8. $\csc \left(x + \dfrac{\pi}{2} \right)$

Find the exact value of $\sin 2\theta$, $\cos 2\theta$, $\tan 2\theta$, *and the quadrant in which* 2θ *lies.*

9. $\sin \theta = \dfrac{4}{5}$, θ in quadrant I

10. $\cos \theta = \dfrac{5}{13}$, θ in quadrant I

11. $\cos \theta = -\dfrac{3}{5}$, θ in quadrant III

12. $\tan \theta = -\dfrac{15}{8}$, θ in quadrant II

13. $\tan \theta = -\dfrac{5}{12}$, θ in quadrant II

14. $\sin \theta = -\dfrac{\sqrt{10}}{10}$, θ in quadrant IV

15. Find an equivalent expression for $\cos 4x$ in terms of function values of x.

16. Find an equivalent expression for $\sin^4 \theta$ in terms of function values of θ, 2θ, or 4θ, raised only to the first power.

Use the half-angle identities to evaluate exactly.

17. $\cos 15°$ **18.** $\tan 67.5°$

19. $\sin 112.5°$ **20.** $\cos \dfrac{\pi}{8}$

21. $\tan 75°$ **22.** $\sin \dfrac{5\pi}{12}$

Given that $\sin \theta = 0.3416$ *and* θ *is in quadrant I, find each of the following using identities.*

23. $\sin 2\theta$ **24.** $\cos \dfrac{\theta}{2}$

25. $\sin \dfrac{\theta}{2}$ **26.** $\sin 4\theta$

Simplify.

27. $2 \cos^2 \dfrac{x}{2} - 1$

28. $\cos^4 x - \sin^4 x$

29. $(\sin x - \cos x)^2 + \sin 2x$

30. $(\sin x + \cos x)^2$

31. $\dfrac{2 - \sec^2 x}{\sec^2 x}$

32. $\dfrac{1 + \sin 2x + \cos 2x}{1 + \sin 2x - \cos 2x}$

33. $(-4 \cos x \sin x + 2 \cos 2x)^2 +$ $(2 \cos 2x + 4 \sin x \cos x)^2$

34. $2 \sin x \cos^3 x - 2 \sin^3 x \cos x$

Technology Connection

In Exercises 35–38, use a graphing calculator to determine which of the following expressions asserts an identity. Then derive the identity algebraically.

35. $\dfrac{\cos 2x}{\cos x - \sin x} = \cdots$

  a) $1 + \cos x$ **b)** $\cos x - \sin x$

$y = 1 + \cos x$ $y = \cos x - \sin x$

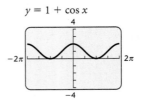   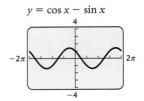

  c) $-\cot x$ **d)** $\sin x (\cot x + 1)$

$y = -\cot x$ $y = \sin x (\cot x + 1)$

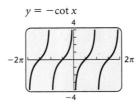   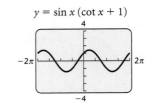

36. $2 \cos^2 \dfrac{x}{2} = \cdots$

  a) $\sin x (\csc x + \tan x)$ **b)** $\sin x - 2 \cos x$
  c) $2(\cos^2 x - \sin^2 x)$ **d)** $1 + \cos x$

37. $\dfrac{\sin 2x}{2 \cos x} = \cdots$

  a) $\cos x$ **b)** $\tan x$
  c) $\cos x + \sin x$ **d)** $\sin x$

38. $2 \sin \dfrac{\theta}{2} \cos \dfrac{\theta}{2} = \cdots$

  a) $\cos^2 \theta$ **b)** $\sin \dfrac{\theta}{2}$

  c) $\sin \theta$ **d)** $\sin \theta - \cos \theta$

Collaborative Discussion and Writing

39. Discuss and compare the graphs of $y = \sin x$, $y = \sin 2x$, and $y = \sin (x/2)$.

40. Find all errors in the following:
$$2 \sin^2 2x + \cos 4x$$
$$= 2(2 \sin x \cos x)^2 + 2 \cos 2x$$
$$= 8 \sin^2 x \cos^2 x + 2(\cos^2 x + \sin^2 x)$$
$$= 8 \sin^2 x \cos^2 x + 2.$$

Skill Maintenance

Complete the identity.

41. $1 - \cos^2 x =$

42. $\sec^2 x - \tan^2 x =$

43. $\sin^2 x - 1 =$

44. $1 + \cot^2 x =$

45. $\csc^2 x - \cot^2 x$

46. $1 + \tan^2 x =$

47. $1 - \sin^2 x$

48. $\sec^2 x - 1$

Consider the following functions (a)–(f). Without graphing them, answer questions 49–52 below.

a) $f(x) = 2 \sin \left(\dfrac{1}{2}x - \dfrac{\pi}{2} \right)$

b) $f(x) = \dfrac{1}{2} \cos \left(2x - \dfrac{\pi}{4} \right) + 2$

c) $f(x) = -\sin \left[2 \left(x - \dfrac{\pi}{2} \right) \right] + 2$

d) $f(x) = \sin (x + \pi) - \dfrac{1}{2}$

e) $f(x) = -2 \cos (4x - \pi)$

f) $f(x) = -\cos 2 \left(x - \dfrac{\pi}{8} \right)$

49. Which functions have a graph with an amplitude of 2?

50. Which functions have a graph with a period of π?

51. Which functions have a graph with a period of 2π?

52. Which functions have a graph with a phase shift of $\dfrac{\pi}{4}$?

Synthesis

53. Given that $\cos 51° \approx 0.6293$, find the six function values for $141°$.

Simplify.

54. $\sin \left(\dfrac{\pi}{2} - x \right) [\sec x - \cos x]$

55. $\cos (\pi - x) + \cot x \sin \left(x - \dfrac{\pi}{2} \right)$

56. $\dfrac{\cos x - \sin \left(\dfrac{\pi}{2} - x \right) \sin x}{\cos x - \cos (\pi - x) \tan x}$

57. $\dfrac{\cos^2 y \sin \left(y + \dfrac{\pi}{2} \right)}{\sin^2 y \sin \left(\dfrac{\pi}{2} - y \right)}$

Find $\sin \theta$, $\cos \theta$, and $\tan \theta$ under the given conditions.

58. $\cos 2\theta = \dfrac{7}{12}$, $\dfrac{3\pi}{2} \leq 2\theta \leq 2\pi$

59. $\tan \dfrac{\theta}{2} = -\dfrac{5}{3}$, $\pi < \theta \leq \dfrac{3\pi}{2}$

60. *Nautical Mile.* Latitude is used to measure north–south location on the Earth between the equator and the poles. For example, Chicago has latitude 42°N. (See the figure.) In Great Britain, the *nautical mile* is defined as the length of a minute of arc of the Earth's radius. Since the Earth is flattened slightly at the poles, a British nautical mile varies with latitude. In fact, it is given, in feet, by the function
$$N(\phi) = 6066 - 31 \cos 2\phi,$$
where ϕ is the latitude in degrees.

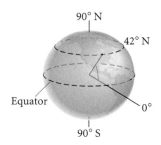

a) What is the length of a British nautical mile at Chicago?

b) What is the length of a British nautical mile at the North Pole?

c) Express $N(\phi)$ in terms of $\cos \phi$ only. That is, do not use the double angle.

61. *Acceleration due to Gravity.*   The acceleration due to gravity is often denoted by g in a formula such as $S = \frac{1}{2}gt^2$, where S is the distance that an object falls in time t. The number g relates to motion near the earth's surface and is usually considered constant. In fact, however, g is not constant, but varies slightly with latitude. If ϕ stands for latitude, in degrees, g is given with good approximation by the formula

$$g = 9.78049(1 + 0.005288 \sin^2 \phi - 0.000006 \sin^2 2\phi),$$

where g is measured in meters per second per second at sea level.

a) Chicago has latitude 42°N. Find g.
b) Philadelphia has latitude 40°N. Find g.

c) Express g in terms of $\sin \phi$ only. That is, eliminate the double angle.

2.3   Proving Trigonometric Identities

◆ Prove identities using other identities.
◆ Use the product-to-sum identities and the sum-to-product identities to derive other identities.

◆ The Logic of Proving Identities

We outline two algebraic methods for proving identities.

Method 1. Start with either the left or the right side of the equation and obtain the other side. For example, suppose you are trying to prove that the equation $P = Q$ is an identity. You might try to produce a string of statements $(R_1, R_2, \ldots$ or $T_1, T_2, \ldots)$ like the following, which start with P and end with Q or start with Q and end with P:

$$
\begin{aligned}
P &= R_1 &\quad \text{or} \quad Q &= T_1 \\
&= R_2 & &= T_2 \\
&\;\;\vdots & &\;\;\vdots \\
&= Q & &= P.
\end{aligned}
$$

Method 2. Work with each side separately until you obtain the same expression. For example, suppose you are trying to prove that $P = Q$ is an identity.

You might be able to produce two strings of statements like the following, each ending with the same statement S.

$$P = R_1 \qquad Q = T_1$$
$$= R_2 \qquad \;\;= T_2$$
$$\quad\;\;\; \vdots \qquad\qquad \vdots$$
$$= S \qquad\;\; = S.$$

The number of steps in each string might be different, but in each case the result is S.

A first step in learning to prove identities is to have at hand a list of the identities that you have already learned. Such a list is on the inside back cover of this text. Ask your instructor which ones you are expected to memorize. The more identities you prove, the easier it will be to prove new ones. A list of helpful hints follows.

Hints for Proving Identities

1. Use method 1 or 2, starting on page 124.
2. Work with the more complex side first.
3. Carry out any algebraic manipulations, such as adding, subtracting, multiplying, or factoring.
4. Multiplying by 1 can be helpful when rational expressions are involved.
5. Converting all expressions to sines and cosines is often helpful.
6. Try something! Put your pencil to work and get involved. You will be amazed at how often this leads to success.

✦ Proving Identities

In what follows, method 1 is used in Examples 1–3 and method 2 is used in Examples 4 and 5.

EXAMPLE 1 Prove the identity $1 + \sin 2\theta = (\sin \theta + \cos \theta)^2$.

Solution Let's use method 1. We begin with the right side and obtain the left side:

$$(\sin \theta + \cos \theta)^2 = \sin^2 \theta + 2 \sin \theta \cos \theta + \cos^2 \theta \qquad \text{Squaring}$$
$$= 1 + 2 \sin \theta \cos \theta \qquad \text{Recalling the identity } \sin^2 x + \cos^2 x = 1 \text{ and substituting}$$
$$= 1 + \sin 2\theta. \qquad \text{Using } \sin 2x = 2 \sin x \cos x$$

We could also begin with the left side and obtain the right side:

$$1 + \sin 2\theta = 1 + 2\sin\theta\cos\theta \qquad \text{Using } \sin 2x = 2\sin x\cos x$$

$$= \sin^2\theta + 2\sin\theta\cos\theta + \cos^2\theta \qquad \begin{array}{l}\text{Replacing 1 with}\\ \sin^2\theta + \cos^2\theta\end{array}$$

$$= (\sin\theta + \cos\theta)^2. \qquad \text{Factoring}$$

▶ **Now Try Exercise 19.**

EXAMPLE 2 Prove the identity

$$\frac{\sec t - 1}{t \sec t} = \frac{1 - \cos t}{t}.$$

Solution We use method 1, starting with the left side. Note that the left side involves $\sec t$, whereas the right side involves $\cos t$, so it might be wise to make use of a basic identity that involves these two expressions: $\sec t = 1/\cos t$.

$$\frac{\sec t - 1}{t \sec t} = \frac{\dfrac{1}{\cos t} - 1}{t\,\dfrac{1}{\cos t}} \qquad \text{Substituting } 1/\cos t \text{ for } \sec t$$

$$= \left(\frac{1}{\cos t} - 1\right) \cdot \frac{\cos t}{t}$$

$$= \frac{1}{t} - \frac{\cos t}{t} \qquad \text{Multiplying}$$

$$= \frac{1 - \cos t}{t}$$

We started with the left side and obtained the right side, so the proof is complete.

▶ **Now Try Exercise 5.**

EXAMPLE 3 Prove the identity

$$\frac{\sin 2x}{\sin x} - \frac{\cos 2x}{\cos x} = \sec x.$$

Solution

$$\frac{\sin 2x}{\sin x} - \frac{\cos 2x}{\cos x} = \frac{2\sin x\cos x}{\sin x} - \frac{\cos^2 x - \sin^2 x}{\cos x} \qquad \begin{array}{l}\text{Using double-angle}\\ \text{identities}\end{array}$$

$$= 2\cos x - \frac{\cos^2 x - \sin^2 x}{\cos x} \qquad \text{Simplifying}$$

$$= \frac{2\cos^2 x}{\cos x} - \frac{\cos^2 x - \sin^2 x}{\cos x} \qquad \begin{array}{l}\text{Multiplying } 2\cos x\\ \text{by 1, or } \cos x/\cos x\end{array}$$

$$= \frac{2\cos^2 x - \cos^2 x + \sin^2 x}{\cos x} \qquad \text{Subtracting}$$

$$= \frac{\cos^2 x + \sin^2 x}{\cos x}$$

Then $\qquad\qquad = \dfrac{1}{\cos x}$ Using a Pythagorean identity

$\qquad\qquad\qquad = \sec x$ Recalling a basic identity

▶ **Now Try Exercise 15.**

EXAMPLE 4   Prove the identity

$$\sin^2 x \tan^2 x = \tan^2 x - \sin^2 x.$$

Solution   For this proof, we are going to work with each side separately using method 2. We try to obtain the same expression on each side. In actual practice, you might work on one side for awhile, then work on the other side, and then go back to the first side. In other words, you work back and forth until you arrive at the same expression. Let's start with the right side:

$$\tan^2 x - \sin^2 x = \dfrac{\sin^2 x}{\cos^2 x} - \sin^2 x \qquad \text{Recalling the identity } \tan x = \dfrac{\sin x}{\cos x} \text{ and substituting}$$

$$= \dfrac{\sin^2 x}{\cos^2 x} - \sin^2 x \cdot \dfrac{\cos^2 x}{\cos^2 x} \qquad \text{Multiplying by 1 in order to subtract}$$

$$= \dfrac{\sin^2 x - \sin^2 x \cos^2 x}{\cos^2 x} \qquad \text{Carrying out the subtraction}$$

$$= \dfrac{\sin^2 x\,(1 - \cos^2 x)}{\cos^2 x} \qquad \text{Factoring}$$

$$= \dfrac{\sin^2 x \sin^2 x}{\cos^2 x} \qquad \text{Recalling the identity } 1 - \cos^2 x = \sin^2 x \text{ and substituting}$$

$$= \dfrac{\sin^4 x}{\cos^2 x}.$$

At this point, we stop and work with the left side, $\sin^2 x \tan^2 x$, of the original identity and try to end with the same expression that we ended with on the right side:

$$\sin^2 x \tan^2 x = \sin^2 x \, \dfrac{\sin^2 x}{\cos^2 x} \qquad \text{Recalling the identity } \tan x = \dfrac{\sin x}{\cos x} \text{ and substituting}$$

$$= \dfrac{\sin^4 x}{\cos^2 x}.$$

We have obtained the same expression from each side, so the proof is complete.

▶ **Now Try Exercise 25.**

EXAMPLE 5 Prove the identity

$$\cot \phi + \csc \phi = \frac{\sin \phi}{1 - \cos \phi}.$$

Solution We are again using method 2, beginning with the left side:

$$\cot \phi + \csc \phi = \frac{\cos \phi}{\sin \phi} + \frac{1}{\sin \phi} \qquad \text{Using basic identities}$$

$$= \frac{1 + \cos \phi}{\sin \phi}. \qquad \text{Adding}$$

At this point, we stop and work with the right side of the original identity:

$$\frac{\sin \phi}{1 - \cos \phi} = \frac{\sin \phi}{1 - \cos \phi} \cdot \frac{1 + \cos \phi}{1 + \cos \phi} \qquad \text{Multiplying by 1}$$

$$= \frac{\sin \phi\,(1 + \cos \phi)}{1 - \cos^2 \phi}$$

$$= \frac{\sin \phi\,(1 + \cos \phi)}{\sin^2 \phi} \qquad \text{Using } \sin^2 x = 1 - \cos^2 x$$

$$= \frac{1 + \cos \phi}{\sin \phi}. \qquad \text{Simplifying}$$

The proof is complete since we obtained the same expression from each side.

▶ Now Try Exercise 29.

◆ Product-to-Sum and Sum-to-Product Identities

On occasion, it is convenient to convert a product of trigonometric expressions to a sum, or the reverse. The following identities are useful in this connection.

Product-to-Sum Identities

$$\sin x \cdot \sin y = \frac{1}{2}[\cos (x - y) - \cos (x + y)] \qquad (1)$$

$$\cos x \cdot \cos y = \frac{1}{2}[\cos (x - y) + \cos (x + y)] \qquad (2)$$

$$\sin x \cdot \cos y = \frac{1}{2}[\sin (x + y) + \sin (x - y)] \qquad (3)$$

$$\cos x \cdot \sin y = \frac{1}{2}[\sin (x + y) - \sin (x - y)] \qquad (4)$$

We can derive product-to-sum identities (1) and (2) using the sum and difference identities for the cosine function:

$$\cos(x + y) = \cos x \cos y - \sin x \sin y, \qquad \text{Sum identity}$$
$$\cos(x - y) = \cos x \cos y + \sin x \sin y. \qquad \text{Difference identity}$$

Subtracting the sum identity from the difference identity, we have

$$\cos(x - y) - \cos(x + y) = 2 \sin x \sin y \qquad \text{Subtracting}$$
$$\frac{1}{2}[\cos(x - y) - \cos(x + y)] = \sin x \sin y. \qquad \text{Multiplying by } \tfrac{1}{2}$$

Thus, $\sin x \sin y = \tfrac{1}{2}[\cos(x - y) - \cos(x + y)]$.

Adding the cosine sum and difference identities, we have

$$\cos(x - y) + \cos(x + y) = 2 \cos x \cos y \qquad \text{Adding}$$
$$\frac{1}{2}[\cos(x - y) + \cos(x + y)] = \cos x \cos y. \qquad \text{Multiplying by } \tfrac{1}{2}$$

Thus, $\cos x \cos y = \tfrac{1}{2}[\cos(x - y) + \cos(x + y)]$.

Identities (3) and (4) can be derived in a similar manner using the sum and difference identities for the sine function.

EXAMPLE 6 Find an identity for $2 \sin 3\theta \cos 7\theta$.

Solution We will use the identity

$$\sin x \cdot \cos y = \frac{1}{2}[\sin(x + y) + \sin(x - y)].$$

Here $x = 3\theta$ and $y = 7\theta$. Thus,

$$2 \sin 3\theta \cos 7\theta = 2 \cdot \frac{1}{2}[\sin(3\theta + 7\theta) + \sin(3\theta - 7\theta)]$$
$$= \sin 10\theta + \sin(-4\theta)$$
$$= \sin 10\theta - \sin 4\theta. \qquad \text{Using } \sin(-\theta) = -\sin\theta$$

▶ **Now Try Exercise 37.**

Sum-to-Product Identities

$$\sin x + \sin y = 2 \sin \frac{x + y}{2} \cos \frac{x - y}{2} \tag{5}$$

$$\sin x - \sin y = 2 \cos \frac{x + y}{2} \sin \frac{x - y}{2} \tag{6}$$

$$\cos y + \cos x = 2 \cos \frac{x + y}{2} \cos \frac{x - y}{2} \tag{7}$$

$$\cos y - \cos x = 2 \sin \frac{x + y}{2} \sin \frac{x - y}{2} \tag{8}$$

The sum-to-product identities (5)–(8) can be derived using the product-to-sum identities. Proofs are left to the exercises.

EXAMPLE 7   Find an identity for $\cos \theta + \cos 5\theta$.

Solution   We will use the identity

$$\cos y + \cos x = 2 \cos \frac{x + y}{2} \cos \frac{x - y}{2}.$$

Here $x = 5\theta$ and $y = \theta$. Thus,

$$\cos \theta + \cos 5\theta = 2 \cos \frac{5\theta + \theta}{2} \cos \frac{5\theta - \theta}{2}$$

$$= 2 \cos 3\theta \cos 2\theta. \qquad \blacktriangleright \text{ Now Try Exercise 35.}$$

2.3   EXERCISE SET

Prove each of the following identities.

1. $\sec x - \sin x \tan x = \cos x$

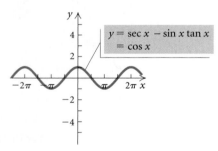

2. $\dfrac{1 + \cos \theta}{\sin \theta} + \dfrac{\sin \theta}{\cos \theta} = \dfrac{\cos \theta + 1}{\sin \theta \cos \theta}$

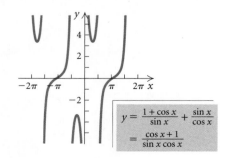

3. $\dfrac{1 - \cos x}{\sin x} = \dfrac{\sin x}{1 + \cos x}$

4. $\dfrac{1 + \tan y}{1 + \cot y} = \dfrac{\sec y}{\csc y}$

5. $\dfrac{1 + \tan \theta}{1 - \tan \theta} + \dfrac{1 + \cot \theta}{1 - \cot \theta} = 0$

6. $\dfrac{\sin x + \cos x}{\sec x + \csc x} = \dfrac{\sin x}{\sec x}$

7. $\dfrac{\cos^2 \alpha + \cot \alpha}{\cos^2 \alpha - \cot \alpha} = \dfrac{\cos^2 \alpha \tan \alpha + 1}{\cos^2 \alpha \tan \alpha - 1}$

8. $\sec 2\theta = \dfrac{\sec^2 \theta}{2 - \sec^2 \theta}$

9. $\dfrac{2 \tan \theta}{1 + \tan^2 \theta} = \sin 2\theta$

10. $\dfrac{\cos (u - v)}{\cos u \sin v} = \tan u + \cot v$

11. $1 - \cos 5\theta \cos 3\theta - \sin 5\theta \sin 3\theta = 2 \sin^2 \theta$

12. $\cos^4 x - \sin^4 x = \cos 2x$

13. $2 \sin \theta \cos^3 \theta + 2 \sin^3 \theta \cos \theta = \sin 2\theta$

14. $\dfrac{\tan 3t - \tan t}{1 + \tan 3t \tan t} = \dfrac{2 \tan t}{1 - \tan^2 t}$

15. $\dfrac{\tan x - \sin x}{2 \tan x} = \sin^2 \dfrac{x}{2}$

16. $\dfrac{\cos^3 \beta - \sin^3 \beta}{\cos \beta - \sin \beta} = \dfrac{2 + \sin 2\beta}{2}$

17. $\sin (\alpha + \beta) \sin (\alpha - \beta) = \sin^2 \alpha - \sin^2 \beta$

18. $\cos^2 x \, (1 - \sec^2 x) = -\sin^2 x$

19. $\tan \theta \, (\tan \theta + \cot \theta) = \sec^2 \theta$

20. $\dfrac{\cos \theta + \sin \theta}{\cos \theta} = 1 + \tan \theta$

21. $\dfrac{1 + \cos^2 x}{\sin^2 x} = 2 \csc^2 x - 1$

22. $\dfrac{\tan y + \cot y}{\csc y} = \sec y$

23. $\dfrac{1 + \sin x}{1 - \sin x} + \dfrac{\sin x - 1}{1 + \sin x} = 4 \sec x \tan x$

24. $\tan \theta - \cot \theta = (\sec \theta - \csc \theta)(\sin \theta + \cos \theta)$

25. $\cos^2 \alpha \cot^2 \alpha = \cot^2 \alpha - \cos^2 \alpha$

26. $\dfrac{\tan x + \cot x}{\sec x + \csc x} = \dfrac{1}{\cos x + \sin x}$

27. $2 \sin^2 \theta \cos^2 \theta + \cos^4 \theta = 1 - \sin^4 \theta$

28. $\dfrac{\cot \theta}{\csc \theta - 1} = \dfrac{\csc \theta + 1}{\cot \theta}$

29. $\dfrac{1 + \sin x}{1 - \sin x} = (\sec x + \tan x)^2$

30. $\sec^4 s - \tan^2 s = \tan^4 s + \sec^2 s$

31. Verify the product-to-sum identities (3) and (4) using the sine sum and difference identities.

32. Verify the sum-to-product identities (5)–(8) using the product-to-sum identities (1)–(4).

Use the product-to-sum and the sum-to-product identities to find identities for each of the following.

33. $\sin 3\theta - \sin 5\theta$

34. $\sin 7x - \sin 4x$

35. $\sin 8\theta + \sin 5\theta$

36. $\cos \theta - \cos 7\theta$

37. $\sin 7u \sin 5u$

38. $2 \sin 7\theta \cos 3\theta$

39. $7 \cos \theta \sin 7\theta$

40. $\cos 2t \sin t$

41. $\cos 55° \sin 25°$

42. $7 \cos 5\theta \cos 7\theta$

Use the product-to-sum and the sum-to-product identities to prove each of the following.

43. $\sin 4\theta + \sin 6\theta = \cot \theta \, (\cos 4\theta - \cos 6\theta)$

44. $\tan 2x \, (\cos x + \cos 3x) = \sin x + \sin 3x$

45. $\cot 4x \, (\sin x + \sin 4x + \sin 7x)$
$\qquad = \cos x + \cos 4x + \cos 7x$

46. $\tan \dfrac{x + y}{2} = \dfrac{\sin x + \sin y}{\cos x + \cos y}$

47. $\cot \dfrac{x + y}{2} = \dfrac{\sin y - \sin x}{\cos x - \cos y}$

48. $\tan \dfrac{\theta + \phi}{2} \tan \dfrac{\phi - \theta}{2} = \dfrac{\cos \theta - \cos \phi}{\cos \theta + \cos \phi}$

49. $\tan \dfrac{\theta + \phi}{2} (\sin \theta - \sin \phi)$
$\qquad = \tan \dfrac{\theta - \phi}{2} (\sin \theta + \sin \phi)$

50. $\sin 2\theta + \sin 4\theta + \sin 6\theta = 4 \cos \theta \cos 2\theta \sin 3\theta$

Technology Connection

In Exercises 51–56, use a graphing calculator to determine which expression (A)–(F) on the right can be used to complete the identity. Then try to prove that identity algebraically.

51. $\dfrac{\cos x + \cot x}{1 + \csc x}$

52. $\cot x + \csc x$

53. $\sin x \cos x + 1$

54. $2 \cos^2 x - 1$

55. $\dfrac{1}{\cot x \sin^2 x}$

56. $(\cos x + \sin x)(1 - \sin x \cos x)$

A. $\dfrac{\sin^3 x - \cos^3 x}{\sin x - \cos x}$

B. $\cos x$

C. $\tan x + \cot x$

D. $\cos^3 x + \sin^3 x$

E. $\dfrac{\sin x}{1 - \cos x}$

F. $\cos^4 x - \sin^4 x$

Collaborative Discussion and Writing

57. What restrictions must be placed on the variable in each of the following identities? Why?

a) $\sin 2x = \dfrac{2 \tan x}{1 + \tan^2 x}$

b) $\dfrac{1 - \cos x}{\sin x} = \dfrac{\sin x}{1 + \cos x}$

c) $2 \sin x \cos^3 x + 2 \sin^3 x \cos x = \sin 2x$

58. Explain why $\tan (x + 450°)$ cannot be simplified using the tangent sum formula, but can be simplified using the sine and cosine sum formulas.

Skill Maintenance

For each function:

a) *Graph the function.*
b) *Determine whether the function is one-to-one.*
c) *If the function is one-to-one, find an equation for its inverse.*
d) *Graph the inverse of the function.*

59. $f(x) = 3x - 2$ **60.** $f(x) = x^3 + 1$

61. $f(x) = x^2 - 4, \ x \geq 0$ **62.** $f(x) = \sqrt{x + 2}$

Solve.

63. $2x^2 = 5x$

64. $3x^2 + 5x - 10 = 18$

65. $x^4 + 5x^2 - 36 = 0$

66. $x^2 - 10x + 1 = 0$

67. $\sqrt{x - 2} = 5$

68. $x = \sqrt{x + 7} + 5$

Synthesis

Prove the identity.

69. $\ln |\tan x| = -\ln |\cot x|$

70. $\ln |\sec \theta + \tan \theta| = -\ln |\sec \theta - \tan \theta|$

71. Prove the identity
$$\log (\cos x - \sin x) + \log (\cos x + \sin x) = \log \cos 2x.$$

72. *Mechanics.* The following equation occurs in the study of mechanics:
$$\sin \theta = \dfrac{I_1 \cos \phi}{\sqrt{(I_1 \cos \phi)^2 + (I_2 \sin \phi)^2}}.$$
It can happen that $I_1 = I_2$. Assuming that this happens, simplify the equation.

73. *Alternating Current.* In the theory of alternating current, the following equation occurs:
$$R = \dfrac{1}{\omega C (\tan \theta + \tan \phi)}.$$
Show that this equation is equivalent to
$$R = \dfrac{\cos \theta \cos \phi}{\omega C \sin (\theta + \phi)}.$$

74. *Electrical Theory.* In electrical theory, the following equations occur:
$$E_1 = \sqrt{2} E_t \cos \left(\theta + \dfrac{\pi}{P} \right)$$
and
$$E_2 = \sqrt{2} E_t \cos \left(\theta - \dfrac{\pi}{P} \right).$$
Assuming that these equations hold, show that
$$\dfrac{E_1 + E_2}{2} = \sqrt{2} E_t \cos \theta \cos \dfrac{\pi}{P}$$
and
$$\dfrac{E_1 - E_2}{2} = -\sqrt{2} E_t \sin \theta \sin \dfrac{\pi}{P}.$$

2.4

Inverses of the Trigonometric Functions

◆ Find values of the inverse trigonometric functions.

◆ Simplify expressions such as $\sin(\sin^{-1}x)$ and $\sin^{-1}(\sin x)$.

◆ Simplify expressions involving compositions such as $\sin\left(\cos^{-1}\frac{1}{2}\right)$ without using a calculator.

◆ Simplify expressions such as $\sin \arctan (a/b)$ by making a drawing and reading off appropriate ratios.

In this section, we develop inverse trigonometric functions. The graphs of the sine, cosine, and tangent functions follow. Do these functions have inverses that are functions? They do have inverses if they are one-to-one, which means that they pass the horizontal-line test.

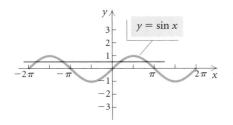
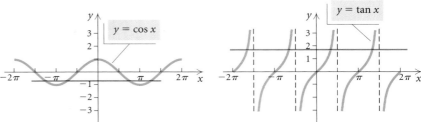

Note that for each function, a horizontal line (shown in red) crosses the graph more than once. Therefore, none of them has an inverse that is a function.

The graphs of an equation and its inverse are reflections of each other across the line $y = x$. Let's examine the graphs of the inverses of each of the three functions graphed above.

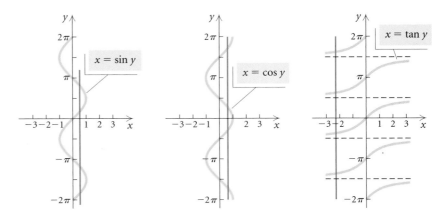

We can check again to see whether these are graphs of functions by using the vertical-line test. In each case, there is a vertical line (shown in red) that crosses the graph more than once, so each *fails* to be a function.

◆ Restricting Ranges to Define Inverse Functions

Recall that a function like $f(x) = x^2$ does not have an inverse that is a function, but by restricting the domain of f to nonnegative numbers, we have a new squaring function, $f(x) = x^2, x \geq 0$, that has an inverse, $f^{-1}(x) = \sqrt{x}$. This is equivalent to restricting the range of the inverse relation to exclude ordered pairs that contain negative numbers.

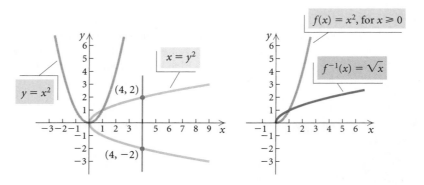

In a similar manner, we can define new trigonometric functions whose inverses are functions. We can do this by restricting either the domains of the basic trigonometric functions or the ranges of their inverse relations. This can be done in many ways, but the restrictions illustrated below with solid red curves are fairly standard in mathematics.

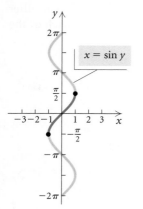

Figure 1

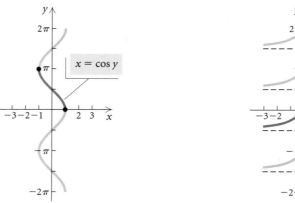

Figure 2

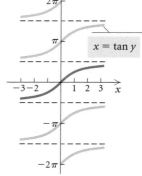

Figure 3

For the inverse sine function, we choose a range close to the origin that allows all inputs on the interval $[-1, 1]$ to have function values. Thus we choose the interval $[-\pi/2, \pi/2]$ for the range (Fig. 1). For the inverse cosine function, we choose a range close to the origin that allows all inputs on the interval $[-1, 1]$ to have function values. We choose the interval $[0, \pi]$ (Fig. 2). For the inverse tangent function, we choose a range close to the origin that allows all real numbers to have function values. The interval $(-\pi/2, \pi/2)$ satisfies this requirement (Fig. 3).

Inverse Trigonometric Functions

FUNCTION	DOMAIN	RANGE
$y = \sin^{-1} x$ $= \arcsin x$, where $x = \sin y$	$[-1, 1]$	$[-\pi/2, \pi/2]$
$y = \cos^{-1} x$ $= \arccos x$, where $x = \cos y$	$[-1, 1]$	$[0, \pi]$
$y = \tan^{-1} x$ $= \arctan x$, where $x = \tan y$	$(-\infty, \infty)$	$(-\pi/2, \pi/2)$

The notation arcsin x arises because the function value, y, is the length of an arc on the unit circle for which the sine is x. Either of the two kinds of notation above can be read "the inverse sine of x" or "the arc sine of x" or "the number (or angle) whose sine is x."

> The notation $\sin^{-1} x$ is *not* exponential notation.
>
> It does *not* mean $\dfrac{1}{\sin x}$!

The graphs of the inverse trigonometric functions are as follows.

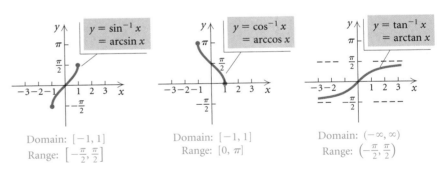

Domain: $[-1, 1]$ Range: $\left[-\frac{\pi}{2}, \frac{\pi}{2}\right]$ Domain: $[-1, 1]$ Range: $[0, \pi]$ Domain: $(-\infty, \infty)$ Range: $\left(-\frac{\pi}{2}, \frac{\pi}{2}\right)$

The following diagrams show the restricted ranges for the inverse trigonometric functions on a unit circle. Compare these graphs with the graphs above. The ranges of these functions should be memorized. The missing endpoints in the graph of the arctangent function indicate inputs that are not in the domain of the original function.

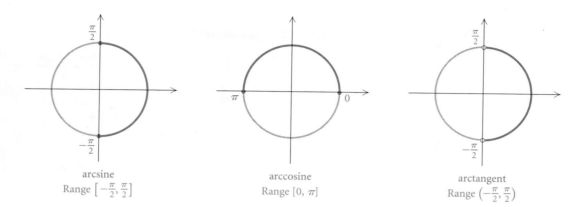

arcsine
Range $\left[-\frac{\pi}{2}, \frac{\pi}{2}\right]$

arccosine
Range $[0, \pi]$

arctangent
Range $\left(-\frac{\pi}{2}, \frac{\pi}{2}\right)$

EXAMPLE 1 Find each of the following function values.

a) $\sin^{-1} \dfrac{\sqrt{2}}{2}$ **b)** $\cos^{-1}\left(-\dfrac{1}{2}\right)$ **c)** $\tan^{-1}\left(-\dfrac{\sqrt{3}}{3}\right)$

Solution

a) Another way to state "find $\sin^{-1}\sqrt{2}/2$" is to say "find β such that $\sin \beta = \sqrt{2}/2$." In the restricted range $[-\pi/2, \pi/2]$, the only number with a sine of $\sqrt{2}/2$ is $\pi/4$. Thus, $\sin^{-1}\left(\sqrt{2}/2\right) = \pi/4$, or $45°$. (See Fig. 4 below.)

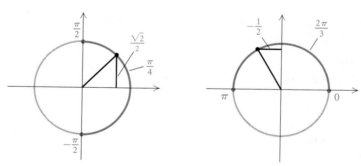

Figure 4   **Figure 5**

b) The only number with a cosine of $-\frac{1}{2}$ in the restricted range $[0, \pi]$ is $2\pi/3$. Thus, $\cos^{-1}\left(-\frac{1}{2}\right) = 2\pi/3$, or $120°$. (See Fig. 5 above.)

c) The only number in the restricted range $(-\pi/2, \pi/2)$ with a tangent of $-\sqrt{3}/3$ is $-\pi/6$. Thus, $\tan^{-1}\left(-\sqrt{3}/3\right)$ is $-\pi/6$, or $-30°$. (See Fig. 6 at left.)

▶ Now Try Exercises 1 and 5.

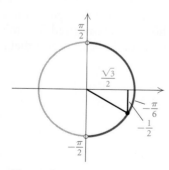

Figure 6

We can also use a calculator to find inverse trigonometric function values. On most graphing calculators, we can find inverse function values in either radians or degrees simply by selecting the appropriate mode. The key strokes involved in finding inverse function values vary with the calculator. Be sure to read the instructions for the particular calculator that you are using.

EXAMPLE 2   Approximate each of the following function values in both radians and degrees. Round radian measure to four decimal places and degree measure to the nearest tenth of a degree.

a) $\cos^{-1}(-0.2689)$
b) $\tan^{-1}(-0.2623)$
c) $\sin^{-1} 0.20345$
d) $\cos^{-1} 1.318$
e) $\csc^{-1} 8.205$

Solution

FUNCTION VALUE	MODE	READOUT	ROUNDED
a) $\cos^{-1}(-0.2689)$	Radian	1.843047111	1.8430
	Degree	105.5988209	105.6°
b) $\tan^{-1}(-0.2623)$	Radian	−.2565212141	−0.2565
	Degree	−14.69758292	−14.7°
c) $\sin^{-1} 0.20345$	Radian	.2048803359	0.2049
	Degree	11.73877855	11.7°
d) $\cos^{-1} 1.318$	Radian	ERR:DOMAIN	
	Degree	ERR:DOMAIN	

The value 1.318 is not in $[-1, 1]$, the domain of the arccosine function.

e) The cosecant function is the reciprocal of the sine function:

$\csc^{-1} 8.205 =$ $\sin^{-1}(1/8.205)$	Radian	.1221806653	0.1222
	Degree	7.000436462	7.0°

▶ Now Try Exercises 21 and 25.

CONNECTING THE CONCEPTS

Domains and Ranges

The following is a summary of the domains and ranges of the trigonometric functions together with a summary of the domains and ranges of the inverse trigonometric functions. For completeness, we have included the arccosecant, the arcsecant, and the arccotangent, though there is a lack of uniformity in their definitions in mathematical literature.

FUNCTION	DOMAIN	RANGE
sin	All reals, $(-\infty, \infty)$	$[-1, 1]$
cos	All reals, $(-\infty, \infty)$	$[-1, 1]$
tan	All reals except $k\pi/2,\ k$ odd	All reals, $(-\infty, \infty)$
csc	All reals except $k\pi$	$(-\infty, -1] \cup [1, \infty)$
sec	All reals except $k\pi/2,\ k$ odd	$(-\infty, -1] \cup [1, \infty)$
cot	All reals except $k\pi$	All reals, $(-\infty, \infty)$

INVERSE FUNCTION	DOMAIN	RANGE
$\sin^{-1}$	$[-1, 1]$	$\left[-\dfrac{\pi}{2}, \dfrac{\pi}{2}\right]$
$\cos^{-1}$	$[-1, 1]$	$[0, \pi]$
$\tan^{-1}$	All reals, or $(-\infty, \infty)$	$\left(-\dfrac{\pi}{2}, \dfrac{\pi}{2}\right)$
$\csc^{-1}$	$(-\infty, -1] \cup [1, \infty)$	$\left[-\dfrac{\pi}{2}, 0\right) \cup \left(0, \dfrac{\pi}{2}\right]$
$\sec^{-1}$	$(-\infty, -1] \cup [1, \infty)$	$\left[0, \dfrac{\pi}{2}\right) \cup \left(\dfrac{\pi}{2}, \pi\right]$
$\cot^{-1}$	All reals, or $(-\infty, \infty)$	$(0, \pi)$

● ● ●

✦ Composition of Trigonometric Functions and Their Inverses

Various compositions of trigonometric functions and their inverses often occur in practice. For example, we might want to try to simplify an expression such as

$$\sin\left(\sin^{-1} x\right) \quad \text{or} \quad \sin\left(\cot^{-1} \frac{x}{2}\right).$$

In the expression on the left, we are finding "the sine of a number whose sine is x." If a function f has an inverse that is also a function, then

$$f(f^{-1}(x)) = x, \quad \text{for all } x \text{ in the } \textit{domain} \text{ of } f^{-1},$$

and

$$f^{-1}(f(x)) = x, \quad \text{for all } x \text{ in the } \textit{domain} \text{ of } f.$$

Thus, if $f(x) = \sin x$ and $f^{-1}(x) = \sin^{-1} x$, then

$$\mathbf{\sin\left(\sin^{-1} x\right) = x, \quad \text{for all } x \text{ in the } \textit{domain} \text{ of } \sin^{-1},}$$

which is any number on the interval $[-1, 1]$. Similar results hold for the other trigonometric functions.

Composition of Trigonometric Functions

$\sin\left(\sin^{-1} x\right) = x,$ for all x in the domain of $\sin^{-1}$.

$\cos\left(\cos^{-1} x\right) = x,$ for all x in the domain of $\cos^{-1}$.

$\tan\left(\tan^{-1} x\right) = x,$ for all x in the domain of $\tan^{-1}$.

EXAMPLE 3   Simplify each of the following.

a) $\cos\left(\cos^{-1} \dfrac{\sqrt{3}}{2}\right)$ **b)** $\sin\left(\sin^{-1} 1.8\right)$

Solution

a) Since $\sqrt{3}/2$ is in $[-1, 1]$, the domain of $\cos^{-1}$, it follows that

$$\cos\left(\cos^{-1} \frac{\sqrt{3}}{2}\right) = \frac{\sqrt{3}}{2}.$$

b) Since 1.8 is not in $[-1, 1]$, the domain of $\sin^{-1}$, we cannot evaluate this expression. We know that there is no number with a sine of 1.8. Since we cannot find $\sin^{-1} 1.8$, we state that $\sin\left(\sin^{-1} 1.8\right)$ does not exist.   ▶ Now Try Exercise 37.

Now let's consider an expression like $\sin^{-1}(\sin x)$. We might also suspect that this is equal to x for any x in the domain of $\sin x$, but this is not true unless x is in the range of the $\sin^{-1}$ function. Note that in order to define $\sin^{-1}$, we had to restrict the domain of the sine function. In doing so, we restricted the range of the inverse sine function. Thus,

$$\sin^{-1}(\sin x) = x, \quad \text{for all } x \text{ in the } \textit{range} \text{ of } \sin^{-1}.$$

Similar results hold for the other trigonometric functions.

Special Cases

$\sin^{-1}(\sin x) = x, \quad$ for all x in the range of $\sin^{-1}$.

$\cos^{-1}(\cos x) = x, \quad$ for all x in the range of $\cos^{-1}$.

$\tan^{-1}(\tan x) = x, \quad$ for all x in the range of $\tan^{-1}$.

EXAMPLE 4 Simplify each of the following.

a) $\tan^{-1}\left(\tan \dfrac{\pi}{6}\right)$ 　　　　　　　　**b)** $\sin^{-1}\left(\sin \dfrac{3\pi}{4}\right)$

Solution

a) Since $\pi/6$ is in $(-\pi/2, \pi/2)$, the range of the $\tan^{-1}$ function, we can use $\tan^{-1}(\tan x) = x$. Thus,

$$\tan^{-1}\left(\tan \frac{\pi}{6}\right) = \frac{\pi}{6}.$$

b) Note that $3\pi/4$ is not in $[-\pi/2, \pi/2]$, the range of the $\sin^{-1}$ function. Thus we *cannot* apply $\sin^{-1}(\sin x) = x$. Instead we first find $\sin(3\pi/4)$, which is $\sqrt{2}/2$, and substitute:

$$\sin^{-1}\left(\sin \frac{3\pi}{4}\right) = \sin^{-1}\left(\frac{\sqrt{2}}{2}\right) = \frac{\pi}{4}.$$

▶ Now Try Exercise 43.

Now we find some other function compositions.

EXAMPLE 5 Simplify each of the following.

a) $\sin[\tan^{-1}(-1)]$ 　　　　　　　　**b)** $\cos^{-1}\left(\sin \dfrac{\pi}{2}\right)$

Solution

a) $\text{Tan}^{-1}(-1)$ is the number (or angle) θ in $(-\pi/2, \pi/2)$ whose tangent is -1. That is, $\tan \theta = -1$. Thus, $\theta = -\pi/4$ and

$$\sin[\tan^{-1}(-1)] = \sin\left[-\frac{\pi}{4}\right] = -\frac{\sqrt{2}}{2}.$$

b) $\cos^{-1}\left(\sin \dfrac{\pi}{2}\right) = \cos^{-1}(1) = 0 \qquad \sin \dfrac{\pi}{2} = 1$

▶ Now Try Exercises 47 and 49.

Next, let's consider

$$\cos\left(\sin^{-1}\frac{3}{5}\right).$$

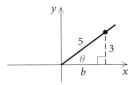

Without using a calculator, we cannot find $\sin^{-1}\frac{3}{5}$. However, we can still evaluate the entire expression by sketching a reference triangle. We are looking for angle θ such that $\sin^{-1}\frac{3}{5} = \theta$, or $\sin\theta = \frac{3}{5}$. Since $\sin^{-1}$ is defined in $[-\pi/2, \pi/2]$ and $\frac{3}{5} > 0$, we know that θ is in quadrant I. We sketch a reference right triangle, as shown at left. The angle θ in this triangle is an angle whose sine is $\frac{3}{5}$. We wish to find the cosine of this angle. Since the triangle is a right triangle, we can find the length of the base, b. It is 4. Thus we know that $\cos\theta = b/5$, or $\frac{4}{5}$. Therefore,

$$\cos\left(\sin^{-1}\frac{3}{5}\right) = \frac{4}{5}.$$

EXAMPLE 6 Find $\sin\left(\cot^{-1}\dfrac{x}{2}\right)$.

Solution Considering all values of x, we draw right triangles, as shown below, whose legs have lengths x and 2, so that $\cot\theta = x/2$.

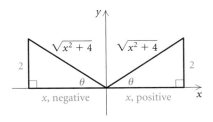

In each, we find the length of the hypotenuse and then read off the sine ratio. We get

$$\sin\left(\cot^{-1}\frac{x}{2}\right) = \frac{2}{\sqrt{x^2+4}}.$$

▶ Now Try Exercise 55.

In the following example, we use a sum identity to evaluate an expression.

EXAMPLE 7 Evaluate:

$$\sin\left(\sin^{-1}\frac{1}{2} + \cos^{-1}\frac{5}{13}\right).$$

Solution Since $\sin^{-1}\frac{1}{2}$ and $\cos^{-1}\frac{5}{13}$ are both angles, the expression is the sine of a sum of two angles, so we use the identity

$$\sin(u+v) = \sin u\cos v + \cos u\sin v.$$

Thus,

$$\sin\left(\sin^{-1}\frac{1}{2} + \cos^{-1}\frac{5}{13}\right)$$

$$= \sin\left(\sin^{-1}\frac{1}{2}\right) \cdot \cos\left(\cos^{-1}\frac{5}{13}\right) + \cos\left(\sin^{-1}\frac{1}{2}\right) \cdot \sin\left(\cos^{-1}\frac{5}{13}\right)$$

$$= \frac{1}{2} \cdot \frac{5}{13} + \cos\left(\sin^{-1}\frac{1}{2}\right) \cdot \sin\left(\cos^{-1}\frac{5}{13}\right). \qquad \text{Using composition identities}$$

Now since $\sin^{-1}\frac{1}{2} = \pi/6$, $\cos\left(\sin^{-1}\frac{1}{2}\right)$ simplifies to $\cos \pi/6$, or $\sqrt{3}/2$. We can illustrate this with a reference triangle in quadrant I.

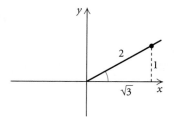

To find $\sin\left(\cos^{-1}\frac{5}{13}\right)$, we use a reference triangle in quadrant I and determine that the sine of the angle whose cosine is $\frac{5}{13}$ is $\frac{12}{13}$.

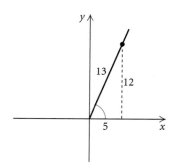

Our expression now simplifies to

$$\frac{1}{2} \cdot \frac{5}{13} + \frac{\sqrt{3}}{2} \cdot \frac{12}{13}, \quad \text{or} \quad \frac{5 + 12\sqrt{3}}{26}.$$

Thus,

$$\sin\left(\sin^{-1}\frac{1}{2} + \cos^{-1}\frac{5}{13}\right) = \frac{5 + 12\sqrt{3}}{26}.$$

▶ Now Try Exercise 63.

2.4 EXERCISE SET

Find each of the following exactly in radians and in degrees.

1. $\sin^{-1}\left(-\dfrac{\sqrt{3}}{2}\right)$

2. $\cos^{-1}\dfrac{1}{2}$

3. $\tan^{-1} 1$

4. $\sin^{-1} 0$

5. $\cos^{-1}\dfrac{\sqrt{2}}{2}$

6. $\sec^{-1}\sqrt{2}$

7. $\tan^{-1} 0$

8. $\tan^{-1}\dfrac{\sqrt{3}}{3}$

9. $\cos^{-1}\dfrac{\sqrt{3}}{2}$

10. $\cot^{-1}\left(-\dfrac{\sqrt{3}}{3}\right)$

11. $\csc^{-1} 2$

12. $\sin^{-1}\dfrac{1}{2}$

13. $\cot^{-1}\left(-\sqrt{3}\right)$

14. $\tan^{-1}(-1)$

15. $\sin^{-1}\left(-\dfrac{1}{2}\right)$

16. $\cos^{-1}\left(-\dfrac{\sqrt{2}}{2}\right)$

17. $\cos^{-1} 0$

18. $\sin^{-1}\dfrac{\sqrt{3}}{2}$

19. $\sec^{-1} 2$

20. $\csc^{-1}(-1)$

Use a calculator to find each of the following in radians, rounded to four decimal places, and in degrees, rounded to the nearest tenth of a degree.

21. $\tan^{-1} 0.3673$

22. $\cos^{-1}(-0.2935)$

23. $\sin^{-1} 0.9613$

24. $\sin^{-1}(-0.6199)$

25. $\cos^{-1}(-0.9810)$

26. $\tan^{-1} 158$

27. $\csc^{-1}(-6.2774)$

28. $\sec^{-1} 1.1677$

29. $\tan^{-1}(1.091)$

30. $\cot^{-1} 1.265$

31. $\sin^{-1}(-0.8192)$

32. $\cos^{-1}(-0.2716)$

33. State the domains of the inverse sine, inverse cosine, and inverse tangent functions.

34. State the ranges of the inverse sine, inverse cosine, and inverse tangent functions.

35. *Angle of Depression.*   An airplane is flying at an altitude of 2000 ft toward an island. The straight-line distance from the airplane to the island is *d* feet. Express θ, the angle of depression, as a function of *d*.

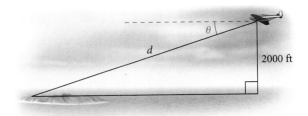

36. *Angle of Inclination.*   A guy wire is attached to the top of a 50-ft pole and stretched to a point that is *d* feet from the bottom of the pole. Express β, the angle of inclination, as a function of *d*.

Evaluate.

37. $\sin(\sin^{-1} 0.3)$

38. $\tan\left[\tan^{-1}(-4.2)\right]$

39. $\cos^{-1}\left[\cos\left(-\dfrac{\pi}{4}\right)\right]$

40. $\sin^{-1}\left(\sin\dfrac{2\pi}{3}\right)$

41. $\sin^{-1}\left(\sin\dfrac{\pi}{5}\right)$

42. $\cot^{-1}\left(\cot\dfrac{2\pi}{3}\right)$

43. $\tan^{-1}\left(\tan\dfrac{2\pi}{3}\right)$

44. $\cos^{-1}\left(\cos\dfrac{\pi}{7}\right)$

45. $\sin\left(\tan^{-1}\dfrac{\sqrt{3}}{3}\right)$

46. $\cos\left(\sin^{-1}\dfrac{\sqrt{3}}{2}\right)$

47. $\tan\left(\cos^{-1}\dfrac{\sqrt{2}}{2}\right)$

48. $\cos^{-1}(\sin\pi)$

49. $\sin^{-1}\left(\cos\dfrac{\pi}{6}\right)$

50. $\sin^{-1}\left[\tan\left(-\dfrac{\pi}{4}\right)\right]$

51. $\tan(\sin^{-1}0.1)$

52. $\cos\left(\tan^{-1}\dfrac{\sqrt{3}}{4}\right)$

53. $\sin^{-1}\left(\sin\dfrac{7\pi}{6}\right)$

54. $\tan^{-1}\left(\tan-\dfrac{3\pi}{4}\right)$

Find.

55. $\sin\left(\tan^{-1}\dfrac{a}{3}\right)$

56. $\tan\left(\cos^{-1}\dfrac{3}{x}\right)$

57. $\cot\left(\sin^{-1}\dfrac{p}{q}\right)$

58. $\sin(\cos^{-1}x)$

59. $\tan\left(\sin^{-1}\dfrac{p}{\sqrt{p^2+9}}\right)$

60. $\tan\left(\dfrac{1}{2}\sin^{-1}\dfrac{1}{2}\right)$

61. $\cos\left(\dfrac{1}{2}\sin^{-1}\dfrac{\sqrt{3}}{2}\right)$

62. $\sin\left(2\cos^{-1}\dfrac{3}{5}\right)$

Evaluate.

63. $\cos\left(\sin^{-1}\dfrac{\sqrt{2}}{2}+\cos^{-1}\dfrac{3}{5}\right)$

64. $\sin\left(\sin^{-1}\dfrac{1}{2}+\cos^{-1}\dfrac{3}{5}\right)$

65. $\sin(\sin^{-1}x+\cos^{-1}y)$

66. $\cos(\sin^{-1}x-\cos^{-1}y)$

67. $\sin(\sin^{-1}0.6032+\cos^{-1}0.4621)$

68. $\cos(\sin^{-1}0.7325-\cos^{-1}0.4838)$

Collaborative Discussion and Writing

69. Explain in your own words why the ranges of the inverse trigonometric functions are restricted.

70. How does the graph of $y=\sin^{-1}x$ differ from the graph of $y=\sin x$?

71. Why is it that

$$\sin\dfrac{5\pi}{6}=\dfrac{1}{2},$$

but

$$\sin^{-1}\left(\dfrac{1}{2}\right)\neq\dfrac{5\pi}{6}?$$

Skill Maintenance

In each of Exercises 72–80, fill in the blank with the correct term. Some of the given choices will not be used.

- linear speed
- angular speed
- angle of elevation
- angle of depression
- complementary
- supplementary
- similar
- congruent
- circular
- periodic
- period
- amplitude
- acute
- obtuse
- quadrantal
- radian measure

72. A function f is said to be _____ if there exists a positive constant p such that $f(s+p)=f(s)$ for all s in the domain of f.

73. The _____ of a rotation is the ratio of the distance s traveled by a point at a radius r from the center of rotation to the length of the radius r.

74. Triangles are _____ if their corresponding angles have the same measure.

75. The angle between the horizontal and a line of sight below the horizontal is called a(n) _____ .

76. _____ is the amount of rotation per unit of time.

77. Two positive angles are _____ if their sum is 180°.

78. The _____ of a periodic function is one half of the distance between its maximum and minimum function values.

79. A(n) _____ angle is an angle with measure greater than 0° and less than 90°.

80. Trigonometric functions with domains composed of real numbers are called _____ functions.

Synthesis

Prove the identity.

81. $\sin^{-1} x + \cos^{-1} x = \dfrac{\pi}{2}$

82. $\tan^{-1} x + \cot^{-1} x = \dfrac{\pi}{2}$

83. $\sin^{-1} x = \tan^{-1} \dfrac{x}{\sqrt{1 - x^2}}$

84. $\tan^{-1} x = \sin^{-1} \dfrac{x}{\sqrt{x^2 + 1}}$

85. $\sin^{-1} x = \cos^{-1} \sqrt{1 - x^2}$, for $x \geq 0$

86. $\cos^{-1} x = \tan^{-1} \dfrac{\sqrt{1 - x^2}}{x}$, for $x > 0$

87. *Height of a Mural.*   An art student's eye is at a point A, looking at a mural of height h, with the bottom of the mural y feet above the eye. The eye is x feet from the wall. Write an expression for θ in terms of x, y, and h. Then evaluate the expression when $x = 20$ ft, $y = 7$ ft, and $h = 25$ ft.

88. Use a calculator to approximate the following expression:

$$16 \tan^{-1} \dfrac{1}{5} - 4 \tan^{-1} \dfrac{1}{239}.$$

What number does this expression seem to approximate?

2.5 ◆ Solving Trigonometric Equations

◆ Solve trigonometric equations.

When an equation contains a trigonometric expression with a variable, such as $\cos x$, it is called a trigonometric equation. Some trigonometric equations are identities, such as $\sin^2 x + \cos^2 x = 1$. Now we consider equations, such as $2 \cos x = -1$, that are usually not identities. As we have done for other types of equations, we will solve such equations by finding all values for x that make the equation true.

EXAMPLE 1 Solve: $2 \cos x = -1$.

ALGEBRAIC SOLUTION

We first solve for $\cos x$:

$$2 \cos x = -1$$

$$\cos x = -\frac{1}{2}.$$

The solutions are numbers that have a cosine of $-\frac{1}{2}$. To find them, we use the unit circle (see Section 1.5).

There are just two points on the unit circle for which the cosine is $-\frac{1}{2}$, as shown in the following figure.

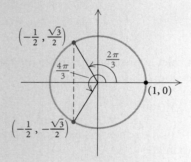

They are the points corresponding to $2\pi/3$ and $4\pi/3$. These numbers, plus any multiple of 2π, are the solutions:

$$\frac{2\pi}{3} + 2k\pi \quad \text{and} \quad \frac{4\pi}{3} + 2k\pi,$$

where k is any integer. In degrees, the solutions are

$$120° + k \cdot 360° \quad \text{and} \quad 240° + k \cdot 360°,$$

where k is any integer.

VISUALIZING THE SOLUTION

We graph $y = 2 \cos x$ and $y = -1$. The first coordinates of the points of intersection of the graphs are the values of x for which $2 \cos x = -1$.

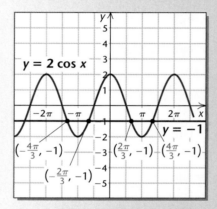

The only solutions in $[-2\pi, 2\pi]$ are

$$-\frac{4\pi}{3}, \quad -\frac{2\pi}{3}, \quad \frac{2\pi}{3}, \quad \text{and} \quad \frac{4\pi}{3}.$$

Since the cosine is periodic, there is an infinite number of solutions. Thus the entire set of solutions is

$$\frac{2\pi}{3} + 2k\pi \quad \text{and} \quad \frac{4\pi}{3} + 2k\pi,$$

where k is any integer.

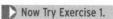 Now Try Exercise 1.

EXAMPLE 2 Solve: $4 \sin^2 x = 1$.

ALGEBRAIC SOLUTION

We begin by solving for $\sin x$:

$$4 \sin^2 x = 1$$

$$\sin^2 x = \frac{1}{4}$$

$$\sin x = \pm \frac{1}{2}.$$

Again, we use the unit circle to find those numbers having a sine of $\frac{1}{2}$ or $-\frac{1}{2}$.

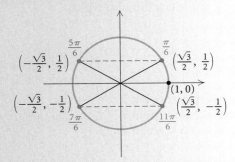

The solutions are

$$\frac{\pi}{6} + 2k\pi, \quad \frac{5\pi}{6} + 2k\pi, \quad \frac{7\pi}{6} + 2k\pi,$$

and

$$\frac{11\pi}{6} + 2k\pi,$$

where k is any integer. In degrees, the solutions are

$$30° + k \cdot 360°, \quad 150° + k \cdot 360°,$$
$$210° + k \cdot 360°, \quad \text{and} \quad 330° + k \cdot 360°,$$

where k is any integer.

The general solutions listed above could be condensed using odd as well as even multiples of π:

$$\frac{\pi}{6} + k\pi \quad \text{and} \quad \frac{5\pi}{6} + k\pi,$$

or, in degrees,

$$30° + k \cdot 180° \quad \text{and} \quad 150° + k \cdot 180°,$$

where k is any integer.

VISUALIZING THE SOLUTION

From the graph shown here, we see that the first coordinates of the points of intersection of the graphs of

$$y = 4 \sin^2 x \quad \text{and} \quad y = 1$$

in $[0, 2\pi)$ are

$$\frac{\pi}{6}, \quad \frac{5\pi}{6}, \quad \frac{7\pi}{6}, \quad \text{and} \quad \frac{11\pi}{6}.$$

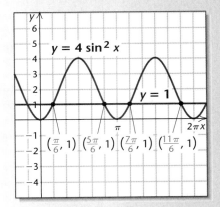

Thus, since the sine function is periodic, the general solutions are

$$\frac{\pi}{6} + k\pi \quad \text{and} \quad \frac{5\pi}{6} + k\pi,$$

or, in degrees,

$$30° + k \cdot 180° \quad \text{and} \quad 150° + k \cdot 180°,$$

where k is any integer.

▷ Now Try Exercise 13.

> In most applications, it is sufficient to find just the solutions from 0 to 2π or from 0° to 360°. We then remember that any multiple of 2π, or 360°, can be added to obtain the rest of the solutions.

We must be careful to find all solutions in $[0, 2\pi)$ when solving trigonometric equations involving double angles.

EXAMPLE 3   Solve $3 \tan 2x = -3$ in the interval $[0, 2\pi)$.

Solution   We first solve for $\tan 2x$:

$$3 \tan 2x = -3$$
$$\tan 2x = -1.$$

We are looking for solutions x to the equation for which

$$0 \le x < 2\pi.$$

Multiplying by 2, we get

$$0 \le 2x < 4\pi,$$

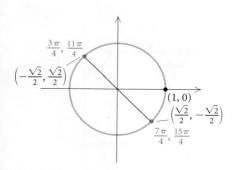

which is the interval we use when solving $\tan 2x = -1$.

Using the unit circle, we find points $2x$ in $[0, 4\pi)$ for which $\tan 2x = -1$. These values of $2x$ are as follows:

$$2x = \frac{3\pi}{4}, \quad \frac{7\pi}{4}, \quad \frac{11\pi}{4}, \quad \text{and} \quad \frac{15\pi}{4}.$$

Thus the desired values of x in $[0, 2\pi)$ are each of these values divided by 2. Therefore,

$$x = \frac{3\pi}{8}, \quad \frac{7\pi}{8}, \quad \frac{11\pi}{8}, \quad \text{and} \quad \frac{15\pi}{8}.$$

▶ Now Try Exercise 21.

Calculators are needed to solve some trigonometric equations. Answers can be found in radians or degrees, depending on the mode setting.

EXAMPLE 4   Solve $\dfrac{1}{2} \cos \phi + 1 = 1.2108$ in $[0, 360°)$.

Solution   We have

$$\frac{1}{2} \cos \phi + 1 = 1.2108$$

$$\frac{1}{2} \cos \phi = 0.2108$$

$$\cos \phi = 0.4216.$$

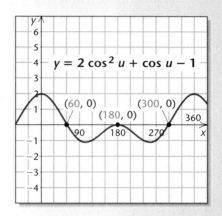

Using a calculator set in DEGREE mode, we find that the reference angle, $\cos^{-1} 0.4216$, is

$$\phi \approx 65.06°.$$

Since $\cos \phi$ is positive, the solutions are in quadrants I and IV. The solutions in $[0, 360°)$ are

$$65.06° \quad \text{and} \quad 360° - 65.06° = 294.94°. \qquad \blacktriangleright \text{ Now Try Exercise 9.}$$

EXAMPLE 5 Solve $2 \cos^2 u = 1 - \cos u$ in $[0°, 360°)$.

ALGEBRAIC SOLUTION

We use the principle of zero products:

$$2 \cos^2 u = 1 - \cos u$$
$$2 \cos^2 u + \cos u - 1 = 0$$
$$(2 \cos u - 1)(\cos u + 1) = 0$$
$$2 \cos u - 1 = 0 \quad \text{or} \quad \cos u + 1 = 0$$
$$2 \cos u = 1 \quad \text{or} \quad \cos u = -1$$
$$\cos u = \frac{1}{2} \quad \text{or} \quad \cos u = -1.$$

Thus,

$$u = 60°, 300° \quad \text{or} \quad u = 180°.$$

The solutions in $[0°, 360°)$ are $60°$, $180°$, and $300°$.

VISUALIZING THE SOLUTION

The solutions of the equation are the zeros of the function

$$y = 2 \cos^2 u + \cos u - 1.$$

Note that they are also the first coordinates of the x-intercepts of the graph.

The zeros in $[0°, 360°)$ are $60°$, $180°$, and $300°$. Thus the solutions of the equation in $[0°, 360°)$ are $60°$, $180°$, and $300°$.

$\blacktriangleright$ **Now Try Exercise 15.**

TECHNOLOGY CONNECTION

We can use either the Intersect method or the Zero method to solve trigonometric equations. Here we illustrate by solving the equation in Example 5 using both methods.

Intersect Method. We graph the equations

$$y_1 = 2\cos^2 x \quad \text{and} \quad y_2 = 1 - \cos x$$

and use the INTERSECT feature to find the first coordinates of the points of intersection.

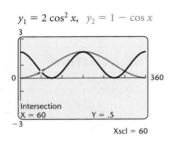

$y_1 = 2\cos^2 x, \quad y_2 = 1 - \cos x$

The leftmost solution is 60°. Using the INTERSECT feature two more times, we find the other solutions, 180° and 300°.

Zero Method. We write the equation in the form

$$2\cos^2 u + \cos u - 1 = 0.$$

Then we graph

$$y = 2\cos^2 x + \cos x - 1$$

and use the ZERO feature to determine the zeros of the function.

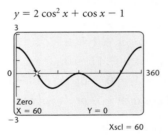

$y = 2\cos^2 x + \cos x - 1$

The leftmost zero is 60°. Using the ZERO feature two more times, we find the other zeros, 180° and 300°. The solutions in $[0°, 360°)$ are 60°, 180°, and 300°.

EXAMPLE 6 Solve $\sin^2 \beta - \sin \beta = 0$ in $[0, 2\pi)$.

ALGEBRAIC SOLUTION

We factor and use the principle of zero products:

$$\sin^2 \beta - \sin \beta = 0$$
$$\sin \beta \,(\sin \beta - 1) = 0 \qquad \text{Factoring}$$
$$\sin \beta = 0 \quad or \quad \sin \beta - 1 = 0$$
$$\sin \beta = 0 \quad or \quad \sin \beta = 1$$
$$\beta = 0, \pi \quad or \qquad \beta = \frac{\pi}{2}.$$

The solutions in $[0, 2\pi)$ are 0, $\pi/2$, and π.

VISUALIZING THE SOLUTION

The solutions of the equation

$$\sin^2 \beta - \sin \beta = 0$$

are the zeros of the function

$$f(\beta) = \sin^2 \beta - \sin \beta.$$

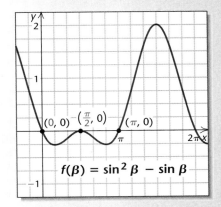

The zeros in $[0, 2\pi)$ are 0, $\pi/2$, and π. Thus the solutions of $\sin^2 \beta - \sin \beta = 0$ are 0, $\pi/2$, and π.

▶ Now Try Exercise 17.

If a trigonometric equation is quadratic but difficult or impossible to factor, we use the *quadratic formula*.

EXAMPLE 7   Solve $10 \sin^2 x - 12 \sin x - 7 = 0$ in $[0°, 360°)$.

Solution   This equation is quadratic in $\sin x$ with $a = 10$, $b = -12$, and $c = -7$. Substituting into the quadratic formula, we get

$$\sin x = \frac{-b \pm \sqrt{b^2 - 4ac}}{2a} \qquad \text{Using the quadratic formula}$$

$$= \frac{-(-12) \pm \sqrt{(-12)^2 - 4(10)(-7)}}{2 \cdot 10} \qquad \text{Substituting}$$

$$= \frac{12 \pm \sqrt{144 + 280}}{20}$$

$$= \frac{12 \pm \sqrt{424}}{20}$$

$$\approx \frac{12 \pm 20.5913}{20}$$

$$\sin x \approx 1.6296 \quad or \quad \sin x \approx -0.4296.$$

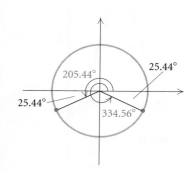

Since sine values are never greater than 1, the first of the equations has no solution. Using the other equation, we find the reference angle to be 25.44°. Since $\sin x$ is negative, the solutions are in quadrants III and IV.
Thus the solutions in $[0°, 360°)$ are

$$180° + 25.44° = 205.44° \quad \text{and} \quad 360° - 25.44° = 334.56°.$$

▶ Now Try Exercise 23.

Trigonometric equations can involve more than one function.

EXAMPLE 8   Solve $2 \cos^2 x \tan x = \tan x$ in $[0, 2\pi)$.

Solution   We have

$$2 \cos^2 x \tan x = \tan x$$

$$2 \cos^2 x \tan x - \tan x = 0$$

$$\tan x (2 \cos^2 x - 1) = 0$$

$$\tan x = 0 \qquad or \qquad 2 \cos^2 x - 1 = 0$$

$$\cos^2 x = \frac{1}{2}$$

$$\cos x = \pm \frac{\sqrt{2}}{2}$$

$$x = 0, \pi \quad or \qquad x = \frac{\pi}{4}, \frac{3\pi}{4}, \frac{5\pi}{4}, \frac{7\pi}{4}.$$

Thus, $x = 0, \pi/4, 3\pi/4, \pi, 5\pi/4$, and $7\pi/4$.

▶ Now Try Exercise 21.

TECHNOLOGY CONNECTION

In Example 9, we can graph the left side and then the right side of the equation as seen in the first window below. Then we look for points of intersection. We could also rewrite the equation as $\sin x + \cos x - 1 = 0$, graph the left side, and look for the zeros of the function, as illustrated in the second window below. In each window, we see the solutions in $[0, 2\pi)$ as 0 and $\pi/2$.

This example illustrates a valuable advantage of the calculator—that is, with a graphing calculator, extraneous solutions do not appear.

$y_1 = \sin x + \cos x, \quad y_2 = 1$

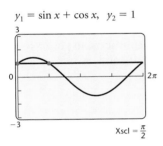

$y = \sin x + \cos x - 1$

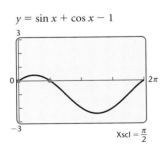

When a trigonometric equation involves more than one function, it is sometimes helpful to use identities to rewrite the equation in terms of a single function.

EXAMPLE 9 Solve $\sin x + \cos x = 1$ in $[0, 2\pi)$.

Solution We have

$$\sin x + \cos x = 1$$
$$(\sin x + \cos x)^2 = 1^2 \qquad \text{Squaring both sides}$$
$$\sin^2 x + 2 \sin x \cos x + \cos^2 x = 1$$
$$2 \sin x \cos x + 1 = 1 \qquad \text{Using } \sin^2 x + \cos^2 x = 1$$
$$2 \sin x \cos x = 0$$
$$\sin 2x = 0. \qquad \text{Using } 2 \sin x \cos x = \sin 2x$$

We are looking for solutions x to the equation for which $0 \leq x < 2\pi$. Multiplying by 2, we get $0 \leq 2x < 4\pi$, which is the interval we consider to solve $\sin 2x = 0$. These values of $2x$ are 0, π, 2π, and 3π. Thus the desired values of x in $[0, 2\pi)$ satisfying this equation are 0, $\pi/2$, π, and $3\pi/2$. Now we check these in the original equation $\sin x + \cos x = 1$:

$$\sin 0 + \cos 0 = 0 + 1 = 1,$$

$$\sin \frac{\pi}{2} + \cos \frac{\pi}{2} = 1 + 0 = 1,$$

$$\sin \pi + \cos \pi = 0 + (-1) = -1,$$

$$\sin \frac{3\pi}{2} + \cos \frac{3\pi}{2} = (-1) + 0 = -1.$$

We find that π and $3\pi/2$ do not check, but the other values do. Thus the solutions in $[0, 2\pi)$ are

$$0 \quad \text{and} \quad \frac{\pi}{2}.$$

When the solution process involves squaring both sides, values are sometimes obtained that are not solutions of the original equation. As we saw in this example, it is important to check the possible solutions.

▶ **Now Try Exercise 39.**

EXAMPLE 10 Solve $\cos 2x + \sin x = 1$ in $[0, 2\pi)$.

ALGEBRAIC SOLUTION

We have

$$\cos 2x + \sin x = 1$$
$$1 - 2 \sin^2 x + \sin x = 1 \qquad \text{Using the identity } \cos 2x = 1 - 2 \sin^2 x$$
$$-2 \sin^2 x + \sin x = 0$$
$$\sin x \,(-2 \sin x + 1) = 0 \qquad \text{Factoring}$$
$$\sin x = 0 \quad or \quad -2 \sin x + 1 = 0 \qquad \text{Principle of zero products}$$
$$\sin x = 0 \quad or \quad \sin x = \frac{1}{2}$$
$$x = 0, \pi \quad or \quad x = \frac{\pi}{6}, \frac{5\pi}{6}.$$

All four values check. The solutions in $[0, 2\pi)$ are 0, $\pi/6$, $5\pi/6$, and π.

VISUALIZING THE SOLUTION

We graph the function
$y = \cos 2x + \sin x - 1$ and look for the zeros of the function.

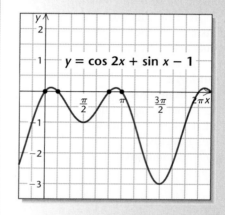

The zeros, or solutions, in $[0, 2\pi)$ are 0, $\pi/6, 5\pi/6$, and π.

◀

EXAMPLE 11 Solve $\tan^2 x + \sec x - 1 = 0$ in $[0, 2\pi)$.

Solution We have

$$\tan^2 x + \sec x - 1 = 0$$
$$\sec^2 x - 1 + \sec x - 1 = 0 \qquad \text{Using the identity } 1 + \tan^2 x = \sec^2 x, \text{ or } \tan^2 x = \sec^2 x - 1$$
$$\sec^2 x + \sec x - 2 = 0$$
$$(\sec x + 2)(\sec x - 1) = 0 \qquad \text{Factoring}$$
$$\sec x = -2 \quad or \quad \sec x = 1 \qquad \text{Principle of zero products}$$
$$\cos x = -\frac{1}{2} \quad or \quad \cos x = 1 \qquad \text{Using the identity } \cos x = 1/\sec x$$
$$x = \frac{2\pi}{3}, \frac{4\pi}{3} \quad or \quad x = 0.$$

All these values check. The solutions in $[0, 2\pi)$ are 0, $2\pi/3$, and $4\pi/3$.

▶ Now Try Exercise 27.

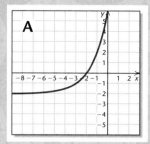

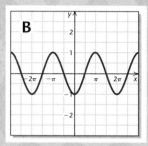

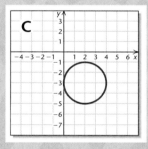

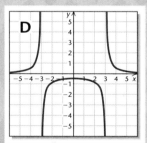

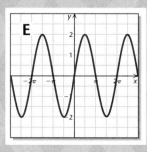

Visualizing the Graph

Match the equation with its graph.

1. $f(x) = \dfrac{4}{x^2 - 9}$

2. $f(x) = \dfrac{1}{2} \sin x - 1$

3. $(x - 2)^2 + (y + 3)^2 = 4$

4. $y = \sin^2 x + \cos^2 x$

5. $f(x) = 3 - \log x$

6. $f(x) = 2^{x+3} - 2$

7. $y = 2 \cos\left(x - \dfrac{\pi}{2}\right)$

8. $y = -x^3 + 3x^2$

9. $f(x) = (x - 3)^2 + 2$

10. $f(x) = -\cos x$

Answers on page A-14

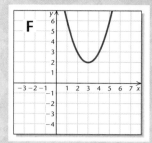

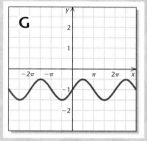

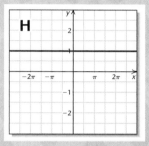

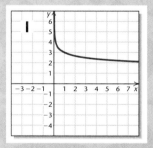

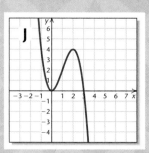

2.5 EXERCISE SET

Solve, finding all solutions. Express the solutions in both radians and degrees.

1. $\cos x = \dfrac{\sqrt{3}}{2}$

2. $\sin x = -\dfrac{\sqrt{2}}{2}$

3. $\tan x = -\sqrt{3}$

4. $\cos x = -\dfrac{1}{2}$

5. $\sin x = \dfrac{1}{2}$

6. $\tan x = -1$

7. $\cos x = -\dfrac{\sqrt{2}}{2}$

8. $\sin x = \dfrac{\sqrt{3}}{2}$

Solve, finding all solutions in $[0, 2\pi)$ or $[0°, 360°)$.

9. $2 \cos x - 1 = -1.2814$

10. $\sin x + 3 = 2.0816$

11. $2 \sin x + \sqrt{3} = 0$

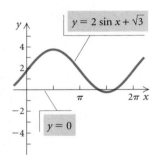

12. $2 \tan x - 4 = 1$

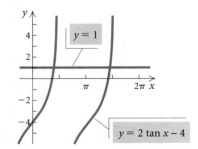

13. $2 \cos^2 x = 1$

14. $\csc^2 x - 4 = 0$

15. $2 \sin^2 x + \sin x = 1$

16. $\cos^2 x + 2 \cos x = 3$

17. $2 \cos^2 x - \sqrt{3} \cos x = 0$

18. $2 \sin^2 \theta + 7 \sin \theta = 4$

19. $6 \cos^2 \phi + 5 \cos \phi + 1 = 0$

20. $2 \sin t \cos t + 2 \sin t - \cos t - 1 = 0$

21. $\sin 2x \cos x - \sin x = 0$

22. $5 \sin^2 x - 8 \sin x = 3$

23. $\cos^2 x + 6 \cos x + 4 = 0$

24. $2 \tan^2 x = 3 \tan x + 7$

25. $7 = \cot^2 x + 4 \cot x$

26. $3 \sin^2 x = 3 \sin x + 2$

Solve, finding all solutions in $[0, 2\pi)$.

27. $\cos 2x - \sin x = 1$

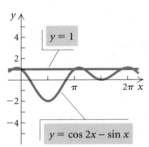

28. $2 \sin x \cos x + \sin x = 0$

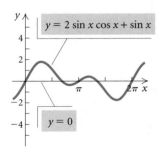

29. $\sin 4x - 2 \sin 2x = 0$

30. $\tan x \sin x - \tan x = 0$

31. $\sin 2x \cos x + \sin x = 0$

32. $\cos 2x \sin x + \sin x = 0$

33. $2 \sec x \tan x + 2 \sec x + \tan x + 1 = 0$

34. $\sin 2x \sin x - \cos 2x \cos x = -\cos x$

35. $\sin 2x + \sin x + 2 \cos x + 1 = 0$

36. $\tan^2 x + 4 = 2 \sec^2 x + \tan x$

37. $\sec^2 x - 2 \tan^2 x = 0$

38. $\cot x = \tan (2x - 3\pi)$

39. $2 \cos x + 2 \sin x = \sqrt{6}$

40. $\sqrt{3} \cos x - \sin x = 1$

41. $\sec^2 x + 2 \tan x = 6$

42. $5 \cos 2x + \sin x = 4$

43. $\cos (\pi - x) + \sin \left(x - \dfrac{\pi}{2} \right) = 1$

44. $\dfrac{\sin^2 x - 1}{\cos \left(\dfrac{\pi}{2} - x \right) + 1} = \dfrac{\sqrt{2}}{2} - 1$

Technology Connection

45. Check the solutions of Exercises 14, 19, 20, and 21 using the Zero method.

46. Check the solutions of Exercises 15, 18, 24, and 25 using the Intersect method.

Solve using a calculator, finding all solutions in $[0, 2\pi)$.

47. $x \sin x = 1$

48. $x^2 + 2 = \sin x$

49. $2 \cos^2 x = x + 1$

50. $x \cos x - 2 = 0$

51. $\cos x - 2 = x^2 - 3x$

52. $\sin x = \tan \dfrac{x}{2}$

Some graphing calculators can use regression to fit a trigonometric function to a set of data.

53. *Sales.* Sales of certain products fluctuate in cycles. The data in the following table show the total sales of skis per month for a business in a northern climate.

Month, x		Total Sales, y (in thousands)
August,	8	$ 0
November,	11	7
February,	2	14
May,	5	7
August,	8	0

a) Using the SINE REGRESSION feature on a graphing calculator, fit a sine function of the form $y = A \sin (Bx - C) + D$ to this set of data.

b) Approximate the total sales for December and for July.

54. *Daylight Hours.*   The data in the following table show the number of daylight hours for certain days in Kajaani, Finland.

Day, x		Number of Daylight Hours, y
January 10,	10	5.0
February 19,	50	9.1
March 3,	62	10.4
April 28,	118	16.4
May 14,	134	18.2
June 11,	162	20.7
July 17,	198	19.5
August 22,	234	15.7
September 19,	262	12.7
October 1,	274	11.4
November 14,	318	6.7
December 28,	362	4.3

Source: *The Astronomical Almanac*, 1995, Washington: U.S. Government Printing Office.

a) Using the SINE REGRESSION feature on a graphing calculator, model these data with an equation of the form $y = A \sin (Bx - C) + D$.

b) Approximate the number of daylight hours in Kajaani for April 22 ($x = 112$), July 4 ($x = 185$), and December 15 ($x = 349$).

Collaborative Discussion and Writing

55. Jan lists her answer to a problem as $\pi/6 + k\pi$, for any integer k, while Jacob lists his answer as $\pi/6 + 2k\pi$ and $7\pi/6 + 2\pi k$, for any integer k. Are their answers equivalent? Why or why not?

56. An identity is an equation that is true for all possible replacements of the variables. Explain the meaning of "possible" in this definition.

Skill Maintenance

Solve the right triangle.

57.

58.

Solve.

59. $\dfrac{x}{27} = \dfrac{4}{3}$

60. $\dfrac{0.01}{0.7} = \dfrac{0.2}{h}$

Synthesis

Solve in $[0, 2\pi)$.

61. $|\sin x| = \dfrac{\sqrt{3}}{2}$

62. $|\cos x| = \dfrac{1}{2}$

63. $\sqrt{\tan x} = \sqrt[4]{3}$

64. $12 \sin x - 7\sqrt{\sin x} + 1 = 0$

65. $\ln (\cos x) = 0$

66. $e^{\sin x} = 1$

67. $\sin (\ln x) = -1$

68. $e^{\ln (\sin x)} = 1$

69. *Temperature during an Illness.* The temperature T, in degrees Fahrenheit, of a patient t days into a 12-day illness is given by

$$T(t) = 101.6° + 3° \sin \left(\frac{\pi}{8}t\right).$$

Find the times t during the illness at which the patient's temperature was 103°.

70. *Satellite Location.* A satellite circles the earth in such a manner that it is y miles from the equator (north or south, height from the surface not considered) t minutes after its launch, where

$$y = 5000 \left[\cos \frac{\pi}{45}(t - 10) \right].$$

At what times t in the interval $[0, 240]$, the first 4 hr, is the satellite 3000 mi north of the equator?

71. *Nautical Mile.* (See Exercise 60 in Exercise Set 2.2.) In Great Britain, the *nautical mile* is defined as the length of a minute of arc of the earth's radius. Since the earth is flattened at the poles, a British nautical mile varies with latitude. In fact, it is given, in feet, by the function

$$N(\phi) = 6066 - 31 \cos 2\phi,$$

where ϕ is the latitude in degrees. At what latitude north is the length of a British nautical mile found to be 6040 ft?

72. *Acceleration due to Gravity.* (See Exercise 61 in Exercise Set 2.2.) The acceleration due to gravity is often denoted by g in a formula such as $S = \frac{1}{2}gt^2$, where S is the distance that an object falls in t seconds. The number g is generally considered constant, but in fact it varies slightly with latitude. If ϕ stands for latitude, in degrees, an excellent approximation of g is given by the formula

$$g = 9.78049(1 + 0.005288 \sin^2 \phi - 0.000006 \sin^2 2\phi),$$

where g is measured in meters per second per second at sea level. At what latitude north does $g = 9.8$?

Solve.

73. $\cos^{-1} x = \cos^{-1} \frac{3}{5} - \sin^{-1} \frac{4}{5}$

74. $\sin^{-1} x = \tan^{-1} \frac{1}{3} + \tan^{-1} \frac{1}{2}$

75. Suppose that $\sin x = 5 \cos x$. Find $\sin x \cos x$.

CHAPTER 2 SUMMARY AND REVIEW

Important Properties and Formulas

Basic Identities

$$\sin x = \frac{1}{\csc x}, \qquad \tan x = \frac{\sin x}{\cos x},$$

$$\cos x = \frac{1}{\sec x}, \qquad \cot x = \frac{\cos x}{\sin x}$$

$$\tan x = \frac{1}{\cot x}$$

$$\sin (-x) = -\sin x,$$

$$\cos (-x) = \cos x,$$

$$\tan (-x) = -\tan x$$

Pythagorean Identities

$$\sin^2 x + \cos^2 x = 1,$$

$$1 + \cot^2 x = \csc^2 x,$$

$$1 + \tan^2 x = \sec^2 x$$

Sum and Difference Identities

$$\sin (u \pm v) = \sin u \cos v \pm \cos u \sin v,$$

$$\cos (u \pm v) = \cos u \cos v \mp \sin u \sin v,$$

$$\tan (u \pm v) = \frac{\tan u \pm \tan v}{1 \mp \tan u \tan v}$$

Double-Angle Identities

$$\sin 2x = 2 \sin x \cos x,$$

$$\cos 2x = \cos^2 x - \sin^2 x$$

$$= 1 - 2 \sin^2 x$$

$$= 2 \cos^2 x - 1,$$

$$\tan 2x = \frac{2 \tan x}{1 - \tan^2 x}$$

Half-Angle Identities

$$\sin \frac{x}{2} = \pm \sqrt{\frac{1 - \cos x}{2}},$$

$$\cos \frac{x}{2} = \pm \sqrt{\frac{1 + \cos x}{2}}$$

$$\tan \frac{x}{2} = \pm \sqrt{\frac{1 - \cos x}{1 + \cos x}}$$

$$= \frac{\sin x}{1 + \cos x}$$

$$= \frac{1 - \cos x}{\sin x}$$

Cofunction Identities

$$\sin \left(\frac{\pi}{2} - x \right) = \cos x, \qquad \cos \left(\frac{\pi}{2} - x \right) = \sin x, \qquad \sin \left(x \pm \frac{\pi}{2} \right) = \mp \cos x,$$

$$\tan \left(\frac{\pi}{2} - x \right) = \cot x, \qquad \cot \left(\frac{\pi}{2} - x \right) = \tan x, \qquad \cos \left(x \pm \frac{\pi}{2} \right) = \pm \sin x$$

$$\sec \left(\frac{\pi}{2} - x \right) = \csc x, \qquad \csc \left(\frac{\pi}{2} - x \right) = \sec x,$$

Product-to-Sum Identities

$$\sin x \cdot \sin y = \frac{1}{2}[\cos (x - y) - \cos (x + y)]$$

$$\cos x \cdot \cos y = \frac{1}{2}[\cos (x - y) + \cos (x + y)]$$

$$\sin x \cdot \cos y = \frac{1}{2}[\sin (x + y) + \sin (x - y)]$$

$$\cos x \cdot \sin y = \frac{1}{2}[\sin (x + y) - \sin (x - y)]$$

Sum-to-Product Identities

$$\sin x + \sin y = 2 \sin \frac{x + y}{2} \cos \frac{x - y}{2}$$

$$\sin x - \sin y = 2 \cos \frac{x + y}{2} \sin \frac{x - y}{2}$$

$$\cos y + \cos x = 2 \cos \frac{x + y}{2} \cos \frac{x - y}{2}$$

$$\cos y - \cos x = 2 \sin \frac{x + y}{2} \sin \frac{x - y}{2}$$

Inverse Trigonometric Functions

Function	Domain	Range
$y = \sin^{-1} x$	$[-1, 1]$	$\left[-\dfrac{\pi}{2}, \dfrac{\pi}{2}\right]$
$y = \cos^{-1} x$	$[-1, 1]$	$[0, \pi]$
$y = \tan^{-1} x$	$(-\infty, \infty)$	$\left(-\dfrac{\pi}{2}, \dfrac{\pi}{2}\right)$

Composition of Trigonometric Functions

The following are true for any x in the domain of the inverse function:

$$\sin (\sin^{-1} x) = x,$$
$$\cos (\cos^{-1} x) = x,$$
$$\tan (\tan^{-1} x) = x.$$

The following are true for any x in the range of the inverse function:

$$\sin^{-1} (\sin x) = x,$$
$$\cos^{-1} (\cos x) = x,$$
$$\tan^{-1} (\tan x) = x.$$

REVIEW EXERCISES

Determine whether the statement is true or false.

1. $\sin^2 s \neq \sin s^2$.   [2.1]

2. Given $0 < \alpha < \pi/2$ and $0 < \beta < \pi/2$ and that $\sin (\alpha + \beta) = 1$ and $\sin (\alpha - \beta) = 0$, then $\alpha = \pi/4$.   [2.1]

3. If the terminal side of θ is in quadrant IV, then $\tan \theta < \cos \theta$.   [2.1]

4. $\cos 5\pi/12 = \cos 7\pi/12$.   [2.2]

5. Given that $\sin \theta = -\dfrac{2}{5}$, $\tan \theta < \cos \theta$.   [2.1]

Complete the Pythagorean identity.   [2.1]

6. $1 + \cot^2 x =$

7. $\sin^2 x + \cos^2 x =$

Multiply and simplify.   [2.1]

8. $(\tan y - \cot y)(\tan y + \cot y)$

9. $(\cos x + \sec x)^2$

Factor and simplify.   [2.1]

10. $\sec x \csc x - \csc^2 x$

11. $3 \sin^2 y - 7 \sin y - 20$

12. $1000 - \cos^3 u$

Simplify.   [2.1]

13. $\dfrac{\sec^4 x - \tan^4 x}{\sec^2 x + \tan^2 x}$

14. $\dfrac{2 \sin^2 x}{\cos^3 x} \cdot \left(\dfrac{\cos x}{2 \sin x}\right)^2$

15. $\dfrac{3 \sin x}{\cos^2 x} \cdot \dfrac{\cos^2 x + \cos x \sin x}{\sin^2 x - \cos^2 x}$

16. $\dfrac{3}{\cos y - \sin y} - \dfrac{2}{\sin^2 y - \cos^2 y}$

17. $\left(\dfrac{\cot x}{\csc x}\right)^2 + \dfrac{1}{\csc^2 x}$

18. $\dfrac{4 \sin x \cos^2 x}{16 \sin^2 x \cos x}$

19. Simplify. Assume the radicand is nonnegative. [2.1]
$$\sqrt{\sin^2 x + 2 \cos x \sin x + \cos^2 x}$$

20. Rationalize the denominator: $\sqrt{\dfrac{1 + \sin x}{1 - \sin x}}$. [2.1]

21. Rationalize the numerator: $\sqrt{\dfrac{\cos x}{\tan x}}$. [2.1]

22. Given that $x = 3 \tan \theta$, express $\sqrt{9 + x^2}$ as a trigonometric function without radicals. Assume that $0 < \theta < \pi/2$. [2.1]

Use the sum and difference formulas to write equivalent expressions. You need not simplify. [2.1]

23. $\cos\left(x + \dfrac{3\pi}{2}\right)$ **24.** $\tan(45° - 30°)$

25. Simplify: $\cos 27° \cos 16° + \sin 27° \sin 16°$. [2.1]

26. Find $\cos 165°$ exactly. [2.1]

27. Given that $\tan \alpha = \sqrt{3}$ and $\sin \beta = \sqrt{2}/2$ and that α and β are between 0 and $\pi/2$, evaluate $\tan(\alpha - \beta)$ exactly. [2.1]

28. Assume that $\sin \theta = 0.5812$ and $\cos \phi = 0.2341$ and that both θ and ϕ are first-quadrant angles. Evaluate $\cos(\theta + \phi)$. [2.1]

Complete the cofunction identity. [2.2]

29. $\cos\left(x + \dfrac{\pi}{2}\right) =$ **30.** $\cos\left(\dfrac{\pi}{2} - x\right) =$

31. $\sin\left(x - \dfrac{\pi}{2}\right) =$

32. Given that $\cos \alpha = -\frac{3}{5}$ and that the terminal side is in quadrant III: [2.2]
  a) Find the other function values for α.
  b) Find the six function values for $\pi/2 - \alpha$.
  c) Find the six function values for $\alpha + \pi/2$.

33. Find an equivalent expression for $\csc\left(x - \dfrac{\pi}{2}\right)$. [2.2]

34. Find $\tan 2\theta$, $\cos 2\theta$, and $\sin 2\theta$ and the quadrant in which 2θ lies, where $\cos \theta = -\frac{4}{5}$ and θ is in quadrant III. [2.2]

35. Find $\sin \dfrac{\pi}{8}$ exactly. [2.2]

36. Given that $\sin \beta = 0.2183$ and β is in quadrant I, find $\sin 2\beta$, $\cos \dfrac{\beta}{2}$, and $\cos 4\beta$. [2.2]

Simplify. [2.2]

37. $1 - 2 \sin^2 \dfrac{x}{2}$

38. $(\sin x + \cos x)^2 - \sin 2x$

39. $2 \sin x \cos^3 x + 2 \sin^3 x \cos x$

40. $\dfrac{2 \cot x}{\cot^2 x - 1}$

Prove the identity. [2.3]

41. $\dfrac{1 - \sin x}{\cos x} = \dfrac{\cos x}{1 + \sin x}$

42. $\dfrac{1 + \cos 2\theta}{\sin 2\theta} = \cot \theta$

43. $\dfrac{\tan y + \sin y}{2 \tan \theta} = \cos^2 \dfrac{y}{2}$

44. $\dfrac{\sin x - \cos x}{\cos^2 x} = \dfrac{\tan^2 x - 1}{\sin x + \cos x}$

Use the product-to-sum and the sum-to-product identities to find identities for each of the following. [2.3]

45. $3 \cos 2\theta \sin\theta$

46. $\sin \theta - \sin 4\theta$

Find each of the following exactly in both radians and degrees. [2.4]

47. $\sin^{-1}\left(-\dfrac{1}{2}\right)$

48. $\cos^{-1} \dfrac{\sqrt{3}}{2}$

49. $\tan^{-1} 1$

50. $\sin^{-1} 0$

Use a calculator to find each of the following in radians, rounded to four decimal places, and in degrees, rounded to the nearest tenth of a degree.   [2.4]

51. $\cos^{-1}(-0.2194)$ **52.** $\cot^{-1} 2.381$

Evaluate.   [2.4]

53. $\cos\left(\cos^{-1}\dfrac{1}{2}\right)$ **54.** $\tan^{-1}\left(\tan\dfrac{\sqrt{3}}{3}\right)$

55. $\sin^{-1}\left(\sin\dfrac{\pi}{7}\right)$ **56.** $\cos\left(\sin^{-1}\dfrac{\sqrt{2}}{2}\right)$

Find.   [2.4]

57. $\cos\left(\tan^{-1}\dfrac{b}{3}\right)$ **58.** $\cos\left(2\sin^{-1}\dfrac{4}{5}\right)$

Solve, finding all solutions. Express the solutions in both radians and degrees.   [2.5]

59. $\cos x = -\dfrac{\sqrt{2}}{2}$ **60.** $\tan x = \sqrt{3}$

Solve, finding all solutions in $[0, 2\pi)$.   [2.5]

61. $4\sin^2 x = 1$

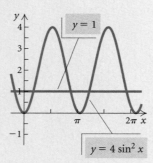

62. $\sin 2x \sin x - \cos x = 0$

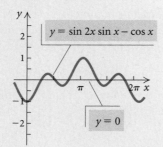

63. $2\cos^2 x + 3\cos x = -1$

64. $\sin^2 x - 7\sin x = 0$

65. $\csc^2 x - 2\cot^2 x = 0$

66. $\sin 4x + 2\sin 2x = 0$

67. $2\cos x + 2\sin x = \sqrt{2}$

68. $6\tan^2 x = 5\tan x + \sec^2 x$

69. Determine the domain of the function $\cos^{-1} x$.   [2.4]

   A. $(0, \pi)$ **B.** $[-1, 1]$
   C. $[-\pi/2, \pi/2]$ **D.** $(-\infty, \infty)$

70. Simplify $\sin^{-1}\left(\sin\dfrac{7\pi}{6}\right)$.   [2.4]

   A. $-\pi/6$ **B.** $7\pi/6$
   C. $-1/2$ **D.** $11\pi/6$

Technology Connection

In Exercises 71–74, use a graphing calculator to determine which expression (A)–(D) on the right can be used to complete the identity. Then prove the identity algebraically.   [2.3]

71. $\csc x - \cos x \cot x$ **A.** $\dfrac{\csc x}{\sec x}$

72. $\dfrac{1}{\sin x \cos x} - \dfrac{\cos x}{\sin x}$ **B.** $\sin x$

73. $\dfrac{\cot x - 1}{1 - \tan x}$ **C.** $\dfrac{2}{\sin x}$

74. $\dfrac{\cos x + 1}{\sin x} + \dfrac{\sin x}{\cos x + 1}$ **D.** $\dfrac{\sin x \cos x}{1 - \sin^2 x}$

Solve using a graphing calculator, finding all solutions in $[0, 2\pi)$.   [2.5]

75. $x \cos x = 1$

76. $2\sin^2 x = x + 1$

Collaborative Discussion and Writing

77. Prove the identity $2\cos^2 x - 1 = \cos^4 x - \sin^4 x$ in three ways:

   a) Start with the left side and deduce the right (method 1).
   b) Start with the right side and deduce the left (method 1).
   c) Work with each side separately until you deduce the same expression (method 2).

   Then determine the most efficient method and explain why you chose that method.   [2.3]

78. Why are the ranges of the inverse trigonometric functions restricted?   [2.4]

Synthesis

79. Find the measure of the angle from l_1 to l_2: [2.1]

$$l_1:\ x + y = 3 \qquad l_2:\ 2x - y = 5.$$

80. Find an identity for $\cos(u + v)$ involving only cosines. [2.1]

81. Simplify: $\cos\left(\dfrac{\pi}{2} - x\right)[\csc x - \sin x]$. [2.2]

82. Find $\sin\theta$, $\cos\theta$, and $\tan\theta$ under the given conditions: [2.2]

$$\sin 2\theta = \frac{1}{5},\ \frac{\pi}{2} \le 2\theta < \pi.$$

83. Prove the following equation to be an identity: [2.3]

$$\ln e^{\sin t} = \sin t.$$

84. Graph: $y = \sec^{-1} x$. [2.4]

85. Show that

$$\tan^{-1} x = \frac{\sin^{-1} x}{\cos^{-1} x}$$

is *not* an identity. [2.4]

86. Solve $e^{\cos x} = 1$ in $[0, 2\pi)$. [2.5]

CHAPTER 2 TEST

Simplify.

1. $\dfrac{2\cos^2 x - \cos x - 1}{\cos x - 1}$

2. $\left(\dfrac{\sec x}{\tan x}\right)^2 - \dfrac{1}{\tan^2 x}$

3. Rationalize the denominator:

$$\sqrt{\frac{1 - \sin\theta}{1 + \sin\theta}}.$$

4. Given that $x = 2\sin\theta$, express $\sqrt{4 - x^2}$ as a trigonometric function without radicals. Assume $0 < \theta < \pi/2$.

Use the sum or difference identities to evaluate exactly.

5. $\sin 75°$

6. $\tan\dfrac{\pi}{12}$

7. Assuming that $\cos u = \frac{5}{13}$ and $\cos v = \frac{12}{13}$ and that u and v are between 0 and $\pi/2$, evaluate $\cos(u - v)$ exactly.

8. Given that $\cos\theta = -\frac{2}{3}$ and that the terminal side is in quadrant II, find $\cos(\pi/2 - \theta)$.

9. Given that $\sin\theta = -\frac{4}{5}$ and θ is in quadrant III, find $\sin 2\theta$ and the quadrant in which 2θ lies.

10. Use a half-angle identity to evaluate $\cos\dfrac{\pi}{12}$ exactly.

11. Given that $\sin\theta = 0.6820$ and that θ is in quadrant I, find $\cos\theta/2$.

12. Simplify: $(\sin x + \cos x)^2 - 1 + 2\sin 2x$.

Prove each of the following identities.

13. $\csc x - \cos x\cot x = \sin x$

14. $(\sin x + \cos x)^2 = 1 + \sin 2x$

15. $(\csc\beta + \cot\beta)^2 = \dfrac{1 + \cos\beta}{1 - \cos\beta}$

16. $\dfrac{1 + \sin\alpha}{1 + \csc\alpha} = \dfrac{\tan\alpha}{\sec\alpha}$

Use the product-to-sum and sum-to-product identities to find identities for each of the following.

17. $\cos 8\alpha - \cos\alpha$

18. $4\sin\beta\cos 3\beta$

19. Find $\sin^{-1}\left(-\dfrac{\sqrt{2}}{2}\right)$ exactly in degrees.

20. Find $\tan^{-1}\sqrt{3}$ exactly in radians.

21. Use a calculator to find $\cos^{-1}(-0.6716)$ in radians, rounded to four decimal places.

22. Evaluate $\cos\left(\sin^{-1}\dfrac{1}{2}\right)$.

23. Find $\tan\left(\sin^{-1}\dfrac{5}{x}\right)$.

24. Evaluate $\cos\left(\sin^{-1}\frac{1}{2}+\cos^{-1}\frac{1}{2}\right)$.

Solve, finding all solutions in $[0, 2\pi)$.

25. $4\cos^2 x = 3$

26. $2\sin^2 x = \sqrt{2}\sin x$

27. $\sqrt{3}\cos x + \sin x = 1$

Synthesis

28. Find $\cos\theta$, given that

$$\cos 2\theta = \frac{5}{6}, \quad \frac{3\pi}{2} < \theta < 2\pi.$$

CHAPTER

3

Applications of Trigonometry

APPLICATION

 musician is constructing an octagonal recording studio in his home and needs to determine two distances for the electrician. The dimensions for the most acoustically perfect studio are shown in the figure on page 182 (*Source*: Tony Medeiros, Indianapolis, IN). Determine the distances from *D* to *F* and from *D* to *B* to the nearest tenth of an inch.

This problem appears as Example 2 in Section 3.2.

3.1 The Law of Sines

✦ Use the law of sines to solve triangles.

✦ Find the area of any triangle given the lengths of two sides and the measure of the included angle.

To **solve a triangle** means to find the lengths of all its sides and the measures of all its angles. We solved right triangles in Section 1.2. For review, let's solve the right triangle shown below. We begin by listing the known measures.

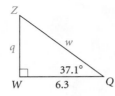

$$Q = 37.1° \qquad q = ?$$
$$W = 90° \qquad w = ?$$
$$Z = ? \qquad z = 6.3$$

Since the sum of the three angle measures of any triangle is 180°, we can immediately find the measure of the third angle:

$$Z = 180° - (90° + 37.1°)$$
$$= 52.9°.$$

Then using the tangent and cosine ratios, respectively, we can find q and w:

$$\tan 37.1° = \frac{q}{6.3}, \quad \text{or}$$

$$q = 6.3 \tan 37.1° \approx 4.8,$$

and $\quad \cos 37.1° = \dfrac{6.3}{w}, \quad$ or

$$w = \frac{6.3}{\cos 37.1°} \approx 7.9.$$

Now all six measures are known and we have solved triangle QWZ.

$$Q = 37.1° \qquad q \approx 4.8$$
$$W = 90° \qquad w \approx 7.9$$
$$Z = 52.9° \qquad z = 6.3$$

✦ Solving Oblique Triangles

The trigonometric functions can also be used to solve triangles that are not right triangles. Such triangles are called **oblique.** Any triangle, right or oblique, can be solved *if at least one side and any other two measures are known.* The five possible situations are illustrated on the next page.

1. **AAS:** Two angles of a triangle and a side opposite one of them are known.

2. **ASA:** Two angles of a triangle and the included side are known.

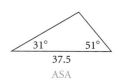

3. **SSA:** Two sides of a triangle and an angle opposite one of them are known. (In this case, there may be no solution, one solution, or two solutions. The latter is known as the ambiguous case.)

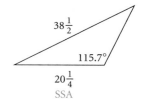

4. **SAS:** Two sides of a triangle and the included angle are known.

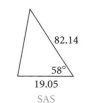

5. **SSS:** All three sides of the triangle are known.

The list above does not include the situation in which only the three angle measures are given. The reason for this lies in the fact that the angle measures determine *only the shape* of the triangle and *not the size*, as shown with the following triangles. Thus we cannot solve a triangle when only the three angle measures are given.

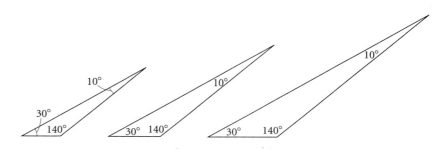

In order to solve oblique triangles, we need to derive the *law of sines* and the *law of cosines*. The law of sines applies to the first three situations listed above. The law of cosines, which we develop in Section 3.2, applies to the last two situations.

✦ The Law of Sines

We consider any oblique triangle. It may or may not have an obtuse angle. Although we look at only the acute-triangle case, the derivation of the obtuse-triangle case is essentially the same.

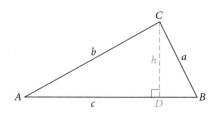

In acute $\triangle ABC$ at left, we have drawn an altitude from vertex C. It has length h. From $\triangle ADC$, we have

$$\sin A = \frac{h}{b}, \quad \text{or} \quad h = b \sin A.$$

From $\triangle BDC$, we have

$$\sin B = \frac{h}{a}, \quad \text{or} \quad h = a \sin B.$$

With $h = b \sin A$ and $h = a \sin B$, we now have

$$a \sin B = b \sin A$$

$$\frac{a \sin B}{\sin A \sin B} = \frac{b \sin A}{\sin A \sin B} \qquad \text{Dividing by } \sin A \sin B$$

$$\frac{a}{\sin A} = \frac{b}{\sin B}. \qquad \text{Simplifying}$$

There is no danger of dividing by 0 here because we are dealing with triangles whose angles are never 0° or 180°. Thus the sine value will never be 0.

If we were to consider altitudes from vertex A and vertex B in the triangle shown above, the same argument would give us

$$\frac{b}{\sin B} = \frac{c}{\sin C} \quad \text{and} \quad \frac{a}{\sin A} = \frac{c}{\sin C}.$$

We combine these results to obtain the law of sines.

The Law of Sines

In any triangle ABC,

$$\frac{a}{\sin A} = \frac{b}{\sin B} = \frac{c}{\sin C}.$$

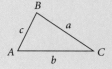

✦ Solving Triangles (AAS and ASA)

When two angles and a side of any triangle are known, the law of sines can be used to solve the triangle.

EXAMPLE 1   In $\triangle EFG$, $e = 4.56$, $E = 43°$, and $G = 57°$. Solve the triangle.

Solution   We first make a drawing. We know three of the six measures.

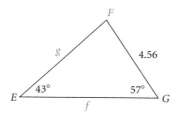

$$
\begin{array}{ll}
E = 43° & e = 4.56 \\
F = \,? & f = \,? \\
G = 57° & g = \,?
\end{array}
$$

From the figure, we see that we have the AAS situation. We begin by finding F:

$$F = 180° - (43° + 57°) = 80°.$$

We can now find the other two sides, using the law of sines:

$$\frac{f}{\sin F} = \frac{e}{\sin E}$$

$$\frac{f}{\sin 80°} = \frac{4.56}{\sin 43°} \qquad \text{Substituting}$$

$$f = \frac{4.56 \sin 80°}{\sin 43°} \qquad \text{Solving for } f$$

$$f \approx 6.58;$$

$$\frac{g}{\sin G} = \frac{e}{\sin E}$$

$$\frac{g}{\sin 57°} = \frac{4.56}{\sin 43°} \qquad \text{Substituting}$$

$$g = \frac{4.56 \sin 57°}{\sin 43°} \qquad \text{Solving for } g$$

$$g \approx 5.61.$$

Thus, we have solved the triangle:

$$
\begin{array}{ll}
E = 43°, & e = 4.56, \\
F = 80°, & f \approx 6.58, \\
G = 57°, & g \approx 5.61.
\end{array}
$$

▶ Now Try Exercise 1.

The law of sines is frequently used in determining distances.

EXAMPLE 2 *Rescue Mission.* During a rescue mission, a Marine fighter pilot receives data on an unidentified aircraft from an AWACS plane and is instructed to intercept the aircraft. The diagram shown below appears on the screen, but before the distance to the point of interception appears on the screen, communications are jammed. Fortunately, the pilot remembers the law of sines. How far must the pilot fly?

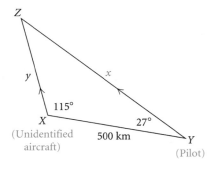

Solution We let x represent the distance that the pilot must fly in order to intercept the aircraft and Z represent the point of interception. We first find angle Z:

$$Z = 180° - (115° + 27°)$$
$$= 38°.$$

Because this application involves the ASA situation, we use the law of sines to determine x:

$$\frac{x}{\sin X} = \frac{z}{\sin Z}$$

$$\frac{x}{\sin 115°} = \frac{500}{\sin 38°} \qquad \text{Substituting}$$

$$x = \frac{500 \sin 115°}{\sin 38°} \qquad \text{Solving for } x$$

$$x \approx 736.$$

Thus the pilot must fly approximately 736 km in order to intercept the unidentified aircraft.   ▶ **Now Try Exercise 23.**

◆ Solving Triangles (SSA)

When two sides of a triangle and an angle opposite one of them are known, the law of sines can be used to solve the triangle.

Suppose for $\triangle ABC$ that b, c, and B are given. The various possibilities are as shown in the eight cases on the following page: 5 cases when B is acute and 3 cases when B is obtuse. Note that $b < c$ in cases 1, 2, 3, and 6; $b = c$ in cases 4 and 7; and $b > c$ in cases 5 and 8.

Angle B Is Acute

Case 1: No solution
$b < c$; side b is too short to reach the base. No triangle is formed.

Case 2: One solution
$b < c$; side b just reaches the base and is perpendicular to it.

Case 3: Two solutions
$b < c$; an arc of radius b meets the base at two points. (This case is called the **ambiguous case.**)

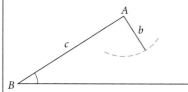

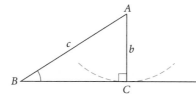

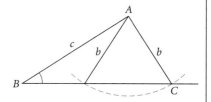

Case 4: One solution
$b = c$; an arc of radius b meets the base at just one point, other than B.

Case 5: One solution
$b > c$; an arc of radius b meets the base at just one point.

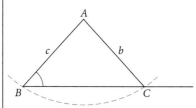

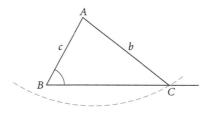

Angle B Is Obtuse

Case 6: No solution
$b < c$; side b is too short to reach the base. No triangle is formed.

Case 7: No solution
$b = c$; an arc of radius b meets the base only at point B. No triangle is formed.

Case 8: One solution
$b > c$; an arc of radius b meets the base at just one point.

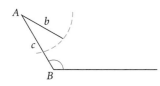

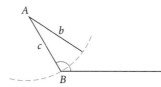

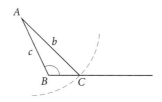

The eight cases above lead us to three possibilities in the SSA situation: *no* solution, *one* solution, or *two* solutions. Let's investigate these possibilities further, looking for ways to recognize the number of solutions.

EXAMPLE 3   *No Solution.*   In △QRS, q = 15, r = 28, and Q = 43.6°. Solve the triangle.

Solution   We make a drawing and list the known measures.

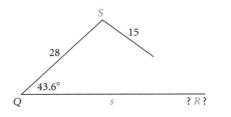

Q = 43.6°	q = 15
R = ?	r = 28
S = ?	s = ?

We observe the SSA situation and use the law of sines to find R:

$$\frac{q}{\sin Q} = \frac{r}{\sin R}$$

$$\frac{15}{\sin 43.6°} = \frac{28}{\sin R} \qquad \text{Substituting}$$

$$\sin R = \frac{28 \sin 43.6°}{15} \qquad \text{Solving for } \sin R$$

$$\sin R \approx 1.2873.$$

Since there is no angle with a sine greater than 1, there is *no solution*.

▶ Now Try Exercise 13.

EXAMPLE 4   *One Solution.*   In △XYZ, x = 23.5, y = 9.8, and X = 39.7°. Solve the triangle.

Solution   We make a drawing and organize the given information.

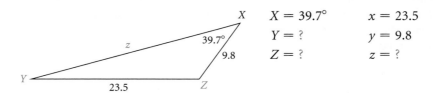

X = 39.7°	x = 23.5
Y = ?	y = 9.8
Z = ?	z = ?

We see the SSA situation and begin by finding Y with the law of sines:

$$\frac{x}{\sin X} = \frac{y}{\sin Y}$$

$$\frac{23.5}{\sin 39.7°} = \frac{9.8}{\sin Y} \qquad \text{Substituting}$$

$$\sin Y = \frac{9.8 \sin 39.7°}{23.5} \qquad \text{Solving for } \sin Y$$

$$\sin Y \approx 0.2664.$$

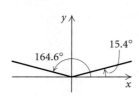

There are two angles less than 180° with a sine of 0.2664. They are 15.4° and 164.6°, to the nearest tenth of a degree. An angle of 164.6° cannot be an angle of this triangle because it already has an angle of 39.7° and these two angles would total more than 180°. Thus, 15.4° is the only possibility for Y. Therefore,

$$Z \approx 180° - (39.7° + 15.4°) \approx 124.9°.$$

We now find z:

$$\frac{z}{\sin Z} = \frac{x}{\sin X}$$

$$\frac{z}{\sin 124.9°} = \frac{23.5}{\sin 39.7°} \qquad \text{Substituting}$$

$$z = \frac{23.5 \sin 124.9°}{\sin 39.7°} \qquad \text{Solving for } z$$

$$z \approx 30.2.$$

We now have solved the triangle:

$X = 39.7°,$	$x = 23.5,$
$Y \approx 15.4°,$	$y = 9.8,$
$Z \approx 124.9°,$	$z \approx 30.2.$

▶ **Now Try Exercise 5.**

The next example illustrates the ambiguous case in which there are two possible solutions.

EXAMPLE 5 *Two Solutions.* In $\triangle ABC$, $b = 15$, $c = 20$, and $B = 29°$. Solve the triangle.

Solution We make a drawing, list the known measures, and see that we again have the SSA situation.

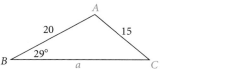

$A = ?$	$a = ?$
$B = 29°$	$b = 15$
$C = ?$	$c = 20$

We first find C:

$$\frac{b}{\sin B} = \frac{c}{\sin C}$$

$$\frac{15}{\sin 29°} = \frac{20}{\sin C} \qquad \text{Substituting}$$

$$\sin C = \frac{20 \sin 29°}{15} \approx 0.6464. \qquad \text{Solving for } \sin C$$

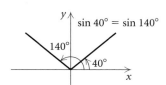

There are two angles less than 180° with a sine of 0.6464. They are 40° and 140°, to the nearest degree. This gives us two possible solutions.

Possible Solution I.

If $C = 40°$, then

$$A = 180° - (29° + 40°) = 111°.$$

Then we find a:

$$\frac{a}{\sin A} = \frac{b}{\sin B}$$

$$\frac{a}{\sin 111°} = \frac{15}{\sin 29°}$$

$$a = \frac{15 \sin 111°}{\sin 29°} \approx 29.$$

These measures make a triangle as shown below; thus we have a solution.

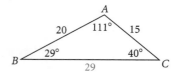

Possible Solution II.

If $C = 140°$, then

$$A = 180° - (29° + 140°) = 11°.$$

Then we find a:

$$\frac{a}{\sin A} = \frac{b}{\sin B}$$

$$\frac{a}{\sin 11°} = \frac{15}{\sin 29°}$$

$$a = \frac{15 \sin 11°}{\sin 29°} \approx 6.$$

These measures make a triangle as shown below; thus we have a second solution.

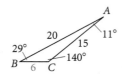

▶ Now Try Exercise 3.

Examples 3–5 illustrate the SSA situation. Note that we need not memorize the eight cases or the procedures in finding no solution, one solution, or two solutions. When we are using the law of sines, the sine value leads us directly to the correct solution or solutions.

✦ The Area of a Triangle

The familiar formula for the area of a triangle, $A = \frac{1}{2}bh$, can be used only when h is known. However, we can use the method used to derive the law of sines to derive an area formula that does not involve the height.

Consider a general triangle $\triangle ABC$, with area K, as shown below.

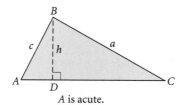

A is acute.

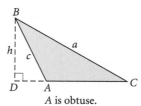

A is obtuse.

Note that in the triangle on the right, $\sin(\angle CAB) = \sin(\angle DAB)$, since $\sin \theta = \sin(180° - \theta)$. Then in each $\triangle ADB$,

$$\sin A = \frac{h}{c}, \quad \text{or} \quad h = c \sin A.$$

Substituting into the formula $K = \frac{1}{2}bh$, we get

$$K = \frac{1}{2}bc \sin A.$$

Any pair of sides and the included angle could have been used. Thus we also have

$$K = \tfrac{1}{2} ab \sin C \quad \text{and} \quad K = \tfrac{1}{2} ac \sin B.$$

The Area of a Triangle

The area K of any $\triangle ABC$ is one half the product of the lengths of two sides and the sine of the included angle:

$$K = \frac{1}{2} bc \sin A = \frac{1}{2} ab \sin C = \frac{1}{2} ac \sin B.$$

EXAMPLE 6 *Area of a Triangular Garden.* A university landscaping architecture department is designing a garden for a triangular area in a dormitory complex. Two sides of the garden, formed by the sidewalks in front of buildings A and B, measure 172 ft and 186 ft, respectively, and together form a 53° angle. The third side of the garden, formed by the sidewalk along Crossroads Avenue, measures 160 ft. What is the area of the garden to the nearest square foot?

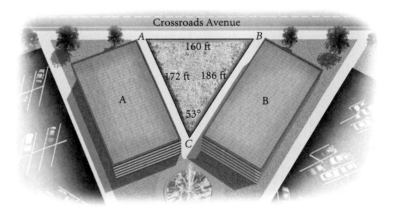

Solution Since we do not know a height of the triangle, we use the area formula:

$$K = \frac{1}{2} ab \sin C$$

$$K = \frac{1}{2} \cdot 186 \text{ ft} \cdot 172 \text{ ft} \cdot \sin 53°$$

$$K \approx 12{,}775 \text{ ft}^2.$$

The area of the garden is approximately 12,775 ft².

▶ Now Try Exercise 25.

3.1 EXERCISE SET

Solve the triangle, if possible.

1. $B = 38°$, $C = 21°$, $b = 24$

2. $A = 131°$, $C = 23°$, $b = 10$

3. $A = 36.5°$, $a = 24$, $b = 34$

4. $B = 118.3°$, $C = 45.6°$, $b = 42.1$

5. $C = 61°10'$, $c = 30.3$, $b = 24.2$

6. $A = 126.5°$, $a = 17.2$, $c = 13.5$

7. $c = 3$ mi, $B = 37.48°$, $C = 32.16°$

8. $a = 2345$ mi, $b = 2345$ mi, $A = 124.67°$

9. $b = 56.78$ yd, $c = 56.78$ yd, $C = 83.78°$

10. $A = 129°32'$, $C = 18°28'$, $b = 1204$ in.

11. $a = 20.01$ cm, $b = 10.07$ cm, $A = 30.3°$

12. $b = 4.157$ km, $c = 3.446$ km, $C = 51°48'$

13. $A = 89°$, $a = 15.6$ in., $b = 18.4$ in.

14. $C = 46°32'$, $a = 56.2$ m, $c = 22.1$ m

15. $a = 200$ m, $A = 32.76°$, $C = 21.97°$

16. $B = 115°$, $c = 45.6$ yd, $b = 23.8$ yd

Find the area of the triangle.

17. $B = 42°$, $a = 7.2$ ft, $c = 3.4$ ft

18. $A = 17°12'$, $b = 10$ in., $c = 13$ in.

19. $C = 82°54'$, $a = 4$ yd, $b = 6$ yd

20. $C = 75.16°$, $a = 1.5$ m, $b = 2.1$ m

21. $B = 135.2°$, $a = 46.12$ ft, $c = 36.74$ ft

22. $A = 113°$, $b = 18.2$ cm, $c = 23.7$ cm

Solve.

23. *Lunar Crater.* Points A and B are on opposite sides of a lunar crater. Point C is 50 m from A. The measure of $\angle BAC$ is determined to be 112° and the measure of $\angle ACB$ is determined to be 42°. What is the width of the crater?

24. *Rock Concert.* In preparation for an outdoor rock concert, a stage crew must determine how far apart to place the two large speaker columns on stage (see the figure below). What generally works best is to place them at 50° angles to the center of the front row. The distance from the center of the front row to each of the speakers is 10 ft. How far apart does the crew need to place the speakers on stage?

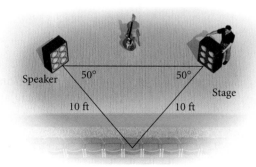

25. *Area of Backyard.* A new homeowner has a triangular-shaped backyard. Two of the three sides measure 53 ft and 42 ft and form an included angle of 135°. To determine the amount of fertilizer and grass seed to be purchased, the owner has to know, or at least approximate,

the area of the yard. Find the area of the yard to the nearest square foot.

26. *Boarding Stable.*   A rancher operates a boarding stable and temporarily needs to make an extra pen. He has a piece of rope 38 ft long and plans to tie the rope to one end of the barn (S) and run the rope around a tree (T) and back to the barn (Q). The tree is 21 ft from where the rope is first tied, and the rope from the barn to the tree makes an angle of $35°$ with the barn. Does the rancher have enough rope if he allows $4\frac{1}{2}$ ft at each end to fasten the rope?

27. *Length of Pole.*   A pole leans away from the sun at an angle of $7°$ to the vertical. When the angle of elevation of the sun is $51°$, the pole casts a shadow 47 ft long on level ground. How long is the pole?

In Exercises 28–31, keep in mind the two types of bearing considered in Sections 1.2 and 1.3.

28. *Reconnaissance Airplane.*   A reconnaissance airplane leaves its airport on the east coast of the United States and flies in a direction of $85°$. Because of bad weather, it returns to another airport 230 km to the north of its home base.

For the return trip, it flies in a direction of $283°$. What is the total distance that the airplane flew?

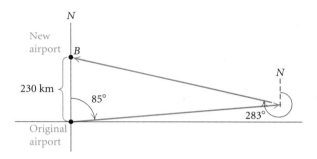

29. *Fire Tower.*   A ranger in fire tower A spots a fire at a direction of $295°$. A ranger in fire tower B, located 45 mi at a direction of $45°$ from tower A, spots the same fire at a direction of $255°$. How far from tower A is the fire? from tower B?

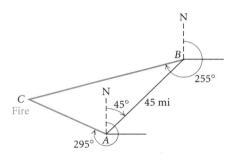

30. *Lighthouse.*   A boat leaves lighthouse A and sails 5.1 km. At this time it is sighted from lighthouse B, 7.2 km west of A. The bearing of the boat from B is N65°10′E. How far is the boat from B?

31. *Mackinac Island.*   Mackinac Island is located 18 mi N31°20′W of Cheboygan, Michigan, where the Coast Guard cutter *Mackinaw* is stationed. A freighter in distress radios the Coast Guard cutter for help. It radios its position as S78°40′E of

Mackinac Island and N64°10′E of Cheboygan. How far is the freighter from Cheboygan?

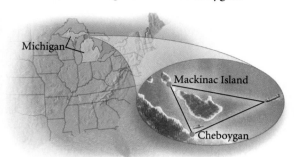

Michigan

Mackinac Island

Cheboygan

32. *Gears.* Three gears are arranged as shown in the figure below. Find the angle ϕ.

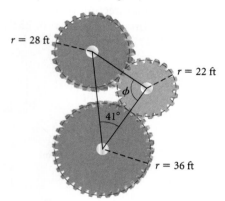

$r = 28$ ft

$r = 22$ ft

ϕ

41°

$r = 36$ ft

Collaborative Discussion and Writing

33. Explain why the law of sines cannot be used to find the first angle when solving a triangle given three sides.

34. We considered eight cases of solving triangles given two sides and an angle opposite one of them. Describe the relationship between side b and the height h in each.

Skill Maintenance

Find the acute angle A, in both radians and degrees, for the given function value.

35. $\cos A = 0.2213$

36. $\cos A = 1.5612$

Convert to decimal degree notation.

37. $18°14′20″$

38. $125°3′42″$

39. Find the absolute value: $|-5|$.

Find the values.

40. $\cos \dfrac{\pi}{6}$

41. $\sin 45°$

42. $\sin 300°$

43. $\cos\left(-\dfrac{2\pi}{3}\right)$

44. Multiply: $(1 - i)(1 + i)$.

Synthesis

45. Prove the following area formulas for a general triangle ABC with area represented by K.

$$K = \frac{a^2 \sin B \sin C}{2 \sin A}$$

$$K = \frac{c^2 \sin A \sin B}{2 \sin C}$$

$$K = \frac{b^2 \sin C \sin A}{2 \sin B}$$

46. *Area of a Parallelogram.* Prove that the area of a parallelogram is the product of two adjacent sides and the sine of the included angle.

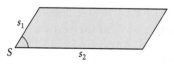

s_1

S

s_2

47. *Area of a Quadrilateral.* Prove that the area of a quadrilateral is one half the product of the lengths of its diagonals and the sine of the angle between the diagonals.

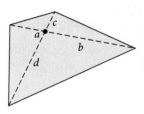

c

a

b

d

48. Find d.

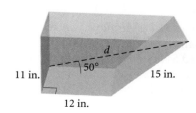

11 in.

50°

d

15 in.

12 in.

49. *Recording Studio.*   A musician is constructing an octagonal recording studio in his home. The studio with dimensions shown to the right is to be built within a rectangular 31′9″ by 29′9″ room. (*Source*: Tony Medeiros, Indianapolis, IN) Point *D* is 9″ from wall 2, and points *C* and *B* are each 9″ from wall 1. Using the law of sines and right triangles, determine to the nearest tenth of an inch how far point *A* is from wall 1 and from wall 4. (For more information on this studio, see Example 2 in Section 3.2.)

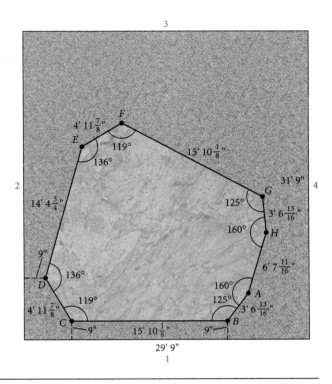

♦ Use the law of cosines to solve triangles.

♦ Determine whether the law of sines or the law of cosines should be applied to solve a triangle.

3.2   The Law of Cosines

The law of sines is used to solve triangles given a side and two angles (AAS and ASA) or given two sides and an angle opposite one of them (SSA). A second law, called the *law of cosines*, is needed to solve triangles given two sides and the included angle (SAS) or given three sides (SSS).

◆ The Law of Cosines

To derive this property, we consider any △*ABC* placed on a coordinate system. We position the origin at one of the vertices—say, *C*—and the positive half of the x-axis along one of the sides—say, CB. Let (x, y) be the coordinates of vertex A. Point B has coordinates $(a, 0)$ and point C has coordinates $(0, 0)$.

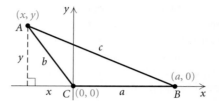

Then $\cos C = \dfrac{x}{b},$ so $x = b \cos C$

and $\sin C = \dfrac{y}{b},$ so $y = b \sin C.$

Thus point A has coordinates

$$(b \cos C, \, b \sin C).$$

Next, we use the distance formula to determine c^2:

$$c^2 = (x - a)^2 + (y - 0)^2,$$

or $c^2 = (b \cos C - a)^2 + (b \sin C - 0)^2.$

Now we multiply and simplify:

$$c^2 = b^2 \cos^2 C - 2ab \cos C + a^2 + b^2 \sin^2 C$$
$$= a^2 + b^2(\sin^2 C + \cos^2 C) - 2ab \cos C$$
$$= a^2 + b^2 - 2ab \cos C. \qquad \text{Using the identity}$$
$$\text{sin}^2 x + \cos^2 x = 1$$

Had we placed the origin at one of the other vertices, we would have obtained

$$a^2 = b^2 + c^2 - 2bc \cos A$$
or $b^2 = a^2 + c^2 - 2ac \cos B.$

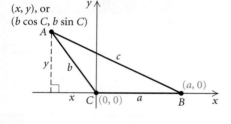

The Law of Cosines

In any triangle ABC,

$$a^2 = b^2 + c^2 - 2bc \cos A,$$
$$b^2 = a^2 + c^2 - 2ac \cos B,$$
or $c^2 = a^2 + b^2 - 2ab \cos C.$

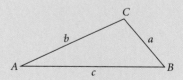

Thus, in any triangle, the square of a side is the sum of the squares of the other two sides, minus twice the product of those sides and the cosine of the included angle. When the included angle is 90°, the law of cosines reduces to the Pythagorean theorem.

✦ Solving Triangles (SAS)

When two sides of a triangle and the included angle are known, we can use the law of cosines to find the third side. The law of cosines or the law of sines can then be used to finish solving the triangle.

EXAMPLE 1   Solve $\triangle ABC$ if $a = 32$, $c = 48$, and $B = 125.2°$.

Solution   We first label a triangle with the known and unknown measures.

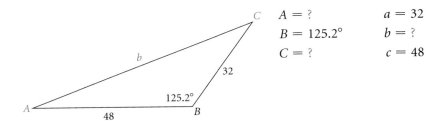

$$A = ? \qquad a = 32$$
$$B = 125.2° \qquad b = ?$$
$$C = ? \qquad c = 48$$

We can find the third side using the law of cosines, as follows:

$$b^2 = a^2 + c^2 - 2ac \cos B$$
$$b^2 = 32^2 + 48^2 - 2 \cdot 32 \cdot 48 \cos 125.2° \qquad \text{Substituting}$$
$$b^2 \approx 5098.8$$
$$b \approx 71.$$

We now have $a = 32$, $b \approx 71$, and $c = 48$, and we need to find the other two angle measures. At this point, we can find them in two ways. One way uses the law of sines. The ambiguous case may arise, however, and we would have to be alert to this possibility. The advantage of using the law of cosines again is that if we solve for the cosine and find that its value is *negative*, then we know that the angle is obtuse. If the value of the cosine is *positive*, then the angle is acute. Thus we use the law of cosines to find a second angle.

Let's find angle A. We select the formula from the law of cosines that contains $\cos A$ and substitute:

$$a^2 = b^2 + c^2 - 2bc \cos A$$
$$32^2 = 71^2 + 48^2 - 2 \cdot 71 \cdot 48 \cos A \qquad \text{Substituting}$$
$$1024 = 5041 + 2304 - 6816 \cos A$$
$$-6321 = -6816 \cos A$$
$$\cos A \approx 0.9273768$$
$$A \approx 22.0°.$$

The third angle is now easy to find:

$$C \approx 180° - (125.2° + 22.0°)$$
$$\approx 32.8°.$$

Thus,

$$A \approx 22.0°, \qquad a = 32,$$
$$B = 125.2°, \qquad b \approx 71,$$
$$C \approx 32.8°, \qquad c = 48.$$

▶ Now Try Exercise 1.

Due to errors created by rounding, answers may vary depending on the order in which they are found. Had we found the measure of angle C first in Example 1, the angle measures would have been $C \approx 34.1°$ and $A \approx 20.7°$. Variances in rounding also change the answers. Had we used 71.4 for b in Example 1, the angle measures would have been $A \approx 21.5°$ and $C \approx 33.3°$.

Suppose we used the law of sines at the outset in Example 1 to find b. We were given only three measures: $a = 32$, $c = 48$, and $B = 125.2°$. When substituting these measures into the proportions, we see that there is not enough information to use the law of sines:

$$\frac{a}{\sin A} = \frac{b}{\sin B} \rightarrow \frac{32}{\sin A} = \frac{b}{\sin 125.2°},$$
$$\frac{b}{\sin B} = \frac{c}{\sin C} \rightarrow \frac{b}{\sin 125.2°} = \frac{48}{\sin C},$$
$$\frac{a}{\sin A} = \frac{c}{\sin C} \rightarrow \frac{32}{\sin A} = \frac{48}{\sin C}.$$

In all three situations, the resulting equation, after the substitutions, still has two unknowns. Thus we cannot use the law of sines to find b.

EXAMPLE 2 *Recording Studio.* A musician is constructing an octagonal recording studio in his home and needs to determine two distances for the electrician. The dimensions for the most acoustically perfect studio are shown in the figure below. (*Source:* Tony Medeiros, Indianapolis, IN) Determine the distances from D to F and from D to B to the nearest tenth of an inch.

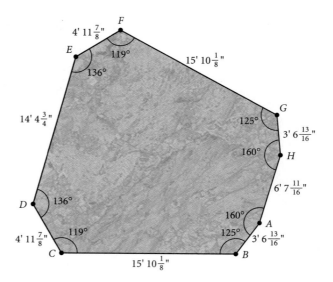

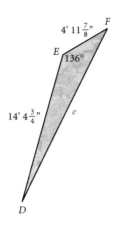

Solution We begin by connecting points D and F and labeling the known measures of $\triangle DEF$. Converting the measures to decimal notation in inches, we have

$$d = 4'11\frac{7}{8}'' = 59.875 \text{ in.},$$

$$f = 14'4\frac{3}{4}'' = 172.75 \text{ in.},$$

$$E = 136°.$$

We can find the measure of the third side, e, using the law of cosines:

$$e^2 = d^2 + f^2 - 2 \cdot d \cdot f \cdot \cos E \qquad \text{Using the law of}$$
$$e^2 = (59.875 \text{ in.})^2 + (172.75 \text{ in.})^2 \qquad \text{cosines}$$
$$\qquad - 2(59.875 \text{ in.})(172.75 \text{ in.}) \cos 136° \qquad \text{Substituting}$$
$$e^2 \approx 48{,}308.4257 \text{ in}^2$$
$$e \approx 219.8 \text{ in.}$$

Thus it is approximately 219.8 in. from D to F.

We continue by connecting points D and B and labeling the known measures of $\triangle DCB$:

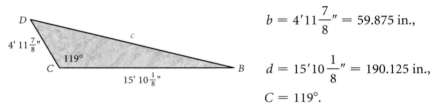

$$b = 4'11\frac{7}{8}'' = 59.875 \text{ in.},$$

$$d = 15'10\frac{1}{8}'' = 190.125 \text{ in.},$$

$$C = 119°.$$

Using the law of cosines, we can determine c, the length of the third side:

$$c^2 = b^2 + d^2 - 2 \cdot b \cdot d \cdot \cos C \qquad \text{Using the law of}$$
$$c^2 = (59.875 \text{ in.})^2 + (190.125 \text{ in.})^2 \qquad \text{cosines}$$
$$\qquad - 2(59.875 \text{ in.})(190.125 \text{ in.}) \cos 119° \qquad \text{Substituting}$$
$$c^2 \approx 50{,}770.4191 \text{ in}^2$$
$$c \approx 225.3 \text{ in.}$$

The distance from D to B is approximately 225.3 in.

▶ Now Try Exercise 25.

✦ Solving Triangles (SSS)

When all three sides of a triangle are known, the law of cosines can be used to solve the triangle.

EXAMPLE 3   Solve $\triangle RST$ if $r = 3.5$, $s = 4.7$, and $t = 2.8$.

Solution   We sketch a triangle and label it with the given measures.

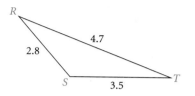

$$R = \text{?} \qquad r = 3.5$$
$$S = \text{?} \qquad s = 4.7$$
$$T = \text{?} \qquad t = 2.8$$

Since we do not know any of the angle measures, we cannot use the law of sines. We begin instead by finding an angle with the law of cosines. We choose to find S first and select the formula that contains $\cos S$:

$$s^2 = r^2 + t^2 - 2rt \cos S$$
$$(4.7)^2 = (3.5)^2 + (2.8)^2 - 2(3.5)(2.8) \cos S \qquad \text{Substituting}$$
$$\cos S = \frac{(3.5)^2 + (2.8)^2 - (4.7)^2}{2(3.5)(2.8)}$$
$$\cos S \approx -0.1020408$$
$$S \approx 95.86°.$$

Similarly, we find angle R:

$$r^2 = s^2 + t^2 - 2st \cos R$$
$$(3.5)^2 = (4.7)^2 + (2.8)^2 - 2(4.7)(2.8) \cos R$$
$$\cos R = \frac{(4.7)^2 + (2.8)^2 - (3.5)^2}{2(4.7)(2.8)}$$
$$\cos R \approx 0.6717325$$
$$R \approx 47.80°.$$

Then

$$T \approx 180° - (95.86° + 47.80°) \approx 36.34°.$$

Thus,

$$R \approx 47.80°, \qquad r = 3.5,$$
$$S \approx 95.86°, \qquad s = 4.7,$$
$$T \approx 36.34°, \qquad t = 2.8.$$

▶ Now Try Exercise 3.

EXAMPLE 4   *Knife Bevel.*   Knifemakers know that the *bevel* of the blade (the angle formed at the cutting edge of the blade) determines the cutting characteristics of the knife. A small bevel like that of a straight razor makes for a keen edge, but is impractical for heavy-duty cutting because the edge dulls quickly and is prone to chipping. A large bevel is suitable for heavy-duty work like chopping wood. Survival knives, being universal in application, are a compromise between small and large bevels. The diagram at left illustrates the blade of a hand-made Randall Model 18 survival knife. What is its bevel? (*Source:* Randall Made Knives, P.O. Box 1988, Orlando, FL 32802)

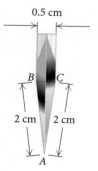

Solution We know three sides of a triangle. We can use the law of cosines to find the bevel, angle A.

$$a^2 = b^2 + c^2 - 2bc \cos A$$
$$(0.5)^2 = 2^2 + 2^2 - 2 \cdot 2 \cdot 2 \cdot \cos A$$
$$0.25 = 4 + 4 - 8 \cos A$$
$$\cos A = \frac{4 + 4 - 0.25}{8}$$
$$\cos A = 0.96875$$
$$A \approx 14.36°.$$

Thus the bevel is approximately $14.36°$.

▶ Now Try Exercise 29.

CONNECTING THE CONCEPTS

Choosing the Appropriate Law

The following summarizes the situations in which to use the law of sines and the law of cosines.

To solve an oblique triangle:

Use the *law of sines* for: Use the *law of cosines* for:

       AAS   SAS
       ASA   SSS
       SSA   ● ● ●

    The law of cosines can also be used for the SSA situation, but since the process involves solving a quadratic equation, we do not include that option in the list above.

EXAMPLE 5 In $\triangle ABC$, three measures are given. Determine which law to use when solving the triangle. You need not solve the triangle.

a) $a = 14, b = 23, c = 10$
b) $a = 207, B = 43.8°, C = 57.6°$
c) $A = 112°, C = 37°, a = 84.7$
d) $B = 101°, a = 960, c = 1042$
e) $b = 17.26, a = 27.29, A = 39°$
f) $A = 61°, B = 39°, C = 80°$

Solution   It is helpful to make a drawing of a triangle with the given information. The triangle need not be drawn to scale. The given parts are shown in color.

FIGURE		SITUATION	LAW TO USE
a)	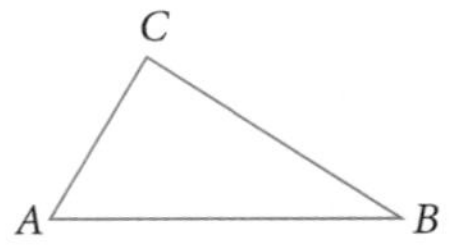	SSS	Law of Cosines
b)	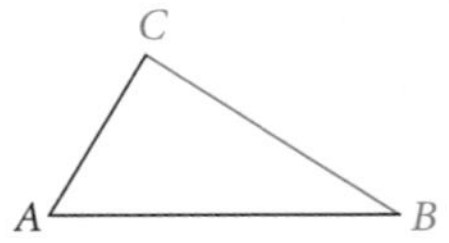	ASA	Law of Sines
c)	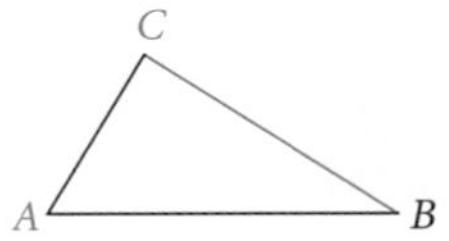	AAS	Law of Sines
d)	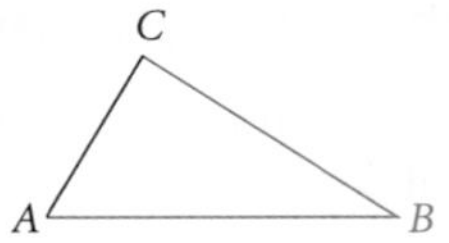	SAS	Law of Cosines
e)	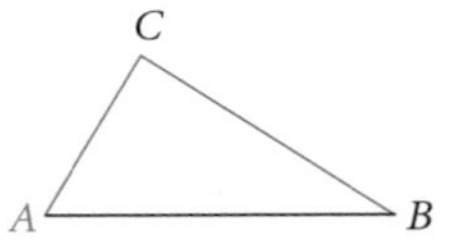	SSA	Law of Sines
f)	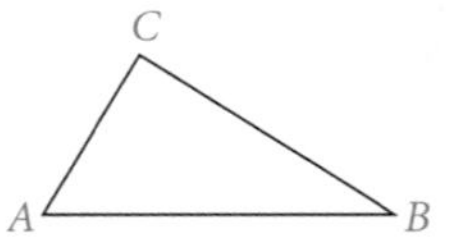	AAA	Cannot be solved

▷ Now Try Exercises 17 and 19.

◆ **3.2   EXERCISE SET**

Solve the triangle, if possible.

1. $A = 30°, b = 12, c = 24$

2. $B = 133°, a = 12, c = 15$

3. $a = 12, b = 14, c = 20$

4. $a = 22.3, b = 22.3, c = 36.1$

5. $B = 72°40', c = 16$ m, $a = 78$ m

6. $C = 22.28°, a = 25.4$ cm, $b = 73.8$ cm

7. $a = 16$ m, $b = 20$ m, $c = 32$ m

8. $B = 72.66°, a = 23.78$ km, $c = 25.74$ km

9. $a = 2$ ft, $b = 3$ ft, $c = 8$ ft

10. $A = 96°13'$, $b = 15.8$ yd, $c = 18.4$ yd

11. $a = 26.12$ km, $b = 21.34$ km, $c = 19.25$ km

12. $C = 28°43'$, $a = 6$ mm, $b = 9$ mm

13. $a = 60.12$ mi, $b = 40.23$ mi, $C = 48.7°$

14. $a = 11.2$ cm, $b = 5.4$ cm, $c = 7$ cm

15. $b = 10.2$ in., $c = 17.3$ in., $A = 53.456°$

16. $a = 17$ yd, $b = 15.4$ yd, $c = 1.5$ yd

Determine which law applies. Then solve the triangle.

17. $A = 70°$, $B = 12°$, $b = 21.4$

18. $a = 15$, $c = 7$, $B = 62°$

19. $a = 3.3$, $b = 2.7$, $c = 2.8$

20. $a = 1.5$, $b = 2.5$, $A = 58°$

21. $A = 40.2°$, $B = 39.8°$, $C = 100°$

22. $a = 60$, $b = 40$, $C = 47°$

23. $a = 3.6$, $b = 6.2$, $c = 4.1$

24. $B = 110°30'$, $C = 8°10'$, $c = 0.912$

Solve.

25. *Poachers.*   A park ranger establishes an observation post from which to watch for poachers. Despite losing her map, the ranger does have a compass and a rangefinder. She observes some poachers, and the rangefinder indicates that they are 500 ft from her position. They are headed toward big game that she knows to be 375 ft from her position. Using her compass, she finds that the poachers' azimuth (the direction measured as an angle from north) is 355° and that of the big game is 42°. What is the distance between the poachers and the game?

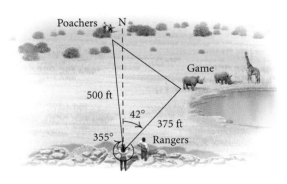

26. *Circus Highwire Act.*   A circus highwire act walks up an approach wire to reach a highwire. The approach wire is 122 ft long and is currently anchored so that it forms the maximum allowable angle of 35° with the ground. A greater approach angle causes the aerialists to slip. However, the aerialists find that there is enough room to anchor the approach wire 30 ft back in order to make the approach angle less severe. When this is done, how much farther will they have to walk up the approach wire, and what will the new approach angle be?

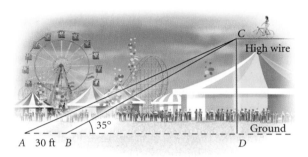

27. *In-line Skater.*   An in-line skater skates on a fitness trail along the Pacific Ocean from point A to point B. As shown below, two streets intersecting at point C also intersect the trail at A and B. In her car, the skater found the lengths of AC and BC to be approximately 0.5 mi and 1.3 mi, respectively. From a map, she estimates the included angle at C to be 110°. How far did she skate from A to B?

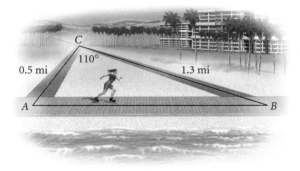

28. *Baseball Bunt.*   A batter in a baseball game drops a bunt down the first-base line. It rolls 34 ft at an angle of 25° with the base path. The pitcher's mound is 60.5 ft from home plate. How far must the pitcher travel to pick up the ball? (*Hint*: A baseball diamond is a square.)

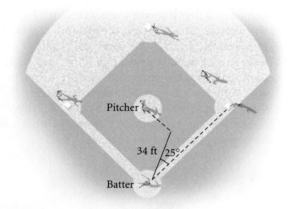

Pitcher
34 ft / 25°
Batter

29. *Survival Trip.*   A group of college students is learning to navigate for an upcoming survival trip. On a map, they have been given three points at which they are to check in. The map also shows the distances between the points. However, to navigate they need to know the angle measurements. Calculate the angles for them.

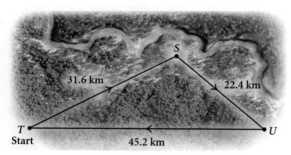

S
31.6 km   22.4 km
T
Start   45.2 km   U

30. *Ships.*   Two ships leave harbor at the same time. The first sails N15°W at 25 knots (a knot is one nautical mile per hour). The second sails N32°E at 20 knots. After 2 hr, how far apart are the ships?

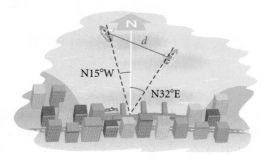

N
d
N15°W
N32°E

31. *Airplanes.*   Two airplanes leave an airport at the same time. The first flies 150 km/h in a direction of 320°. The second flies 200 km/h in a direction of 200°. After 3 hr, how far apart are the planes?

32. *Slow-Pitch Softball.*   A slow-pitch softball diamond is a square 65 ft on a side. The pitcher's mound is 46 ft from home plate. How far is it from the pitcher's mound to first base?

33. *Isosceles Trapezoid.*   The longer base of an isosceles trapezoid measures 14 ft. The nonparallel sides measure 10 ft, and the base angles measure 80°.

a) Find the length of a diagonal.
b) Find the area.

34. *Area of Sail.*   A sail that is in the shape of an isosceles triangle has a vertex angle of 38°. The angle is included by two sides, each measuring 20 ft. Find the area of the sail.

35. Three circles are arranged as shown in the figure below. Find the length *PQ*.

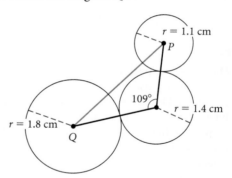

r = 1.1 cm
P
109°
r = 1.4 cm
r = 1.8 cm
Q

36. *Swimming Pool.*   A triangular swimming pool measures 44 ft on one side and 32.8 ft on another side. These sides form an angle that measures 40.8°. How long is the other side?

Collaborative Discussion and Writing

37. Try to solve this triangle using the law of cosines. Then explain why it is easier to solve it using the law of sines.

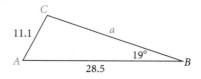

C
11.1   a
A   19°   B
28.5

38. Explain why we cannot solve a triangle given SAS with the law of sines.

Skill Maintenance

Classify the function as linear, quadratic, cubic, quartic, rational, exponential, logarithmic, or trigonometric.

39. $f(x) = -\frac{3}{4}x^4$

40. $y - 3 = 17x$

41. $y = \sin^2 x - 3 \sin x$

42. $f(x) = 2^{x-1/2}$

43. $f(x) = \dfrac{x^2 - 2x + 3}{x - 1}$

44. $f(x) = 27 - x^3$

45. $y = e^x + e^{-x} - 4$

46. $y = \log_2 (x - 2) - \log_2 (x + 3)$

47. $f(x) = -\cos(\pi x - 3)$

48. $y = \frac{1}{2}x^2 - 2x + 2$

Synthesis

49. *Canyon Depth.* A bridge is being built across a canyon. The length of the bridge is 5045 ft. From the deepest point in the canyon, the angles of elevation of the ends of the bridge are 78° and 72°. How deep is the canyon?

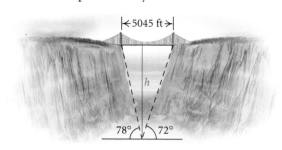

50. *Heron's Formula.* If a, b, and c are the lengths of the sides of a triangle, then the area K of the triangle is given by
$$K = \sqrt{s(s - a)(s - b)(s - c)},$$
where $s = \frac{1}{2}(a + b + c)$. The number s is called the *semiperimeter*. Prove Heron's formula. (*Hint*: Use the area formula $K = \frac{1}{2}bc \sin A$ developed in Section 3.1.) Then use Heron's formula to find the area of the triangular swimming pool described in Exercise 36.

51. *Area of Isosceles Triangle.* Find a formula for the area of an isosceles triangle in terms of the congruent sides and their included angle. Under what conditions will the area of a triangle with fixed congruent sides be maximum?

52. *Reconnaissance Plane.* A reconnaissance plane patrolling at 5000 ft sights a submarine at bearing 35° and at an angle of depression of 25°. A carrier is at bearing 105° and at an angle of depression of 60°. How far is the submarine from the carrier?

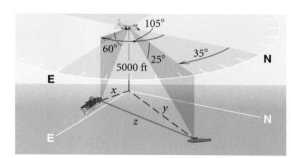

3.3   Complex Numbers: Trigonometric Form

◆ Graph complex numbers.

◆ Given a complex number in standard form, find trigonometric, or polar, notation; and given a complex number in trigonometric form, find standard notation.

◆ Use trigonometric notation to multiply and divide complex numbers.

◆ Use DeMoivre's theorem to raise complex numbers to powers.

◆ Find the nth roots of a complex number.

◆ Graphical Representation

Just as real numbers can be graphed on a line, complex numbers can be graphed on a plane. We graph a complex number $a + bi$ in the same way that we graph an ordered pair of real numbers (a, b). However, in place of an x-axis, we have a real axis, and in place of a y-axis, we have an imaginary axis. Horizontal distances correspond to the real part of a number. Vertical distances correspond to the imaginary part.

EXAMPLE 1   Graph each of the following complex numbers.

a) $3 + 2i$ **b)** $-4 - 5i$ **c)** $-3i$

d) $-1 + 3i$ **e)** 2

Solution

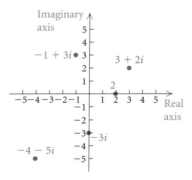

We recall that the absolute value of a real number is its distance from 0 on the number line. The absolute value of a complex number is its distance from the origin in the complex plane. For example, if $z = a + bi$, then using the distance formula, we have

$$|z| = |a + bi| = \sqrt{(a - 0)^2 + (b - 0)^2} = \sqrt{a^2 + b^2}.$$

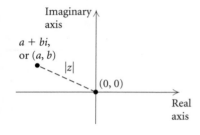

Absolute Value of a Complex Number

The **absolute value** of a complex number $a + bi$ is

$$|a + bi| = \sqrt{a^2 + b^2}.$$

TECHNOLOGY · · · · · · · · · · · · · · · ·
CONNECTION

We can check our work in
Example 2 with a graphing
calculator. Note that
$\sqrt{5} \approx 2.236067977$ and
$\frac{4}{5} = 0.8$.

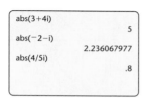

EXAMPLE 2 Find the absolute value of each of the following.

a) $3 + 4i$ 　　　　　b) $-2 - i$ 　　　　　c) $\frac{4}{5}i$

Solution

a) $|3 + 4i| = \sqrt{3^2 + 4^2} = \sqrt{9 + 16} = \sqrt{25} = 5$

b) $|-2 - i| = \sqrt{(-2)^2 + (-1)^2} = \sqrt{5}$

c) $\left|\frac{4}{5}i\right| = \left|0 + \frac{4}{5}i\right| = \sqrt{0^2 + \left(\frac{4}{5}\right)^2} = \frac{4}{5}$

▶ Now Try Exercises 3 and 5.

✦ Trigonometric Notation for Complex Numbers

Now let's consider a nonzero complex number $a + bi$. Suppose that its absolute value is r. If we let θ be an angle in standard position whose terminal side passes through the point (a, b), as shown in the figure, then

$$\cos \theta = \frac{a}{r}, \quad \text{or} \quad a = r \cos \theta$$

and

$$\sin \theta = \frac{b}{r}, \quad \text{or} \quad b = r \sin \theta.$$

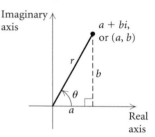

Substituting these values for a and b into the $(a + bi)$ notation, we get

$$a + bi = r \cos \theta + (r \sin \theta)i$$
$$= r(\cos \theta + i \sin \theta).$$

This is **trigonometric notation** for a complex number $a + bi$. The number r is called the **absolute value** of $a + bi$, and θ is called the **argument** of $a + bi$. Trigonometric notation for a complex number is also called **polar notation.**

> ### Trigonometric Notation for Complex Numbers
> $$a + bi = r(\cos \theta + i \sin \theta)$$

　　To find trigonometric notation for a complex number given in **standard notation**, $a + bi$, we must find r and determine the angle θ for which $\sin \theta = b/r$ and $\cos \theta = a/r$.

EXAMPLE 3 Find trigonometric notation for each of the following complex numbers.

a) $1 + i$ **b)** $\sqrt{3} - i$

Solution

a) We note that $a = 1$ and $b = 1$. Then

$$r = \sqrt{a^2 + b^2} = \sqrt{1^2 + 1^2} = \sqrt{2},$$

$$\sin \theta = \frac{b}{r} = \frac{1}{\sqrt{2}}, \quad \text{or} \quad \frac{\sqrt{2}}{2},$$

and

$$\cos \theta = \frac{a}{r} = \frac{1}{\sqrt{2}}, \quad \text{or} \quad \frac{\sqrt{2}}{2}.$$

Since θ is in quadrant I, $\theta = \pi/4$, or $45°$, and we have

$$1 + i = \sqrt{2}\left(\cos \frac{\pi}{4} + i \sin \frac{\pi}{4} \right),$$

or

$$1 + i = \sqrt{2}(\cos 45° + i \sin 45°).$$

b) We see that $a = \sqrt{3}$ and $b = -1$. Then

$$r = \sqrt{(\sqrt{3})^2 + (-1)^2} = 2,$$

$$\sin \theta = \frac{-1}{2} = -\frac{1}{2},$$

and

$$\cos \theta = \frac{\sqrt{3}}{2}.$$

Since θ is in quadrant IV, $\theta = 11\pi/6$, or $330°$, and we have

$$\sqrt{3} - i = 2\left(\cos \frac{11\pi}{6} + i \sin \frac{11\pi}{6} \right),$$

or

$$\sqrt{3} - i = 2(\cos 330° + i \sin 330°). \qquad \blacktriangleright \text{ Now Try Exercise 13.}$$

In changing to trigonometric notation, note that there are many angles satisfying the given conditions. We ordinarily choose the *smallest positive* angle.

To change from trigonometric notation to standard notation, $a + bi$, we recall that $a = r \cos \theta$ and $b = r \sin \theta$.

TECHNOLOGY ·················
CONNECTION

We can perform the computations in Example 4 on a graphing calculator.

Degree Mode

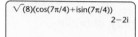

2(cos(120)+isin(120))
                    −1+1.732050808i

Radian Mode

√(8)(cos(7π/4)+isin(7π/4))
                    2−2i

EXAMPLE 4 Find standard notation, $a + bi$, for each of the following complex numbers.

a) $2(\cos 120° + i \sin 120°)$ **b)** $\sqrt{8}\left(\cos \dfrac{7\pi}{4} + i \sin \dfrac{7\pi}{4}\right)$

Solution

a) Rewriting, we have

$$2(\cos 120° + i \sin 120°) = 2 \cos 120° + (2 \sin 120°)i.$$

Thus,

$$a = 2 \cos 120° = 2 \cdot \left(-\frac{1}{2}\right) = -1$$

and

$$b = 2 \sin 120° = 2 \cdot \frac{\sqrt{3}}{2} = \sqrt{3},$$

so

$$2(\cos 120° + i \sin 120°) = -1 + \sqrt{3}i.$$

b) Rewriting, we have

$$\sqrt{8}\left(\cos \frac{7\pi}{4} + i \sin \frac{7\pi}{4}\right) = \sqrt{8} \cos \frac{7\pi}{4} + \left(\sqrt{8} \sin \frac{7\pi}{4}\right)i.$$

Thus,

$$a = \sqrt{8} \cos \frac{7\pi}{4} = \sqrt{8} \cdot \frac{\sqrt{2}}{2} = 2$$

and

$$b = \sqrt{8} \sin \frac{7\pi}{4} = \sqrt{8} \cdot \left(-\frac{\sqrt{2}}{2}\right) = -2,$$

so

$$\sqrt{8}\left(\cos \frac{7\pi}{4} + i \sin \frac{7\pi}{4}\right) = 2 - 2i.$$

▶ Now Try Exercises 23 and 27.

✦ Multiplication and Division with Trigonometric Notation

Multiplication of complex numbers is easier to manage with trigonometric notation than with standard notation. We simply multiply the absolute values and add the arguments. Let's state this in a more formal manner.

> **Complex Numbers: Multiplication**
>
> For any complex numbers $r_1(\cos \theta_1 + i \sin \theta_1)$ and $r_2(\cos \theta_2 + i \sin \theta_2)$,
>
> $$r_1(\cos \theta_1 + i \sin \theta_1) \cdot r_2(\cos \theta_2 + i \sin \theta_2)$$
> $$= r_1 r_2 [\cos (\theta_1 + \theta_2) + i \sin (\theta_1 + \theta_2)].$$

Proof

$$r_1(\cos \theta_1 + i \sin \theta_1) \cdot r_2(\cos \theta_2 + i \sin \theta_2) =$$
$$r_1 r_2(\cos \theta_1 \cos \theta_2 - \sin \theta_1 \sin \theta_2) + r_1 r_2(\sin \theta_1 \cos \theta_2 + \cos \theta_1 \sin \theta_2)i$$

Now, using identities for sums of angles, we simplify, obtaining

$$r_1 r_2 \cos (\theta_1 + \theta_2) + r_1 r_2 \sin (\theta_1 + \theta_2)i,$$

or

$$r_1 r_2 [\cos (\theta_1 + \theta_2) + i \sin (\theta_1 + \theta_2)],$$

which was to be shown.

TECHNOLOGY
CONNECTION

We can multiply complex numbers on a graphing calculator. The products in Example 5 are shown below.

Degree Mode

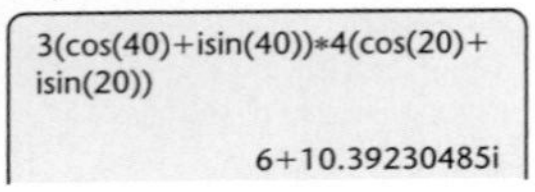

3(cos(40)+isin(40))*4(cos(20)+ isin(20))
6+10.39230485i

Radian Mode

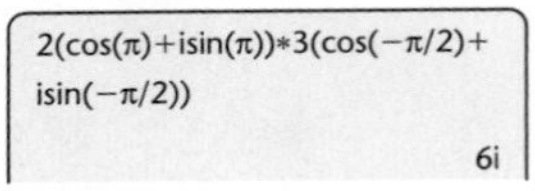

2(cos(π)+isin(π))*3(cos(−π/2)+ isin(−π/2))
6i

EXAMPLE 5 Multiply and express the answer to each of the following in standard notation.

a) $3(\cos 40° + i \sin 40°)$ and $4(\cos 20° + i \sin 20°)$

b) $2(\cos \pi + i \sin \pi)$ and $3\left[\cos \left(-\dfrac{\pi}{2} \right) + i \sin \left(-\dfrac{\pi}{2} \right) \right]$

Solution

a) $3(\cos 40° + i \sin 40°) \cdot 4(\cos 20° + i \sin 20°)$
$$= 3 \cdot 4 \cdot [\cos (40° + 20°) + i \sin (40° + 20°)]$$
$$= 12(\cos 60° + i \sin 60°)$$
$$= 12\left(\frac{1}{2} + \frac{\sqrt{3}}{2}i \right)$$
$$= 6 + 6\sqrt{3}i$$

b) $2(\cos \pi + i \sin \pi) \cdot 3\left[\cos \left(-\dfrac{\pi}{2} \right) + i \sin \left(-\dfrac{\pi}{2} \right) \right]$

$$= 2 \cdot 3 \cdot \left[\cos \left(\pi + \left(-\frac{\pi}{2} \right) \right) + i \sin \left(\pi + \left(-\frac{\pi}{2} \right) \right) \right]$$

$$= 6\left(\cos \frac{\pi}{2} + i \sin \frac{\pi}{2} \right)$$

$$= 6(0 + i \cdot 1)$$

$$= 6i$$

EXAMPLE 6 Convert to trigonometric notation and multiply:
$$(1 + i)(\sqrt{3} - i).$$

Solution We first find trigonometric notation:

$$1 + i = \sqrt{2}(\cos 45° + i \sin 45°), \qquad \text{See Example 3(a).}$$
$$\sqrt{3} - i = 2(\cos 330° + i \sin 330°). \qquad \text{See Example 3(b).}$$

Then we multiply:

$$\sqrt{2}(\cos 45° + i \sin 45°) \cdot 2(\cos 330° + i \sin 330°)$$
$$= 2\sqrt{2}[\cos (45° + 330°) + i \sin (45° + 330°)]$$
$$= 2\sqrt{2}(\cos 375° + i \sin 375°)$$
$$= 2\sqrt{2}(\cos 15° + i \sin 15°). \qquad \text{375° has the same terminal side as 15°.}$$

▶ Now Try Exercise 35.

To divide complex numbers, we divide the absolute values and subtract the arguments. We state this fact below, but omit the proof.

Complex Numbers: Division

For any complex numbers $r_1(\cos \theta_1 + i \sin \theta_1)$ and
$r_2(\cos \theta_2 + i \sin \theta_2)$, $r_2 \neq 0$,

$$\frac{r_1(\cos \theta_1 + i \sin \theta_1)}{r_2(\cos \theta_2 + i \sin \theta_2)} = \frac{r_1}{r_2}[\cos (\theta_1 - \theta_2) + i \sin (\theta_1 - \theta_2)].$$

EXAMPLE 7 Divide

$$2\left(\cos \frac{3\pi}{2} + i \sin \frac{3\pi}{2}\right) \quad \text{by} \quad 4\left(\cos \frac{\pi}{2} + i \sin \frac{\pi}{2}\right)$$

and express the solution in standard notation.

Solution We have

$$\frac{2\left(\cos \dfrac{3\pi}{2} + i \sin \dfrac{3\pi}{2}\right)}{4\left(\cos \dfrac{\pi}{2} + i \sin \dfrac{\pi}{2}\right)} = \frac{2}{4}\left[\cos \left(\frac{3\pi}{2} - \frac{\pi}{2}\right) + i \sin \left(\frac{3\pi}{2} - \frac{\pi}{2}\right)\right]$$

$$= \frac{1}{2}(\cos \pi + i \sin \pi)$$

$$= \frac{1}{2}(-1 + i \cdot 0)$$

$$= -\frac{1}{2}.$$

EXAMPLE 8 Convert to trigonometric notation and divide:

$$\frac{1 + i}{1 - i}.$$

Solution We first convert to trigonometric notation:

$$1 + i = \sqrt{2}(\cos 45° + i \sin 45°), \qquad \text{See Example 3(a).}$$
$$1 - i = \sqrt{2}(\cos 315° + i \sin 315°).$$

We now divide:

$$\frac{\sqrt{2}(\cos 45° + i \sin 45°)}{\sqrt{2}(\cos 315° + i \sin 315°)}$$

$$= 1[\cos (45° - 315°) + i \sin (45° - 315°)]$$
$$= \cos (-270°) + i \sin (-270°)$$
$$= 0 + i \cdot 1$$
$$= i. \qquad \blacktriangleright \text{ Now Try Exercise 39.}$$

✦ Powers of Complex Numbers

An important theorem about powers and roots of complex numbers is
named for the French mathematician Abraham DeMoivre (1667–1754).
Let's consider the square of a complex number $r(\cos \theta + i \sin \theta)$:

$$[r(\cos \theta + i \sin \theta)]^2 = [r(\cos \theta + i \sin \theta)] \cdot [r(\cos \theta + i \sin \theta)]$$
$$= r \cdot r \cdot [\cos (\theta + \theta) + i \sin (\theta + \theta)]$$
$$= r^2(\cos 2\theta + i \sin 2\theta).$$

Similarly, we see that

$$[r(\cos \theta + i \sin \theta)]^3$$
$$= r \cdot r \cdot r \cdot [\cos (\theta + \theta + \theta) + i \sin (\theta + \theta + \theta)]$$
$$= r^3(\cos 3\theta + i \sin 3\theta).$$

DeMoivre's theorem is the generalization of these results.

> ### DeMoivre's Theorem
> For any complex number $r(\cos \theta + i \sin \theta)$ and any natural
> number n,
>
> $$[r(\cos \theta + i \sin \theta)]^n = r^n(\cos n\theta + i \sin n\theta).$$

TECHNOLOGY ·················
CONNECTION

We can find powers of complex numbers, like those in Example 9, on a graphing calculator.

```
(1+i)^9
                16+16i
(√(3)-i)^10
        512+886.8100135i
```

EXAMPLE 9 Find each of the following.

a) $(1 + i)^9$ **b)** $(\sqrt{3} - i)^{10}$

Solution

a) We first find trigonometric notation:

$$1 + i = \sqrt{2}(\cos 45° + i \sin 45°). \text{See Example 3(a).}$$

Then

$$(1 + i)^9 = \left[\sqrt{2}(\cos 45° + i \sin 45°)\right]^9$$
$$= (\sqrt{2})^9[\cos (9 \cdot 45°) + i \sin (9 \cdot 45°)] \text{DeMoivre's theorem}$$
$$= 2^{9/2}(\cos 405° + i \sin 405°)$$
$$= 16\sqrt{2}(\cos 45° + i \sin 45°) \text{405° has the same terminal side as 45°.}$$
$$= 16\sqrt{2}\left(\frac{\sqrt{2}}{2} + i\frac{\sqrt{2}}{2}\right)$$
$$= 16 + 16i.$$

b) We first convert to trigonometric notation:

$$\sqrt{3} - i = 2(\cos 330° + i \sin 330°). \text{See Example 3(b).}$$

Then

$$(\sqrt{3} - i)^{10} = [2(\cos 330° + i \sin 330°)]^{10}$$
$$= 2^{10}(\cos 3300° + i \sin 3300°)$$
$$= 1024(\cos 60° + i \sin 60°) \text{3300° has the same terminal side as 60°.}$$
$$= 1024\left(\frac{1}{2} + i\frac{\sqrt{3}}{2}\right)$$
$$= 512 + 512\sqrt{3}i.$$

▶ Now Try Exercise 47.

✦ Roots of Complex Numbers

As we will see, every nonzero complex number has two square roots. A nonzero complex number has three cube roots, four fourth roots, and so on. In general, a nonzero complex number has n different nth roots. They can be found using the formula that we now state but do not prove.

Roots of Complex Numbers

The nth roots of a complex number $r(\cos \theta + i \sin \theta)$, $r \neq 0$, are given by

$$r^{1/n}\left[\cos\left(\frac{\theta}{n} + k \cdot \frac{360°}{n}\right) + i \sin\left(\frac{\theta}{n} + k \cdot \frac{360°}{n}\right)\right],$$

where $k = 0, 1, 2, \ldots, n - 1$.

EXAMPLE 10   Find the square roots of $2 + 2\sqrt{3}i$.

Solution   We first find trigonometric notation:

$$2 + 2\sqrt{3}i = 4(\cos 60° + i \sin 60°).$$

Then $n = 2$, $1/n = 1/2$, and $k = 0, 1$; and

$$[4(\cos 60° + i \sin 60°)]^{1/2}$$

$$= 4^{1/2}\left[\cos\left(\frac{60°}{2} + k \cdot \frac{360°}{2}\right) + i \sin\left(\frac{60°}{2} + k \cdot \frac{360°}{2}\right)\right], \quad k = 0, 1$$

$$= 2[\cos(30° + k \cdot 180°) + i \sin(30° + k \cdot 180°), \quad k = 0, 1.$$

Thus the roots are

$$2(\cos 30° + i \sin 30°) \text{ for } k = 0$$

and $2(\cos 210° + i \sin 210°)$ for $k = 1$,

or $\sqrt{3} + i$ and $-\sqrt{3} - i$.   ▶ Now Try Exercise 57.

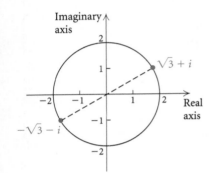

In Example 10, we see that the two square roots of the number are opposites of each other. We can illustrate this graphically. We also note that the roots are equally spaced about a circle of radius r—in this case, $r = 2$. The roots are $360°/2$, or $180°$ apart.

EXAMPLE 11   Find the cube roots of 1. Then locate them on a graph.

Solution   We begin by finding trigonometric notation:

$$1 = 1(\cos 0° + i \sin 0°).$$

Then $n = 3$, $1/n = 1/3$, and $k = 0, 1, 2$; and

$$[1(\cos 0° + i \sin 0°)]^{1/3}$$

$$= 1^{1/3}\left[\cos\left(\frac{0°}{3} + k \cdot \frac{360°}{3}\right) + i \sin\left(\frac{0°}{3} + k \cdot \frac{360°}{3}\right)\right], \quad k = 0, 1, 2.$$

The roots are

$$1(\cos 0° + i \sin 0°), \quad 1(\cos 120° + i \sin 120°),$$

and $1(\cos 240° + i \sin 240°)$,

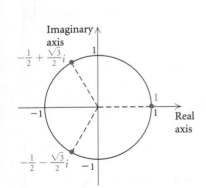

or 1, $-\frac{1}{2} + \frac{\sqrt{3}}{2}i$,   and $-\frac{1}{2} - \frac{\sqrt{3}}{2}i$.

The graphs of the cube roots lie equally spaced about a circle of radius 1. The roots are $360°/3$, or $120°$ apart.   ▶ Now Try Exercise 59.

The nth roots of 1 are often referred to as the **nth roots of unity**. In Example 11, we found the cube roots of unity.

Using a graphing calculator set in PARAMETRIC mode, we can approximate the nth roots of a number p. We use the following window and let

$$X_1T = (p^\wedge(1/n)) \cos T \quad \text{and} \quad Y_1T = (p^\wedge(1/n)) \sin T.$$

WINDOW

 Tmin $= 0$

 Tmax $= 360$, if in degree mode, or
    $= 2\pi$, if in radian mode

 Tstep $= 360/n$, or $2\pi/n$

 Xmin $= -3$, Xmax $= 3$, Xscl $= 1$

 Ymin $= -2$, Ymax $= 2$, Yscl $= 1$

 To find the fifth roots of 8, enter $X_1T = (8^\wedge(1/5)) \cos T$ and $Y_1T = (8^\wedge(1/5)) \sin T$. In this case, use DEGREE mode. After the graph has been generated, use the TRACE feature to locate the fifth roots. The T, X, and Y values appear on the screen. What do they represent?

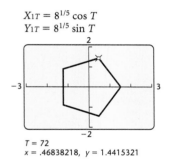

$X_1T = 8^{1/5} \cos T$
$Y_1T = 8^{1/5} \sin T$

$T = 72$
$x = .46838218, \ y = 1.4415321$

 Three of the fifth roots of 8 are approximately

  1.5157, $0.46838 + 1.44153i$, and $-1.22624 + 0.89092i$.

Find the other two. Then use a calculator to approximate the cube roots of unity that were found in Example 11. Also approximate the fourth roots of 5 and the tenth roots of unity.

3.3 EXERCISE SET

Graph the complex number and find its absolute value.

1. $4 + 3i$

2. $-2 - 3i$

3. i

4. $-5 - 2i$

5. $4 - i$

6. $6 + 3i$

7. 3

8. $-2i$

Express the indicated number in both standard notation and trigonometric notation.

9.

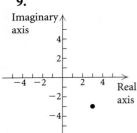

10.

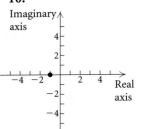

11.

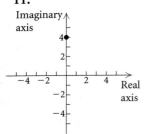

12.

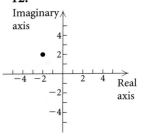

Find trigonometric notation.

13. $1 - i$

14. $-10\sqrt{3} + 10i$

15. $-3i$

16. $-5 + 5i$

17. $\sqrt{3} + i$

18. 4

19. $\dfrac{2}{5}$

20. $7.5i$

21. $-3\sqrt{2} - 3\sqrt{2}i$

22. $-\dfrac{9}{2} - \dfrac{9\sqrt{3}}{2}i$

Find standard notation, $a + bi$.

23. $3(\cos 30° + i \sin 30°)$

24. $6(\cos 120° + i \sin 120°)$

25. $10(\cos 270° + i \sin 270°)$

26. $3(\cos 0° + i \sin 0°)$

27. $\sqrt{8}\left(\cos \dfrac{\pi}{4} + i \sin \dfrac{\pi}{4}\right)$

28. $5\left(\cos \dfrac{\pi}{3} + i \sin \dfrac{\pi}{3}\right)$

29. $2\left(\cos \dfrac{\pi}{2} + i \sin \dfrac{\pi}{2}\right)$

30. $3\left[\cos\left(-\dfrac{3\pi}{4}\right) + i \sin\left(-\dfrac{3\pi}{4}\right)\right]$

31. $\sqrt{2}[\cos(-60°) + i \sin(-60°)]$

32. $4(\cos 135° + i \sin 135°)$

Multiply or divide and leave the answer in trigonometric notation.

33. $\dfrac{12(\cos 48° + i \sin 48°)}{3(\cos 6° + i \sin 6°)}$

34. $5\left(\cos \dfrac{\pi}{3} + i \sin \dfrac{\pi}{3}\right) \cdot 2\left(\cos \dfrac{\pi}{4} + i \sin \dfrac{\pi}{4}\right)$

35. $2.5(\cos 35° + i \sin 35°) \cdot 4.5(\cos 21° + i \sin 21°)$

36. $\dfrac{\dfrac{1}{2}\left(\cos \dfrac{2\pi}{3} + i \sin \dfrac{2\pi}{3}\right)}{\dfrac{3}{8}\left(\cos \dfrac{\pi}{6} + i \sin \dfrac{\pi}{6}\right)}$

Convert to trigonometric notation and then multiply or divide.

37. $(1 - i)(2 + 2i)$

38. $\left(1 + i\sqrt{3}\right)(1 + i)$

39. $\dfrac{1 - i}{1 + i}$

40. $\dfrac{1 - i}{\sqrt{3} - i}$

41. $\left(3\sqrt{3} - 3i\right)(2i)$

42. $\left(2\sqrt{3} + 2i\right)(2i)$

43. $\dfrac{2\sqrt{3} - 2i}{1 + \sqrt{3}i}$

44. $\dfrac{3 - 3\sqrt{3}i}{\sqrt{3} - i}$

Raise the number to the given power and write trigonometric notation for the answer.

45. $\left[2\left(\cos\dfrac{\pi}{3} + i\sin\dfrac{\pi}{3}\right)\right]^3$

46. $[2(\cos 120° + i\sin 120°)]^4$

47. $(1 + i)^6$

48. $\left(-\sqrt{3} + i\right)^5$

Raise the number to the given power and write standard notation for the answer.

49. $[3(\cos 20° + i\sin 20°)]^3$

50. $[2(\cos 10° + i\sin 10°)]^9$

51. $(1 - i)^5$

52. $(2 + 2i)^4$

53. $\left(\dfrac{1}{\sqrt{2}} - \dfrac{1}{\sqrt{2}}i\right)^{12}$

54. $\left(\dfrac{\sqrt{3}}{2} + \dfrac{1}{2}i\right)^{10}$

Find the square roots of the number.

55. $-i$

56. $1 + i$

57. $2\sqrt{2} - 2\sqrt{2}i$

58. $-\sqrt{3} - i$

Find the cube roots of the number.

59. i

60. $-64i$

61. $2\sqrt{3} - 2i$

62. $1 - \sqrt{3}i$

63. Find and graph the fourth roots of 16.

64. Find and graph the fourth roots of i.

65. Find and graph the fifth roots of -1.

66. Find and graph the sixth roots of 1.

67. Find the tenth roots of 8.

68. Find the ninth roots of -4.

69. Find the sixth roots of -1.

70. Find the fourth roots of 12.

Find all the complex solutions of the equation.

71. $x^3 = 1$

72. $x^5 - 1 = 0$

73. $x^4 + i = 0$

74. $x^4 + 81 = 0$

75. $x^6 + 64 = 0$

76. $x^5 + \sqrt{3} + i = 0$

Technology Connection

77. Using a graphing calculator, check your work in each of the odd-numbered Exercises 13–53.

78. Using a graphing calculator, check your work in each of the even-numbered Exercises 14–54.

Collaborative Discussion and Writing

79. Find and graph the square roots of $1 - i$. Explain geometrically why they are the opposites of each other.

80. Explain why trigonometric notation for a complex number is not unique, but rectangular, or standard, notation is unique.

Skill Maintenance

Convert to degree measure.

81. $\dfrac{\pi}{12}$

82. 3π

Convert to radian measure.

83. $330°$

84. $-225°$

85. Find r.

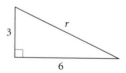

86. Graph these points in the rectangular coordinate system: $(2, -1), (0, 3),$ and $\left(-\dfrac{1}{2}, -4\right)$.

Find the function value using coordinates of points on the unit circle.

87. $\sin\dfrac{2\pi}{3}$

88. $\cos\dfrac{\pi}{6}$

89. $\cos\dfrac{\pi}{4}$

90. $\sin\dfrac{5\pi}{6}$

Synthesis

Solve.

91. $x^2 + (1 - i)x + i = 0$

92. $3x^2 + (1 + 2i)x + 1 - i = 0$

93. Find polar notation for $(\cos\theta + i\sin\theta)^{-1}$.

94. Show that for any complex number z,
$$|z| = |-z|.$$

95. Show that for any complex number z and its conjugate $\bar{z}$,
$$|z| = |\bar{z}|.$$
(*Hint*: Let $z = a + bi$ and $\bar{z} = a - bi$.)

96. Show that for any complex number z and its conjugate $\bar{z}$,
$$|z\bar{z}| = |z^2|.$$
(*Hint*: Let $z = a + bi$ and $\bar{z} = a - bi$.)

97. Show that for any complex number z,
$$|z^2| = |z|^2.$$

98. Show that for any complex numbers z and w,
$$|z \cdot w| = |z| \cdot |w|.$$
(*Hint*: Let $z = r_1(\cos \theta_1 + i \sin \theta_1)$ and $w = r_2(\cos \theta_2 + i \sin \theta_2)$.)

99. Show that for any complex number z and any nonzero, complex number w,
$$\left|\frac{z}{w}\right| = \frac{|z|}{|w|}. \quad \text{(Use the hint for Exercise 98.)}$$

100. On a complex plane, graph $|z| = 1$.

101. On a complex plane, graph $z + \bar{z} = 3$.

3.4 Polar Coordinates and Graphs

◆ Graph points given their polar coordinates.
◆ Convert from rectangular coordinates to polar coordinates and from polar coordinates to rectangular coordinates.
◆ Convert from rectangular equations to polar equations and from polar equations to rectangular equations.
◆ Graph polar equations.

◆ Polar Coordinates

All graphing throughout this text has been done with rectangular coordinates, (x, y), in the Cartesian coordinate system. We now introduce the polar coordinate system. As shown in the diagram at left, any point P has rectangular coordinates (x, y) and polar coordinates (r, θ). On a polar graph, the origin is called the **pole** and the positive half of the x-axis is called the **polar axis**. The point P can be plotted given the directed angle θ from the polar axis to the ray OP and the directed distance r from the pole to the point. The angle θ can be expressed in degrees or radians.

To plot points on a polar graph:

1. Locate the directed angle θ.
2. Move a directed distance r from the pole. If $r > 0$, move along ray OP. If $r < 0$, move in the opposite direction of ray OP.

Polar graph paper, shown below, facilitates plotting. Points B and G illustrate that θ may be in radians. Points E and F illustrate that the polar coordinates of a point are not unique.

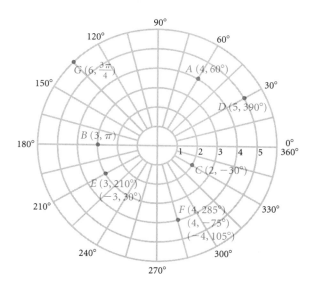

EXAMPLE 1 Graph each of the following points.

a) $A(3, 60°)$ **b)** $B(0, 10°)$

c) $C(-5, 120°)$ **d)** $D(1, -60°)$

e) $E\left(2, \dfrac{3\pi}{2}\right)$ **f)** $F\left(-4, \dfrac{\pi}{3}\right)$

Solution

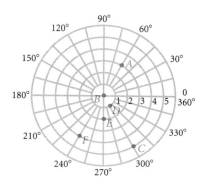

> ▶ Now Try Exercises 3 and 7.

To convert from rectangular to polar coordinates and from polar to rectangular coordinates, we need to recall the following relationships.

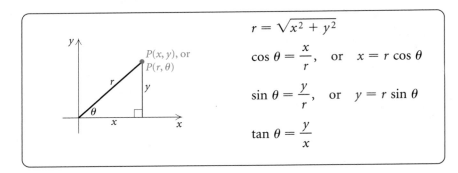

$$r = \sqrt{x^2 + y^2}$$

$$\cos \theta = \frac{x}{r}, \quad \text{or} \quad x = r \cos \theta$$

$$\sin \theta = \frac{y}{r}, \quad \text{or} \quad y = r \sin \theta$$

$$\tan \theta = \frac{y}{x}$$

EXAMPLE 2 Convert each of the following to polar coordinates.

a) $(3, 3)$ **b)** $\left(2\sqrt{3}, -2\right)$

Solution

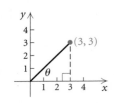

a) We first find r:

$$r = \sqrt{3^2 + 3^2} = \sqrt{18} = 3\sqrt{2}.$$

Then we determine θ:

$$\tan \theta = \frac{3}{3} = 1; \quad \text{therefore,} \quad \theta = 45°, \text{or} \ \frac{\pi}{4}.$$

We know that $\theta = \pi/4$ and not $5\pi/4$ since $(3, 3)$ is in quadrant I. Thus, $(r, \theta) = \left(3\sqrt{2}, 45°\right)$, or $\left(3\sqrt{2}, \pi/4\right)$. Other possibilities for polar coordinates include $\left(3\sqrt{2}, -315°\right)$ and $\left(-3\sqrt{2}, 5\pi/4\right)$.

b) We first find r:

$$r = \sqrt{\left(2\sqrt{3}\right)^2 + (-2)^2} = \sqrt{12 + 4} = \sqrt{16} = 4.$$

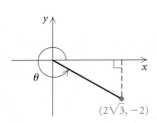

Then we determine θ:

$$\tan \theta = \frac{-2}{2\sqrt{3}} = -\frac{1}{\sqrt{3}}; \quad \text{therefore,} \quad \theta = 330°, \text{or} \ \frac{11\pi}{6}.$$

Thus, $(r, \theta) = (4, 330°)$, or $(4, 11\pi/6)$. Other possibilities for polar coordinates for this point include $(4, -\pi/6)$ and $(-4, 150°)$.

▶ Now Try Exercise 19.

It is easier to convert from polar to rectangular coordinates than from rectangular to polar coordinates.

EXAMPLE 3 Convert each of the following to rectangular coordinates.

a) $\left(10, \dfrac{\pi}{3}\right)$ 　　　　　　　　　　**b)** $(-5, 135°)$

Solution

a) The ordered pair $(10, \pi/3)$ gives us $r = 10$ and $\theta = \pi/3$. We now find x and y:

$$x = r \cos \theta = 10 \cos \frac{\pi}{3} = 10 \cdot \frac{1}{2} = 5$$

and

$$y = r \sin \theta = 10 \sin \frac{\pi}{3} = 10 \cdot \frac{\sqrt{3}}{2} = 5\sqrt{3}.$$

Thus, $(x, y) = \left(5, 5\sqrt{3}\right)$.

b) From the ordered pair $(-5, 135°)$, we know that $r = -5$ and $\theta = 135°$. We now find x and y:

$$x = -5 \cos 135° = -5 \cdot \left(-\frac{\sqrt{2}}{2}\right) = \frac{5\sqrt{2}}{2}$$

and

$$y = -5 \sin 135° = -5 \cdot \left(\frac{\sqrt{2}}{2}\right) = -\frac{5\sqrt{2}}{2}.$$

Thus, $(x, y) = \left(\dfrac{5\sqrt{2}}{2}, -\dfrac{5\sqrt{2}}{2}\right)$.

▶ Now Try Exercises 27 and 33.

✦ Polar and Rectangular Equations

Some curves have simpler equations in polar coordinates than in rectangular coordinates. For others, the reverse is true.

EXAMPLE 4 Convert each of the following to a polar equation.

a) $x^2 + y^2 = 25$ 　　　　　　　　　**b)** $2x - y = 5$

Solution

a) We have

$$x^2 + y^2 = 25$$
$$(r \cos \theta)^2 + (r \sin \theta)^2 = 25 \qquad \text{Substituting for } x \text{ and } y$$
$$r^2 \cos^2 \theta + r^2 \sin^2 \theta = 25$$
$$r^2(\cos^2 \theta + \sin^2 \theta) = 25$$
$$r^2 = 25 \qquad \cos^2 \theta + \sin^2 \theta = 1$$
$$r = 5.$$

This example illustrates that the polar equation of a circle centered at the origin is much simpler than the rectangular equation.

b) We have

$$2x - y = 5$$
$$2(r \cos \theta) - (r \sin \theta) = 5$$
$$r(2 \cos \theta - \sin \theta) = 5.$$

In this example, we see that the rectangular equation is simpler than the polar equation.

▶ Now Try Exercises 39 and 43.

EXAMPLE 5   Convert each of the following to a rectangular equation.

a) $r = 4$
b) $r \cos \theta = 6$
c) $r = 2 \cos \theta + 3 \sin \theta$

Solution

a) We have

$$r = 4$$
$$\sqrt{x^2 + y^2} = 4 \qquad \text{Substituting for } r$$
$$x^2 + y^2 = 16. \qquad \text{Squaring}$$

In squaring, we must be careful not to introduce solutions of the equation that are not already present. In this case, we did not, because the graph of either equation is a circle of radius 4 centered at the origin.

b) We have

$$r \cos \theta = 6$$
$$x = 6. \qquad x = r \cos \theta$$

The graph of $r \cos \theta = 6$, or $x = 6$, is a vertical line.

c) We have

$$r = 2 \cos \theta + 3 \sin \theta$$
$$r^2 = 2r \cos \theta + 3r \sin \theta \qquad \text{Multiplying both sides by } r$$
$$x^2 + y^2 = 2x + 3y. \qquad \text{Substituting } x^2 + y^2 \text{ for } r^2,$$
$$\qquad\qquad\qquad\qquad\qquad x \text{ for } r \cos \theta, \text{ and } y \text{ for } r \sin \theta$$

▶ Now Try Exercises 51 and 55.

✦ Graphing Polar Equations

To graph a polar equation, we can make a table of values, choosing values of θ and calculating corresponding values of r. We plot the points and complete the graph, as we do when graphing a rectangular equation. A difference occurs in the case of a polar equation, however, because as θ increases sufficiently, points may begin to repeat and the curve will be traced again and again. When this happens, the curve is complete.

EXAMPLE 6 Graph: $r = 1 - \sin\theta$.

Solution We first make a table of values. Note that the points begin to repeat at $\theta = 360°$. We plot these points and draw the curve, as shown below.

θ	r
0°	1
15°	0.7412
30°	0.5
45°	0.2929
60°	0.1340
75°	0.0341
90°	0
105°	0.0341
120°	0.1340
135°	0.2929
150°	0.5
165°	0.7412
180°	1

θ	r
195°	1.2588
210°	1.5
225°	1.7071
240°	1.8660
255°	1.9659
270°	2
285°	1.9659
300°	1.8660
315°	1.7071
330°	1.5
345°	1.2588
360°	1
375°	0.7412
390°	0.5

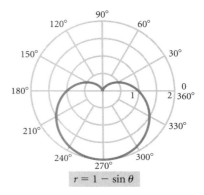

$r = 1 - \sin\theta$

Because of its heart shape, this curve is called a *cardioid*.

▷ Now Try Exercise 69.

We can graph polar equations using a graphing calculator. The equation usually must be written first in the form $r = f(\theta)$. It is necessary to decide on not only the best window dimensions but also the range of values for θ. Typically, we begin with a range of 0 to 2π for θ in radians and 0° to 360° for θ in degrees. Because most polar graphs are curved, it is important to square the window to minimize distortion.

Graph $r = 4 \sin 3\theta$. Begin by setting the calculator in POLAR mode, and use either of the following windows:

WINDOW
(Radians)
 θmin $= 0$
 θmax $= 2\pi$
 θstep $= \pi/24$
 Xmin $= -9$
 Xmax $= 9$
 Xscl $= 1$
 Ymin $= -6$
 Ymax $= 6$
 Yscl $= 1$

WINDOW
(Degrees)
 θmin $= 0$
 θmax $= 360$
 θstep $= 1$
 Xmin $= -9$
 Xmax $= 9$
 Xscl $= 1$
 Ymin $= -6$
 Ymax $= 6$
 Yscl $= 1$

$r = 4 \sin 3\theta$

We observe the same graph in both windows. The calculator allows us to view the curve as it is formed.

Now graph each of the following equations and observe the effect of changing the coefficient of $\sin 3\theta$ and the coefficient of θ:

$r = 2 \sin 3\theta,$ $r = 6 \sin 3\theta,$ $r = 4 \sin \theta,$

$r = 4 \sin 5\theta,$ $r = 4 \sin 2\theta,$ $r = 4 \sin 4\theta.$

Polar equations of the form $r = a \cos n\theta$ and $r = a \sin n\theta$ have rose-shaped curves. The number a determines the length of the petals, and the number n determines the number of petals. If n is odd, there are n petals. If n is even, there are $2n$ petals.

EXAMPLE 7 Graph each of the following polar equations. Try to visualize the shape of the curve before graphing it.

a) $r = 3$

b) $r = 5 \sin \theta$

c) $r = 2 \csc \theta$

Solution

For each graph, we can begin with a table of values. Then we plot points and complete the graph.

a) $r = 3$

For all values of θ, r is 3. Thus the graph of $r = 3$ is a circle of radius 3 centered at the origin.

θ	r
0°	3
60°	3
135°	3
210°	3
300°	3
360°	3

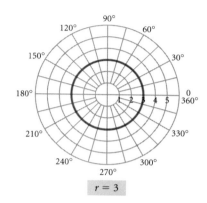

$r = 3$

We can verify our graph by converting to the equivalent rectangular equation. For $r = 3$, we substitute $\sqrt{x^2 + y^2}$ for r and square. The resulting equation,

$$x^2 + y^2 = 3^2,$$

is the equation of a circle with radius 3 centered at the origin.

b) $r = 5 \sin \theta$

θ	r
0°	0
15°	1.2941
30°	2.5
45°	3.5355
60°	4.3301
75°	4.8296
90°	5
105°	4.8296
120°	4.3301
135°	3.5355
150°	2.5
165°	1.2941
180°	0

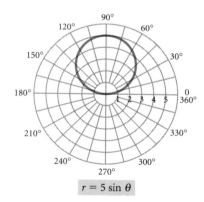

$r = 5 \sin \theta$

c) $r = 2 \csc \theta$

We can rewrite $r = 2 \csc \theta$ as $r = 2/\sin \theta$.

θ	r
0°	Not defined
15°	7.7274
30°	4
45°	2.8284
60°	2.3094
75°	2.0706
90°	2
105°	2.0706
120°	2.3094
135°	2.8284
150°	4
165°	7.7274
180°	Not defined

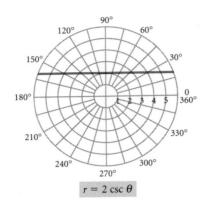

$r = 2 \csc \theta$

▶ Now Try Exercise 63.

We can check our graph in Example 7(c) by converting the polar equation to the equivalent rectangular equation:

$$r = 2 \csc \theta$$

$$r = \frac{2}{\sin \theta}$$

$$r \sin \theta = 2$$

$$y = 2. \qquad \text{Substituting } y \text{ for } r \sin \theta$$

The graph of $y = 2$ is a horizontal line passing through $(0, 2)$ on a rectangular grid.

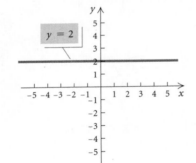

$y = 2$

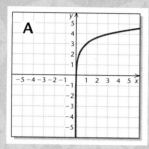

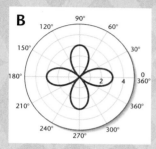

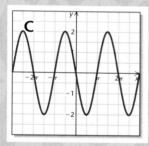

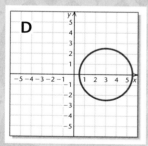

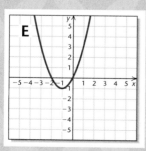

Visualizing the Graph

Match the equation with its graph.

1. $f(x) = 2^{(1/2)x}$

2. $y = -2 \sin x$

3. $y = (x + 1)^2 - 1$

4. $f(x) = \dfrac{x - 3}{x^2 + x - 6}$

5. $r = 1 + \sin \theta$

6. $f(x) = 2 \log x + 3$

7. $(x - 3)^2 + y^2 = \dfrac{25}{4}$

8. $y = -\cos\left(x - \dfrac{\pi}{2}\right)$

9. $r = 3 \cos 2\theta$

10. $f(x) = x^4 - x^3 + x^2 - x$

Answers on page A-20

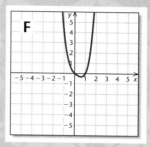

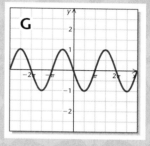

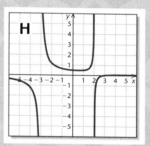

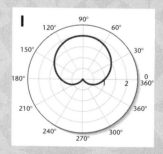

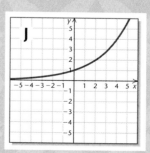

3.4 EXERCISE SET

Graph the point on a polar grid.

1. $(2, 45°)$ **2.** $(4, \pi)$ **3.** $(3.5, 210°)$

4. $(-3, 135°)$ **5.** $\left(1, \dfrac{\pi}{6}\right)$ **6.** $(2.75, 150°)$

7. $\left(-5, \dfrac{\pi}{2}\right)$ **8.** $(0, 15°)$ **9.** $(3, -315°)$

10. $\left(1.2, -\dfrac{2\pi}{3}\right)$ **11.** $(4.3, -60°)$ **12.** $(3, 405°)$

Find polar coordinates of points A, B, C, and D. Give three answers for each point.

13.

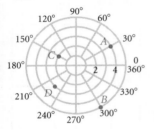

14.

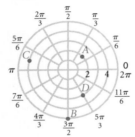

Find the polar coordinates of the point. Express the angle in degrees and then in radians, using the smallest possible positive angle.

15. $(0, -3)$ **16.** $(-4, 4)$

17. $\left(3, -3\sqrt{3}\right)$ **18.** $\left(-\sqrt{3}, 1\right)$

19. $\left(4\sqrt{3}, -4\right)$ **20.** $\left(2\sqrt{3}, 2\right)$

21. $\left(-\sqrt{2}, -\sqrt{2}\right)$ **22.** $\left(-3, 3\sqrt{3}\right)$

23. $\left(1, \sqrt{3}\right)$ **24.** $(0, -1)$

25. $\left(\dfrac{5\sqrt{2}}{2}, -\dfrac{5\sqrt{2}}{2}\right)$ **26.** $\left(-\dfrac{3}{2}, -\dfrac{3\sqrt{3}}{2}\right)$

Find the rectangular coordinates of the point.

27. $(5, 60°)$ **28.** $(0, -23°)$

29. $(-3, 45°)$ **30.** $(6, 30°)$

31. $(3, -120°)$ **32.** $\left(7, \dfrac{\pi}{6}\right)$

33. $\left(-2, \dfrac{5\pi}{3}\right)$ **34.** $(1.4, 225°)$

35. $(2, 210°)$ **36.** $\left(1, \dfrac{7\pi}{4}\right)$

37. $\left(-6, \dfrac{5\pi}{6}\right)$ **38.** $(4, 180°)$

Convert to a polar equation.

39. $3x + 4y = 5$ **40.** $5x + 3y = 4$

41. $x = 5$ **42.** $y = 4$

43. $x^2 + y^2 = 36$ **44.** $x^2 - 4y^2 = 4$

45. $x^2 = 25y$ **46.** $2x - 9y + 3 = 0$

47. $y^2 - 5x - 25 = 0$ **48.** $x^2 + y^2 = 8y$

49. $x^2 - 2x + y^2 = 0$ **50.** $3x^2y = 81$

Convert to a rectangular equation.

51. $r = 5$ **52.** $\theta = \dfrac{3\pi}{4}$

53. $r \sin \theta = 2$ **54.** $r = -3 \sin \theta$

55. $r + r \cos \theta = 3$ **56.** $r = \dfrac{2}{1 - \sin \theta}$

57. $r - 9 \cos \theta = 7 \sin \theta$ **58.** $r + 5 \sin \theta = 7 \cos \theta$

59. $r = 5 \sec \theta$ **60.** $r = 3 \cos \theta$

61. $\theta = \dfrac{5\pi}{3}$ **62.** $r = \cos \theta - \sin \theta$

Graph the equation.

63. $r = \sin \theta$ **64.** $r = 1 - \cos \theta$

65. $r = 4 \cos 2\theta$ **66.** $r = 1 - 2 \sin \theta$

67. $r = \cos \theta$ **68.** $r = 2 \sec \theta$

69. $r = 2 - \cos 3\theta$ **70.** $r = \dfrac{1}{1 + \cos \theta}$

Technology Connection

Use a graphing calculator to convert from rectangular coordinates to polar coordinates. Express the answer in both degrees and radians, using the smallest possible positive angle.

71. $(3, 7)$

72. $\left(-2, -\sqrt{5}\right)$

73. $\left(-\sqrt{10}, 3.4\right)$

74. $(0.9, -6)$

Use a graphing calculator to convert from polar coordinates to rectangular coordinates. Round the coordinates to the nearest hundredth.

75. $(3, -43°)$

76. $\left(-5, \dfrac{\pi}{7}\right)$

77. $\left(-4.2, \dfrac{3\pi}{5}\right)$

78. $(2.8, 166°)$

In Exercises 79–90, use a graphing calculator to match the equation with one of figures (a)–(l), which follow. Try matching the graphs mentally before using a calculator.

a)

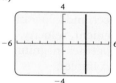

b)

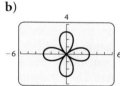

c)

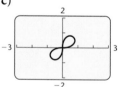

d)

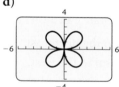

e)

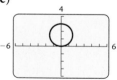

f)

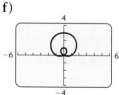

g)

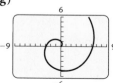

h)

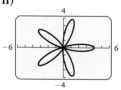

i)

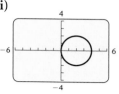

j)

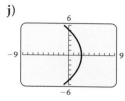

k)

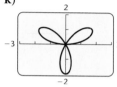

l)

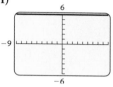

79. $r = 3 \sin 2\theta$

80. $r = 4 \cos \theta$

81. $r = \theta$

82. $r^2 = \sin 2\theta$

83. $r = \dfrac{5}{1 + \cos \theta}$

84. $r = 1 + 2 \sin \theta$

85. $r = 3 \cos 2\theta$

86. $r = 3 \sec \theta$

87. $r = 3 \sin \theta$

88. $r = 4 \cos 5\theta$

89. $r = 2 \sin 3\theta$

90. $r \sin \theta = 6$

Graph.

91. $r = \sin \theta \tan \theta$ (Cissoid)

92. $r = 3\theta$ (Spiral of Archimedes)

93. $r = e^{\theta/10}$ (Logarithmic spiral)

94. $r = 10^{2\theta}$ (Logarithmic spiral)

95. $r = \cos 2\theta \sec \theta$ (Strophoid)

96. $r = \cos 2\theta - 2$ (Peanut)

97. $r = \frac{1}{4} \tan^2 \theta \sec \theta$ (Semicubical parabola)

98. $r = \sin 2\theta + \cos \theta$ (Twisted sister)

Collaborative Discussion and Writing

99. Explain why the rectangular coordinates of a point are unique and the polar coordinates of a point are not unique.

100. Give an example of an equation that is easier to graph in polar notation than in rectangular notation and explain why.

Skill Maintenance

Solve.

101. $2x - 4 = x + 8$ **102.** $4 - 5y = 3$

Graph.

103. $y = 2x - 5$ **104.** $4x - y = 6$

105. $x = -3$ **106.** $y = 0$

Synthesis

107. Convert to a rectangular equation:

$$r = \sec^2 \frac{\theta}{2}.$$

108. The center of a regular hexagon is at the origin, and one vertex is the point $(4, 0°)$. Find the coordinates of the other vertices.

3.5 Vectors and Applications

- ◆ Determine whether two vectors are equivalent.
- ◆ Find the sum, or resultant, of two vectors.
- ◆ Resolve a vector into its horizontal and vertical components.
- ◆ Solve applied problems involving vectors.

We measure some quantities using only their magnitudes. For example, we describe time, length, and mass using units like seconds, feet, and kilograms, respectively. However, to measure quantities like **displacement, velocity,** or **force,** we need to describe a *magnitude* and a *direction*. Together magnitude and direction describe a **vector.** The following are some examples.

Displacement

An object moves a certain distance in a certain direction.

> A surveyor steps 20 yd to the northeast.
> A hiker follows a trail 5 mi to the west.
> A batter hits a ball 100 m along the left-field line.

Velocity

An object travels at a certain speed in a certain direction.

> A breeze is blowing 15 mph from the northwest.
> An airplane is traveling 450 km/h in a direction of 243°.

Force

A push or pull is exerted on an object in a certain direction.

> A force of 200 lb is required to pull a cart up a 30° incline.
> A 25-lb force is required to lift a box upward.
> A force of 15 newtons is exerted downward on the handle of a jack. (A newton, abbreviated N, is a unit of force used in physics, and $1 \text{ N} \approx 0.22 \text{ lb.}$)

✦ Vectors

Vectors can be graphically represented by directed line segments. The length is chosen, according to some scale, to represent the **magnitude of the vector,** and the direction of the directed line segment represents the **direction of the vector.** For example, if we let 1 cm represent 5 km/h, then a 15-km/h wind from the northwest would be represented by a directed line segment 3 cm long, as shown in the figure at left.

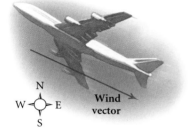

> ### *Vector*
>
> A **vector** in the plane is a directed line segment. Two vectors are **equivalent** if they have the same *magnitude* and *direction*.

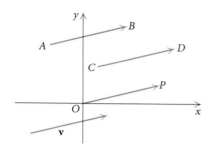

Consider a vector drawn from point A to point B. Point A is called the **initial point** of the vector, and point B is called the **terminal point.** Symbolic notation for this vector is $\overrightarrow{AB}$ (read "vector AB"). Vectors are also denoted by boldface letters such as **u, v,** and **w.** The four vectors in the figure at left have the *same* length and direction. Thus they represent **equivalent** vectors; that is,

$$\overrightarrow{AB} = \overrightarrow{CD} = \overrightarrow{OP} = \mathbf{v}.$$

In the context of vectors, we use $=$ to mean equivalent.

The length, or **magnitude,** of $\overrightarrow{AB}$ is expressed as $|\overrightarrow{AB}|$. In order to determine whether vectors are equivalent, we find their magnitudes and directions.

EXAMPLE 1 The vectors **u,** $\overrightarrow{OR}$, and **w** are shown in the figure below. Show that $\mathbf{u} = \overrightarrow{OR} = \mathbf{w}$.

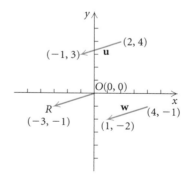

Solution We first find the length of each vector using the distance formula.

$$|\mathbf{u}| = \sqrt{[2 - (-1)]^2 + (4 - 3)^2} = \sqrt{9 + 1} = \sqrt{10},$$
$$|\overrightarrow{OR}| = \sqrt{[0 - (-3)]^2 + [0 - (-1)]^2} = \sqrt{9 + 1} = \sqrt{10},$$
$$|\mathbf{w}| = \sqrt{(4 - 1)^2 + [-1 - (-2)]^2} = \sqrt{9 + 1} = \sqrt{10}.$$

u ≠ **v** (not equivalent)
Different magnitudes;
different directions

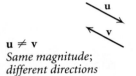

u ≠ **v**
Same magnitude;
different directions

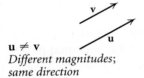

u ≠ **v**
Different magnitudes;
same direction

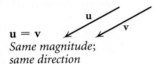

u = **v**
Same magnitude;
same direction

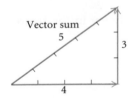

Thus

$$|\mathbf{u}| = |\overrightarrow{OR}| = |\mathbf{w}|.$$

The vectors **u**, $\overrightarrow{OR}$, and **w** appear to go in the same direction so we check their slopes. If the lines that they are on all have the same slope, the vectors have the same direction. We calculate the slopes:

$$\text{Slope} = \underset{\mathbf{u}}{\frac{4-3}{2-(-1)}} = \underset{\overrightarrow{OR}}{\frac{0-(-1)}{0-(-3)}} = \underset{\mathbf{w}}{\frac{-1-(-2)}{4-1}} = \frac{1}{3}.$$

Since **u**, $\overrightarrow{OR}$, and **w** have the *same* magnitude and the *same* direction,

$$\mathbf{u} = \overrightarrow{OR} = \mathbf{w}.$$

▶ **Now Try Exercise 1.**

Keep in mind that the equivalence of vectors requires only the same magnitude and the same direction—not the same location. In the illustrations at left, each of the first three pairs of vectors are not equivalent. The fourth set of vectors is an example of equivalence.

✦ Vector Addition

Suppose a person takes 4 steps east and then 3 steps north. He or she will then be 5 steps from the starting point in the direction shown at left. A vector 4 units long and pointing to the right represents 4 steps east and a vector 3 units long and pointing up represents 3 steps north. The **sum** of the two vectors is the vector 5 steps in magnitude and in the direction shown. The sum is also called the **resultant** of the two vectors.

In general, two nonzero vectors **u** and **v** can be added geometrically by placing the initial point of **v** at the terminal point of **u** and then finding the vector that has the same initial point as **u** and the same terminal point as **v**, as shown in the following figure.

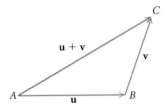

The sum **u** + **v** is the vector represented by the directed line segment from the initial point *A* of **u** to the terminal point *C* of **v**. That is, if

$$\mathbf{u} = \overrightarrow{AB} \quad \text{and} \quad \mathbf{v} = \overrightarrow{BC},$$

then

$$\mathbf{u} + \mathbf{v} = \overrightarrow{AB} + \overrightarrow{BC} = \overrightarrow{AC}.$$

We can also describe vector addition by placing the initial points of the vectors together, completing a parallelogram, and finding the diagonal of the parallelogram. (See the figure on the left on the following page.) This description of addition is sometimes called the **parallelogram law** of vector addition. Vector addition is **commutative.** As shown in the figure on the right below, both **u** + **v** and **v** + **u** are represented by the same directed line segment.

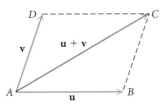 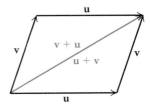

◆ Applications

If two forces F_1 and F_2 act on an object, the *combined* effect is the sum, or resultant, $F_1 + F_2$ of the separate forces.

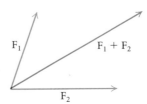

EXAMPLE 2 Forces of 15 newtons and 25 newtons act on an object at right angles to each other. Find their sum, or resultant, giving the magnitude of the resultant and the angle that it makes with the larger force.

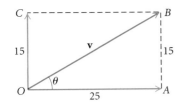

Solution We make a drawing—this time, a rectangle—using **v** or $\overrightarrow{OB}$ to represent the resultant. To find the magnitude, we use the Pythagorean theorem:

$$|\mathbf{v}|^2 = 15^2 + 25^2 \qquad \text{Here } |\mathbf{v}| \text{ denotes the length, or magnitude, of v.}$$
$$|\mathbf{v}| = \sqrt{15^2 + 25^2}$$
$$|\mathbf{v}| \approx 29.2.$$

To find the direction, we note that since OAB is a right triangle,

$$\tan \theta = \tfrac{15}{25} = 0.6.$$

Using a calculator, we find θ, the angle that the resultant makes with the larger force:

$$\theta = \tan^{-1}(0.6) \approx 31°.$$

The resultant $\overrightarrow{OB}$ has a magnitude of 29.2 and makes an angle of 31° with the larger force.

▶ Now Try Exercise 13.

AERIAL BEARINGS

REVIEW SECTION **1.3.**

Pilots must adjust the direction of their flight when there is a crosswind. Both the wind and the aircraft velocities can be described by vectors.

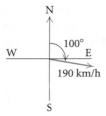

Airplane airspeed

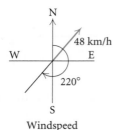

Windspeed

EXAMPLE 3   *Airplane Speed and Direction.*   An airplane travels on a bearing of 100° at an airspeed of 190 km/h while a wind is blowing 48 km/h from 220°. Find the ground speed of the airplane and the direction of its track, or course, over the ground.

Solution   We first make a drawing. The wind is represented by $\overrightarrow{OC}$ and the velocity vector of the airplane by $\overrightarrow{OA}$. The resultant velocity vector is **v**, the sum of the two vectors. The angle θ between **v** and $\overrightarrow{OA}$ is called a **drift angle**.

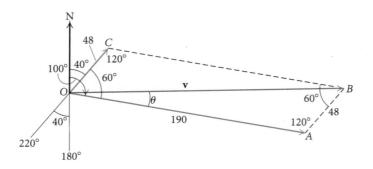

Note that the measure of $\angle COA = 100° - 40° = 60°$. Thus the measure of $\angle CBA$ is also 60° (opposite angles of a parallelogram are equal). Since the sum of all the angles of the parallelogram is 360° and $\angle OCB$ and $\angle OAB$ have the same measure, each must be 120°. By the *law of cosines* in $\triangle OAB$, we have

$$|\mathbf{v}|^2 = 48^2 + 190^2 - 2 \cdot 48 \cdot 190 \cos 120°$$
$$|\mathbf{v}|^2 = 47{,}524$$
$$|\mathbf{v}| = 218.$$

Thus, $|\mathbf{v}|$ is 218 km/h. By the *law of sines* in the same triangle,

$$\frac{48}{\sin \theta} = \frac{218}{\sin 120°},$$

or

$$\sin \theta = \frac{48 \sin 120°}{218} \approx 0.1907$$
$$\theta \approx 11°.$$

Thus, $\theta = 11°$, to the nearest degree. The ground speed of the airplane is 218 km/h, and its track is in the direction of $100° - 11°$, or 89°.

▶ Now Try Exercise 27.

✦ Components

Given a vector **w**, we may want to find two other vectors **u** and **v** whose sum is **w**. The vectors **u** and **v** are called **components** of **w** and the process of finding them is called **resolving**, or **representing**, a vector into its vector components.

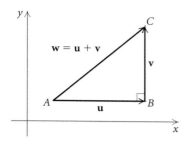

When we resolve a vector, we generally look for perpendicular components. Most often, one component will be parallel to the *x*-axis and the other will be parallel to the *y*-axis. For this reason, they are often called the **horizontal** and **vertical** components of a vector. In the figure at left, the vector $\mathbf{w} = \overrightarrow{AC}$ is resolved as the sum of $\mathbf{u} = \overrightarrow{AB}$ and $\mathbf{v} = \overrightarrow{BC}$. The horizontal component of $\mathbf{w}$ is $\mathbf{u}$ and the vertical component is $\mathbf{v}$.

EXAMPLE 4 A vector $\mathbf{w}$ has a magnitude of 130 and is inclined 40° with the horizontal. Resolve the vector into horizontal and vertical components.

Solution We first make a drawing showing horizontal and vertical vectors $\mathbf{u}$ and $\mathbf{v}$ whose sum is $\mathbf{w}$.

From $\triangle ABC$, we find $|\mathbf{u}|$ and $|\mathbf{v}|$ using the definitions of the cosine and sine functions:

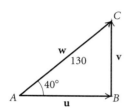

$$\cos 40° = \frac{|\mathbf{u}|}{130}, \quad \text{or} \quad |\mathbf{u}| = 130 \cos 40° \approx 100,$$

$$\sin 40° = \frac{|\mathbf{v}|}{130}, \quad \text{or} \quad |\mathbf{v}| = 130 \sin 40° \approx 84.$$

Thus the horizontal component of $\mathbf{w}$ is 100 right, and the vertical component of $\mathbf{w}$ is 84 up.

▶ Now Try Exercise 31.

EXAMPLE 5 *Shipping Crate.* A wooden shipping crate that weighs 816 lb is placed on a loading ramp that makes an angle of 25° with the horizontal. To keep the crate from sliding, a chain is hooked to the crate and to a pole at the top of the ramp. Find the magnitude of the components of the crate's weight (disregarding friction) perpendicular and parallel to the incline.

Solution We first make a drawing illustrating the forces with a rectangle. We let

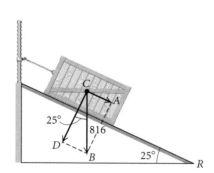

$|\overrightarrow{CB}| =$ the weight of the crate = 816 lb (force of gravity),

$|\overrightarrow{CD}| =$ the magnitude of the component of the crate's weight perpendicular to the incline (force against the ramp), and

$|\overrightarrow{CA}| =$ the magnitude of the component of the crate's weight parallel to the incline (force that pulls the crate down the ramp).

The angle at *R* is given to be 25° and $\angle BCD = \angle R = 25°$ because the sides of these angles are, respectively, perpendicular. Using the cosine and sine functions, we find that

$$\cos 25° = \frac{|\overrightarrow{CD}|}{816}, \quad \text{or} \quad |\overrightarrow{CD}| = 816 \cos 25° \approx 740 \text{ lb}, \quad \text{and}$$

$$\sin 25° = \frac{|\overrightarrow{CA}|}{816}, \quad \text{or} \quad |\overrightarrow{CA}| = 816 \sin 25° \approx 345 \text{ lb}.$$

▶ Now Try Exercise 35.

3.5 EXERCISE SET

Sketch the pair of vectors and determine whether they are equivalent. Use the following ordered pairs for the initial and terminal points.

$A(-2, 2)$ $E(-4, 1)$ $I(-6, -3)$
$B(3, 4)$ $F(2, 1)$ $J(3, 1)$
$C(-2, 5)$ $G(-4, 4)$ $K(-3, -3)$
$D(-1, -1)$ $H(1, 2)$ $O(0, 0)$

1. $\overrightarrow{GE}$, $\overrightarrow{BJ}$ 2. $\overrightarrow{DJ}$, $\overrightarrow{OF}$

3. $\overrightarrow{DJ}$, $\overrightarrow{AB}$ 4. $\overrightarrow{CG}$, $\overrightarrow{FO}$

5. $\overrightarrow{DK}$, $\overrightarrow{BH}$ 6. $\overrightarrow{BA}$, $\overrightarrow{DI}$

7. $\overrightarrow{EG}$, $\overrightarrow{BJ}$ 8. $\overrightarrow{GC}$, $\overrightarrow{FO}$

9. $\overrightarrow{GA}$, $\overrightarrow{BH}$ 10. $\overrightarrow{JD}$, $\overrightarrow{CG}$

11. $\overrightarrow{AB}$, $\overrightarrow{ID}$ 12. $\overrightarrow{OF}$, $\overrightarrow{HB}$

13. Two forces of 32 N (newtons) and 45 N act on an object at right angles. Find the magnitude of the resultant and the angle that it makes with the smaller force.

14. Two forces of 50 N and 60 N act on an object at right angles. Find the magnitude of the resultant and the angle that it makes with the larger force.

15. Two forces of 410 N and 600 N act on an object. The angle between the forces is 47°. Find the magnitude of the resultant and the angle that it makes with the larger force.

16. Two forces of 255 N and 325 N act on an object. The angle between the forces is 64°. Find the magnitude of the resultant and the angle that it makes with the smaller force.

*In Exercises 17–24, magnitudes of vectors **u** and **v** and the angle θ between the vectors are given. Find the sum of **u** + **v**. Give the magnitude to the nearest tenth and give the direction by specifying to the nearest degree the angle that the resultant makes with **u**.*

17. $|\mathbf{u}| = 45$, $|\mathbf{v}| = 35$, $\theta = 90°$

18. $|\mathbf{u}| = 54$, $|\mathbf{v}| = 43$, $\theta = 150°$

19. $|\mathbf{u}| = 10$, $|\mathbf{v}| = 12$, $\theta = 67°$

20. $|\mathbf{u}| = 25$, $|\mathbf{v}| = 30$, $\theta = 75°$

21. $|\mathbf{u}| = 20$, $|\mathbf{v}| = 20$, $\theta = 117°$

22. $|\mathbf{u}| = 30$, $|\mathbf{v}| = 30$, $\theta = 123°$

23. $|\mathbf{u}| = 23$, $|\mathbf{v}| = 47$, $\theta = 27°$

24. $|\mathbf{u}| = 32$, $|\mathbf{v}| = 74$, $\theta = 72°$

25. *Hot-Air Balloon.* A hot-air balloon is rising vertically 10 ft/sec while the wind is blowing horizontally 5 ft/sec. Find the speed **v** of the balloon and the angle θ that it makes with the horizontal.

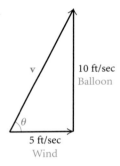

26. *Ship.* A ship sails first N80°E for 120 nautical mi, and then S20°W for 200 nautical mi. How far is the ship, then, from the starting point, and in what direction?

27. *Boat.* A boat heads 35°, propelled by a force of 750 lb. A wind from 320° exerts a force of 150 lb on the boat. How large is the resultant force **F**, and in what direction is the boat moving?

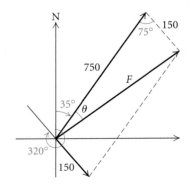

28. *Airplane.* An airplane flies 32° for 210 km, and then 280° for 170 km. How far is the airplane, then, from the starting point, and in what direction?

29. *Airplane.* An airplane has an airspeed of 150 km/h. It is to make a flight in a direction of 70° while there is a 25-km/h wind from 340°. What will the airplane's actual heading be?

30. *Wind.* A wind has an easterly component (*from* the east) of 10 km/h and a southerly component (*from* the south) of 16 km/h. Find the magnitude and the direction of the wind.

31. A vector **w** has magnitude 100 and points southeast. Resolve the vector into easterly and southerly components.

32. A vector **u** with a magnitude of 150 lb is inclined to the right and upward 52° from the horizontal. Resolve the vector into components.

33. *Airplane.* An airplane takes off at a speed **S** of 225 mph at an angle of 17° with the horizontal. Resolve the vector **S** into components.

34. *Wheelbarrow.* A wheelbarrow is pushed by applying a 97-lb force **F** that makes a 38° angle with the horizontal. Resolve **F** into its horizontal and vertical components. (The horizontal component is the effective force in the direction of motion and the vertical component adds weight to the wheelbarrow.)

35. *Luggage Wagon.* A luggage wagon is being pulled with vector force **V**, which has a magnitude of 780 lb at an angle of elevation of 60°. Resolve the vector **V** into components.

36. *Hot-air Balloon.* A hot-air balloon exerts a 1200-lb pull on a tether line at a 45° angle with the horizontal. Resolve the vector **B** into components.

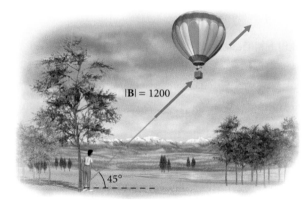

37. *Airplane.* An airplane is flying at 200 km/h in a direction of 305°. Find the westerly and northerly components of its velocity.

38. *Baseball.* A baseball player throws a baseball with a speed **S** of 72 mph at an angle of 45° with the horizontal. Resolve the vector **S** into components.

39. A block weighing 100 lb rests on a 25° incline. Find the magnitude of the components of the block's weight perpendicular and parallel to the incline.

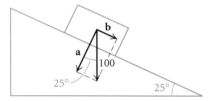

40. A shipping crate that weighs 450 kg is placed on a loading ramp that makes an angle of 30° with the horizontal. Find the magnitude of the components of the crate's weight perpendicular and parallel to the incline.

41. An 80-lb block of ice rests on a 37° incline. What force parallel to the incline is necessary in order to keep the ice from sliding down?

42. What force is necessary to pull a 3500-lb truck up a 9° incline?

Collaborative Discussion and Writing

43. Describe the concept of a vector as though you were explaining it to a classmate. Use the concept of an arrow shot from a bow in the explanation.

44. Explain why vectors $\overrightarrow{QR}$ and $\overrightarrow{RQ}$ are not equivalent.

Skill Maintenance

In each of Exercises 45–54, fill in the blank with the correct term. Some of the given choices will not be used.

angular speed
linear speed
acute
obtuse
secant of θ
cotangent of θ
identity
inverse
absolute value
sines
cosine
common
natural
horizontal line
vertical line
double-angle
half-angle
coterminal
reference angle

45. Logarithms, base e, are called _____ logarithms.

46. _____ identities give trigonometric function values of $x/2$ in terms of function values of x.

47. _____ is distance traveled per unit of time.

48. The sine of an angle is also the _____ of the angle's complement.

49. A(n) _____ is an equation that is true for all possible replacements of the variables.

50. The _____ is the length of the side adjacent to θ divided by the length of the side opposite θ.

51. If two or more angles have the same terminal side, the angles are said to be _____ .

52. In any triangle, the sides are proportional to the _____ of the opposite angles.

53. If it is possible for a(n) _____ to intersect the graph of a function more than once, then the function is not one-to-one and its _____ is not a function.

54. The _____ for an angle is the _____ angle formed by the terminal side of the angle and the x-axis.

Synthesis

55. *Eagle's Flight.*   An eagle flies from its nest 7 mi in the direction northeast, where it stops to rest on a cliff. It then flies 8 mi in the direction S30°W to land on top of a tree. Place an xy-coordinate system so that the origin is the bird's nest, the x-axis points east, and the y-axis points north.

   a) At what point is the cliff located?
   b) At what point is the tree located?

◆ Perform calculations with vectors in component form.
◆ Express a vector as a linear combination of unit vectors.
◆ Express a vector in terms of its magnitude and its direction.
◆ Find the angle between two vectors using the dot product.
◆ Solve applied problems involving forces in equilibrium.

✦ Position Vectors

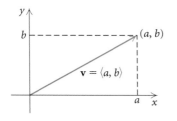

Let's consider a vector **v** whose initial point is the *origin* in an *xy*-coordinate system and whose terminal point is (a, b). We say that the vector is in **standard position** and refer to it as a position vector. Note that the ordered pair (a, b) defines the vector uniquely. Thus we can use (a, b) to denote the vector. To emphasize that we are thinking of a vector and to avoid the confusion of notation with ordered-pair and interval notation, we generally write

$$\mathbf{v} = \langle a, b \rangle.$$

The coordinate a is the *scalar* **horizontal component** of the vector, and the coordinate b is the *scalar* **vertical component** of the vector. By **scalar,** we mean a *numerical* quantity rather than a *vector* quantity. Thus, $\langle a, b \rangle$ is considered to be the *component form* of **v**. Note that a and b are *not* vectors and should not be confused with the vector component definition given in Section 3.5.

Now consider $\overrightarrow{AC}$ with $A = (x_1, y_1)$ and $C = (x_2, y_2)$. Let's see how to find the position vector equivalent to $\overrightarrow{AC}$. As you can see in the figure below, the initial point A is relocated to the origin $(0, 0)$. The coordinates of P are found by subtracting the coordinates of A from the coordinates of C. Thus, $P = (x_2 - x_1, y_2 - y_1)$ and the position vector is $\overrightarrow{OP}$.

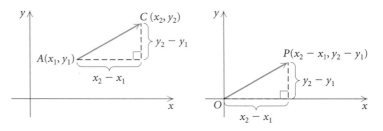

It can be shown that $\overrightarrow{OP}$ and $\overrightarrow{AC}$ have the same magnitude and direction and are therefore equivalent. Thus, $\overrightarrow{AC} = \overrightarrow{OP} = \langle x_2 - x_1, y_2 - y_1 \rangle$.

Component Form of a Vector

The **component form** of $\overrightarrow{AC}$ with $A = (x_1, y_1)$ and $C = (x_2, y_2)$ is

$$\overrightarrow{AC} = \langle x_2 - x_1, y_2 - y_1 \rangle.$$

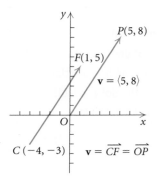

EXAMPLE 1 Find the component form of $\overrightarrow{CF}$ if $C = (-4, -3)$ and $F = (1, 5)$.

Solution We have

$$\overrightarrow{CF} = \langle 1 - (-4), 5 - (-3) \rangle = \langle 5, 8 \rangle.$$

Note that vector $\overrightarrow{CF}$ is equivalent to *position vector* $\overrightarrow{OP}$ with $P = (5, 8)$ as shown in the figure at left. ◀

Now that we know how to write vectors in component form, let's restate some definitions that we first considered in Section 3.5.

The length of a vector $\mathbf{v}$ is easy to determine when the components of the vector are known. For $\mathbf{v} = \langle v_1, v_2 \rangle$, we have

$$|\mathbf{v}|^2 = v_1^2 + v_2^2 \quad \text{Using the Pythagorean theorem}$$
$$|\mathbf{v}| = \sqrt{v_1^2 + v_2^2}.$$

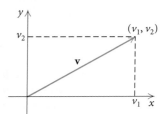

Length of a Vector

The **length**, or **magnitude**, of a vector $\mathbf{v} = \langle v_1, v_2 \rangle$ is given by

$$|\mathbf{v}| = \sqrt{v_1^2 + v_2^2}.$$

EXAMPLE 2 Find the length, or magnitude, of vector $\mathbf{v} = \langle 5, 8 \rangle$, illustrated in Example 1.

Solution

$$|\mathbf{v}| = \sqrt{v_1^2 + v_2^2} \quad \text{Length of vector } v = \langle v_1, v_2 \rangle$$
$$|\mathbf{v}| = \sqrt{5^2 + 8^2} \quad \text{Substituting 5 for } v_1 \text{ and 8 for } v_2$$
$$= \sqrt{25 + 64}$$
$$= \sqrt{89}$$

▶ Now Try Exercises 1 and 7.

Two vectors are **equivalent** if they have the *same* magnitude and the *same* direction.

Equivalent Vectors

Let $\mathbf{u} = \langle u_1, u_2 \rangle$ and $\mathbf{v} = \langle v_1, v_2 \rangle$. Then

$$\langle u_1, u_2 \rangle = \langle v_1, v_2 \rangle \quad \text{if and only if} \quad u_1 = v_1 \quad \text{and} \quad u_2 = v_2.$$

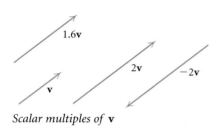

Scalar multiples of **v**

✦ Operations on Vectors

To multiply a vector **v** by a positive real number, we multiply its length by the number. Its direction stays the same. When a vector **v** is multiplied by 2, for instance, its length is doubled and its direction is not changed. When a vector is multiplied by 1.6, its length is increased by 60% and its direction stays the same. To multiply a vector **v** by a negative real number, we multiply its length by the number and reverse its direction. When a vector is multiplied by −2, its length is doubled and its direction is reversed. Since real numbers work like scaling factors in vector multiplication, we call them **scalars** and the products $k\mathbf{v}$ are called **scalar multiples** of **v**.

Scalar Multiplication

For a real number k and a vector $\mathbf{v} = \langle v_1, v_2 \rangle$, the **scalar product** of k and **v** is

$$k\mathbf{v} = k\langle v_1, v_2 \rangle = \langle kv_1, kv_2 \rangle.$$

The vector $k\mathbf{v}$ is a **scalar multiple** of the vector **v**.

EXAMPLE 3 Let $\mathbf{u} = \langle -5, 4 \rangle$ and $\mathbf{w} = \langle 1, -1 \rangle$. Find $-7\mathbf{w}$, $3\mathbf{u}$, and $-1\mathbf{w}$.

Solution

$$-7\mathbf{w} = -7\langle 1, -1 \rangle = \langle -7, 7 \rangle,$$
$$3\mathbf{u} = 3\langle -5, 4 \rangle = \langle -15, 12 \rangle,$$
$$-1\mathbf{w} = -1\langle 1, -1 \rangle = \langle -1, 1 \rangle$$

In Section 3.5, we used the parallelogram law to add two vectors, but now we can add two vectors using components. To add two vectors given in component form, we add the corresponding components. Let $\mathbf{u} = \langle u_1, u_2 \rangle$ and $\mathbf{v} = \langle v_1, v_2 \rangle$. Then

$$\mathbf{u} + \mathbf{v} = \langle u_1 + v_1, u_2 + v_2 \rangle.$$

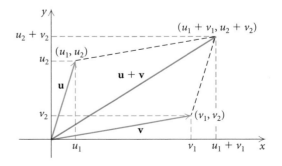

For example, if $\mathbf{v} = \langle -3, 2 \rangle$ and $\mathbf{w} = \langle 5, -9 \rangle$, then

$$\mathbf{v} + \mathbf{w} = \langle -3 + 5, 2 + (-9) \rangle = \langle 2, -7 \rangle.$$

> ### Vector Addition
> If $\mathbf{u} = \langle u_1, u_2 \rangle$ and $\mathbf{v} = \langle v_1, v_2 \rangle$, then
>
> $$\mathbf{u} + \mathbf{v} = \langle u_1 + v_1, u_2 + v_2 \rangle.$$

Before we define vector subtraction, we need to define $-\mathbf{v}$. The opposite of $\mathbf{v} = \langle v_1, v_2 \rangle$, shown below, is

$$-\mathbf{v} = (-1)\mathbf{v} = (-1)\langle v_1, v_2 \rangle$$
$$= \langle -v_1, -v_2 \rangle.$$

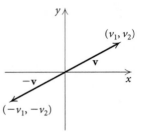

Vector subtraction such as $\mathbf{u} - \mathbf{v}$ involves subtracting corresponding components. We show this by rewriting $\mathbf{u} - \mathbf{v}$ as $\mathbf{u} + (-\mathbf{v})$. If $\mathbf{u} = \langle u_1, u_2 \rangle$ and $\mathbf{v} = \langle v_1, v_2 \rangle$, then

$$\mathbf{u} - \mathbf{v} = \mathbf{u} + (-\mathbf{v}) = \langle u_1, u_2 \rangle + \langle -v_1, -v_2 \rangle$$
$$= \langle u_1 + (-v_1), u_2 + (-v_2) \rangle$$
$$= \langle u_1 - v_1, u_2 - v_2 \rangle.$$

We can illustrate vector subtraction with parallelograms, just as we did vector addition.

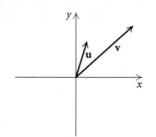

Sketch $\mathbf{u}$ and $\mathbf{v}$.

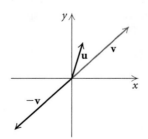

Sketch $-\mathbf{v}$.

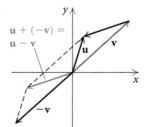

Sketch $\mathbf{u} + (-\mathbf{v})$, or $\mathbf{u} - \mathbf{v}$, using the parallelogram law.

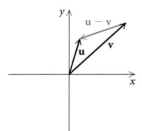

$\mathbf{u} - \mathbf{v}$ is the vector from the terminal point of $\mathbf{v}$ to the terminal point of $\mathbf{u}$.

> **Vector Subtraction**
>
> If $\mathbf{u} = \langle u_1, u_2 \rangle$ and $\mathbf{v} = \langle v_1, v_2 \rangle$, then
>
> $$\mathbf{u} - \mathbf{v} = \langle u_1 - v_1, u_2 - v_2 \rangle.$$

It is interesting to compare the sum of two vectors with the difference of the same two vectors in the same parallelogram. The vectors $\mathbf{u} + \mathbf{v}$ and $\mathbf{u} - \mathbf{v}$ are the diagonals of the parallelogram.

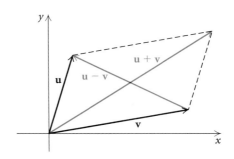

EXAMPLE 4 Do the following calculations, where $\mathbf{u} = \langle 7, 2 \rangle$ and $\mathbf{v} = \langle -3, 5 \rangle$.

a) $\mathbf{u} + \mathbf{v}$ b) $\mathbf{u} - 6\mathbf{v}$

c) $3\mathbf{u} + 4\mathbf{v}$ d) $|5\mathbf{v} - 2\mathbf{u}|$

Solution

a) $\mathbf{u} + \mathbf{v} = \langle 7, 2 \rangle + \langle -3, 5 \rangle = \langle 7 + (-3), 2 + 5 \rangle = \langle 4, 7 \rangle$

b) $\mathbf{u} - 6\mathbf{v} = \langle 7, 2 \rangle - 6\langle -3, 5 \rangle = \langle 7, 2 \rangle - \langle -18, 30 \rangle = \langle 25, -28 \rangle$

c) $3\mathbf{u} + 4\mathbf{v} = 3\langle 7, 2 \rangle + 4\langle -3, 5 \rangle = \langle 21, 6 \rangle + \langle -12, 20 \rangle = \langle 9, 26 \rangle$

d) $|5\mathbf{v} - 2\mathbf{u}| = |5\langle -3, 5 \rangle - 2\langle 7, 2 \rangle| = |\langle -15, 25 \rangle - \langle 14, 4 \rangle|$

$$= |\langle -29, 21 \rangle|$$
$$= \sqrt{(-29)^2 + 21^2}$$
$$= \sqrt{1282}$$
$$\approx 35.8$$

▶ Now Try Exercises 9 and 11.

Before we state the properties of vector addition and scalar multiplication, we need to define another special vector—the zero vector. The vector whose initial and terminal points are both $(0, 0)$ is the **zero vector,** denoted by $\mathbf{O}$, or $\langle 0, 0 \rangle$. Its magnitude is 0. In vector addition, the zero vector is the additive identity vector:

$$\mathbf{v} + \mathbf{O} = \mathbf{v}. \qquad \langle v_1, v_2 \rangle + \langle 0, 0 \rangle = \langle v_1, v_2 \rangle$$

Operations on vectors share many of the same properties as operations on real numbers.

> **Properties of Vector Addition and Scalar Multiplication**
> For all vectors **u**, **v**, and **w**, and for all scalars b and c:
>
> 1. $\mathbf{u} + \mathbf{v} = \mathbf{v} + \mathbf{u}$.
> 2. $\mathbf{u} + (\mathbf{v} + \mathbf{w}) = (\mathbf{u} + \mathbf{v}) + \mathbf{w}$.
> 3. $\mathbf{v} + \mathbf{O} = \mathbf{v}$.
> 4. $1\mathbf{v} = \mathbf{v}$; $0\mathbf{v} = \mathbf{O}$.
> 5. $\mathbf{v} + (-\mathbf{v}) = \mathbf{O}$.
> 6. $b(c\mathbf{v}) = (bc)\mathbf{v}$.
> 7. $(b + c)\mathbf{v} = b\mathbf{v} + c\mathbf{v}$.
> 8. $b(\mathbf{u} + \mathbf{v}) = b\mathbf{u} + b\mathbf{v}$.

◆ Unit Vectors

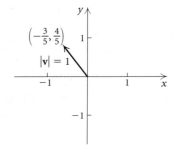

A vector of magnitude, or length, 1 is called a **unit vector.** The vector $\mathbf{v} = \left\langle -\frac{3}{5}, \frac{4}{5} \right\rangle$ is a unit vector because

$$|\mathbf{v}| = \left| \left\langle -\tfrac{3}{5}, \tfrac{4}{5} \right\rangle \right| = \sqrt{\left(-\tfrac{3}{5}\right)^2 + \left(\tfrac{4}{5}\right)^2}$$

$$= \sqrt{\tfrac{9}{25} + \tfrac{16}{25}}$$

$$= \sqrt{\tfrac{25}{25}} = \sqrt{1} = 1.$$

EXAMPLE 5 Find a unit vector that has the same direction as the vector $\mathbf{w} = \langle -3, 5 \rangle$.

Solution   We first find the length of **w**:

$$|\mathbf{w}| = \sqrt{(-3)^2 + 5^2} = \sqrt{34}.$$

Thus we want a vector whose length is $1/\sqrt{34}$ of **w** and whose direction is the same as vector **w**. That vector is

$$\mathbf{u} = \frac{1}{\sqrt{34}}\mathbf{w} = \frac{1}{\sqrt{34}}\langle -3, 5 \rangle = \left\langle \frac{-3}{\sqrt{34}}, \frac{5}{\sqrt{34}} \right\rangle.$$

The vector **u** is a *unit vector* because

$$|\mathbf{u}| = \left| \frac{1}{\sqrt{34}}\mathbf{w} \right| = \sqrt{\left(\frac{-3}{\sqrt{34}}\right)^2 + \left(\frac{5}{\sqrt{34}}\right)^2} = \sqrt{\frac{9}{34} + \frac{25}{34}}$$

$$= \sqrt{\frac{34}{34}} = \sqrt{1} = 1.$$

▶ Now Try Exercise 33.

> **Unit Vector**
> If **v** is a vector and $\mathbf{v} \neq \mathbf{O}$, then
>
> $$\frac{1}{|\mathbf{v}|} \cdot \mathbf{v}, \quad \text{or} \quad \frac{\mathbf{v}}{|\mathbf{v}|},$$
>
> is a **unit vector** in the direction of **v**.

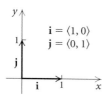

Although unit vectors can have any direction, the unit vectors parallel to the x- and y-axes are particularly useful. They are defined as

$$\mathbf{i} = \langle 1, 0 \rangle \quad \text{and} \quad \mathbf{j} = \langle 0, 1 \rangle.$$

Any vector can be expressed as a **linear combination** of unit vectors $\mathbf{i}$ and $\mathbf{j}$. For example, let $\mathbf{v} = \langle v_1, v_2 \rangle$. Then

$$\mathbf{v} = \langle v_1, v_2 \rangle = \langle v_1, 0 \rangle + \langle 0, v_2 \rangle$$
$$= v_1 \langle 1, 0 \rangle + v_2 \langle 0, 1 \rangle = v_1 \mathbf{i} + v_2 \mathbf{j}.$$

EXAMPLE 6 Express the vector $\mathbf{r} = \langle 2, -6 \rangle$ as a linear combination of $\mathbf{i}$ and $\mathbf{j}$.

Solution

$$\mathbf{r} = \langle 2, -6 \rangle = 2\mathbf{i} + (-6)\mathbf{j} = 2\mathbf{i} - 6\mathbf{j}$$

▶ Now Try Exercise 39.

EXAMPLE 7 Write the vector $\mathbf{q} = -\mathbf{i} + 7\mathbf{j}$ in component form.

Solution

$$\mathbf{q} = -\mathbf{i} + 7\mathbf{j} = -1\mathbf{i} + 7\mathbf{j} = \langle -1, 7 \rangle$$

◀

Vector operations can also be performed when vectors are written as linear combinations of $\mathbf{i}$ and $\mathbf{j}$.

EXAMPLE 8 If $\mathbf{a} = 5\mathbf{i} - 2\mathbf{j}$ and $\mathbf{b} = -\mathbf{i} + 8\mathbf{j}$, find $3\mathbf{a} - \mathbf{b}$.

Solution

$$3\mathbf{a} - \mathbf{b} = 3(5\mathbf{i} - 2\mathbf{j}) - (-\mathbf{i} + 8\mathbf{j})$$
$$= 15\mathbf{i} - 6\mathbf{j} + \mathbf{i} - 8\mathbf{j}$$
$$= 16\mathbf{i} - 14\mathbf{j}$$

▶ Now Try Exercise 45.

◆ Direction Angles

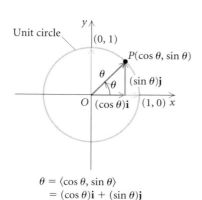

The terminal point P of a unit vector in standard position is a point on the unit circle denoted by $(\cos \theta, \sin \theta)$. Thus the unit vector can be expressed in component form,

$$\mathbf{u} = \langle \cos \theta, \sin \theta \rangle,$$

or as a linear combination of the unit vectors $\mathbf{i}$ and $\mathbf{j}$,

$$\mathbf{u} = (\cos \theta)\mathbf{i} + (\sin \theta)\mathbf{j},$$

where the components of $\mathbf{u}$ are functions of the **direction angle** θ measured counterclockwise from the x-axis to the vector. As θ varies from 0 to 2π, the point P traces the circle $x^2 + y^2 = 1$. This takes in all possible directions for unit vectors so the equation $\mathbf{u} = (\cos \theta)\mathbf{i} + (\sin \theta)\mathbf{j}$ describes every possible unit vector in the plane.

UNIT CIRCLE

REVIEW SECTION 1.5.

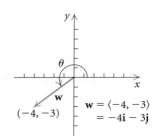

EXAMPLE 9 Calculate and sketch the unit vector $\mathbf{u} = (\cos\theta)\mathbf{i} + (\sin\theta)\mathbf{j}$ for $\theta = 2\pi/3$. Include the unit circle in your sketch.

Solution

$$\mathbf{u} = \left(\cos\frac{2\pi}{3}\right)\mathbf{i} + \left(\sin\frac{2\pi}{3}\right)\mathbf{j}$$

$$= \left(-\frac{1}{2}\right)\mathbf{i} + \left(\frac{\sqrt{3}}{2}\right)\mathbf{j}$$

▶ Now Try Exercise 49.

Let $\mathbf{v} = \langle v_1, v_2 \rangle$ with direction angle θ. Using the definition of the tangent function, we can determine the direction angle from the components of $\mathbf{v}$:

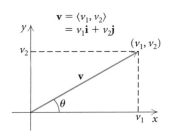

$$\tan\theta = \frac{v_2}{v_1}$$

$$\theta = \tan^{-1}\frac{v_2}{v_1}.$$

EXAMPLE 10 Determine the direction angle θ of the vector $\mathbf{w} = -4\mathbf{i} - 3\mathbf{j}$.

Solution We know that

$$\mathbf{w} = -4\mathbf{i} - 3\mathbf{j} = \langle -4, -3 \rangle.$$

Thus we have

$$\tan\theta = \frac{-3}{-4} = \frac{3}{4} \quad \text{and} \quad \theta = \tan^{-1}\frac{3}{4}.$$

Since $\mathbf{w}$ is in the third quadrant, we know that θ is a third-quadrant angle. The reference angle is

$$\tan^{-1}\frac{3}{4} \approx 37°, \quad \text{and} \quad \theta \approx 180° + 37°, \text{ or } 217°.$$

▶ Now Try Exercise 55.

It is convenient for work with applied problems and in subsequent courses, such as calculus, to have a way to express a vector so that both its magnitude and its direction can be determined, or read, easily. Let $\mathbf{v}$ be a vector. Then $\mathbf{v}/|\mathbf{v}|$ is a unit vector in the same direction as $\mathbf{v}$. Thus we have

$$\frac{\mathbf{v}}{|\mathbf{v}|} = (\cos\theta)\mathbf{i} + (\sin\theta)\mathbf{j}$$

$$\mathbf{v} = |\mathbf{v}|[(\cos\theta)\mathbf{i} + (\sin\theta)\mathbf{j}] \quad \text{Multiplying by } |\mathbf{v}|$$

$$\mathbf{v} = |\mathbf{v}|(\cos\theta)\mathbf{i} + |\mathbf{v}|(\sin\theta)\mathbf{j}.$$

Let's revisit the applied problem in Example 3 of Section 3.5 and use this new notation.

EXAMPLE 11 *Airplane Speed and Direction.* An airplane travels on a bearing of 100° at an airspeed of 190 km/h while a wind is blowing 48 km/h from 220°. Find the ground speed of the airplane and the direction of its track, or course, over the ground.

Solution We first make a drawing. The wind is represented by $\overrightarrow{OC}$ and the velocity vector of the airplane by $\overrightarrow{OA}$. The resultant velocity vector is **v**, the sum of the two vectors:

$$\mathbf{v} = \overrightarrow{OC} + \overrightarrow{OA}.$$

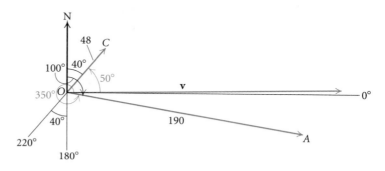

The bearing (measured from north) of the airspeed vector $\overrightarrow{OA}$ is 100°. Its *direction angle* (measured counterclockwise from the positive x-axis) is 350°. The bearing (measured from north) of the wind vector $\overrightarrow{OC}$ is 220°. Its direction angle (measured counterclockwise from the positive x-axis) is 50°. The magnitudes of $\overrightarrow{OA}$ and $\overrightarrow{OC}$ are 190 and 48, respectively. We have

$$\overrightarrow{OA} = 190(\cos 350°)\mathbf{i} + 190(\sin 350°)\mathbf{j}, \quad \text{and}$$
$$\overrightarrow{OC} = 48(\cos 50°)\mathbf{i} + 48(\sin 50°)\mathbf{j}.$$

Thus,

$$\mathbf{v} = \overrightarrow{OA} + \overrightarrow{OC}$$
$$= [190(\cos 350°)\mathbf{i} + 190(\sin 350°)\mathbf{j}] + [48(\cos 50°)\mathbf{i} + 48(\sin 50°)\mathbf{j}]$$
$$= [190(\cos 350°) + 48(\cos 50°)]\mathbf{i} + [190(\sin 350°) + 48(\sin 50°)]\mathbf{j}$$
$$\approx 217.97\mathbf{i} + 3.78\mathbf{j}.$$

From this form, we can determine the ground speed and the course:

$$\text{Ground speed} \approx \sqrt{(217.97)^2 + (3.78)^2}$$
$$\approx 218 \text{ km/h}.$$

We let α be the direction angle of **v**. Then

$$\tan \alpha = \frac{3.78}{217.97}$$

$$\alpha = \tan^{-1} \frac{3.78}{217.97} \approx 1°.$$

Thus the course of the airplane (the direction from north) is 90° − 1°, or 89°.

✦ Angle between Vectors

When a vector is multiplied by a scalar, the result is a vector. When two vectors are added, the result is also a vector. Thus we might expect the product of two vectors to be a vector as well, but it is not. The *dot product* of two vectors is a real number, or scalar. This product is useful in finding the angle between two vectors and in determining whether two vectors are perpendicular.

Dot Product

The **dot product** of two vectors $\mathbf{u} = \langle u_1, u_2 \rangle$ and $\mathbf{v} = \langle v_1, v_2 \rangle$ is

$$\mathbf{u} \cdot \mathbf{v} = u_1 v_1 + u_2 v_2.$$

(Note that $u_1 v_1 + u_2 v_2$ is a *scalar*, not a vector.)

EXAMPLE 12   Find the indicated dot product when

$$\mathbf{u} = \langle 2, -5 \rangle, \quad \mathbf{v} = \langle 0, 4 \rangle, \quad \text{and} \quad \mathbf{w} = \langle -3, 1 \rangle.$$

a) $\mathbf{u} \cdot \mathbf{w}$

b) $\mathbf{w} \cdot \mathbf{v}$

Solution

a) $\mathbf{u} \cdot \mathbf{w} = 2(-3) + (-5)1 = -6 - 5 = -11$

b) $\mathbf{w} \cdot \mathbf{v} = -3(0) + 1(4) = 0 + 4 = 4$

The dot product can be used to find the angle between two vectors. The angle *between* two vectors is the smallest positive angle formed by the two directed line segments. Thus the angle θ between $\mathbf{u}$ and $\mathbf{v}$ is the same angle as between $\mathbf{v}$ and $\mathbf{u}$, and $0 \leq \theta \leq \pi$.

Angle between Two Vectors

If θ is the angle between two *nonzero* vectors $\mathbf{u}$ and $\mathbf{v}$, then

$$\cos \theta = \frac{\mathbf{u} \cdot \mathbf{v}}{|\mathbf{u}| \, |\mathbf{v}|}.$$

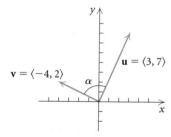

EXAMPLE 13   Find the angle between $\mathbf{u} = \langle 3, 7 \rangle$ and $\mathbf{v} = \langle -4, 2 \rangle$.

Solution   We begin by finding $\mathbf{u} \cdot \mathbf{v}$, $|\mathbf{u}|$, and $|\mathbf{v}|$:

$$\mathbf{u} \cdot \mathbf{v} = 3(-4) + 7(2) = 2,$$
$$|\mathbf{u}| = \sqrt{3^2 + 7^2} = \sqrt{58}, \quad \text{and}$$
$$|\mathbf{v}| = \sqrt{(-4)^2 + 2^2} = \sqrt{20}.$$

Then

$$\cos \alpha = \frac{\mathbf{u} \cdot \mathbf{v}}{|\mathbf{u}| |\mathbf{v}|} = \frac{2}{\sqrt{58}\sqrt{20}}$$

$$\alpha = \cos^{-1} \frac{2}{\sqrt{58}\sqrt{20}} \approx 86.6°.$$

▶ Now Try Exercise 63.

◆ Forces in Equilibrium

When several forces act through the same point on an object, their vector sum must be **O** in order for a balance to occur. When a balance occurs, then the object is either stationary or moving in a straight line without acceleration. The fact that the vector sum must be **O** for a balance, and vice versa, allows us to solve many applied problems involving forces.

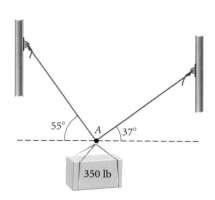

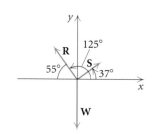

EXAMPLE 14 *Suspended Block.* A 350-lb block is suspended by two cables, as shown at left. At point *A*, there are three forces acting: **W**, the block pulling down, and **R** and **S**, the two cables pulling upward and outward. Find the tension in each cable.

Solution We draw a force diagram with the initial points of each vector at the origin. For there to be a balance, the vector sum must be the vector **O**: **R** + **S** + **W** = **O**. We can express each vector in terms of its magnitude and its direction angle:

$$\mathbf{R} = |\mathbf{R}| [(\cos 125°)\mathbf{i} + (\sin 125°)\mathbf{j}],$$
$$\mathbf{S} = |\mathbf{S}| [(\cos 37°)\mathbf{i} + (\sin 37°)\mathbf{j}], \quad \text{and}$$
$$\mathbf{W} = |\mathbf{W}| [(\cos 270°)\mathbf{i} + (\sin 270°)\mathbf{j}]$$
$$= 350(\cos 270°)\mathbf{i} + 350(\sin 270°)\mathbf{j} = -350\mathbf{j}.$$

Substituting for **R**, **S**, and **W** in **R** + **S** + **W** = **O**, we have

$$[|\mathbf{R}| (\cos 125°) + |\mathbf{S}| (\cos 37°)]\mathbf{i} + [|\mathbf{R}| (\sin 125°) + |\mathbf{S}| (\sin 37°) - 350]\mathbf{j}$$
$$= 0\mathbf{i} + 0\mathbf{j}.$$

This gives us two equations:

$$|\mathbf{R}| (\cos 125°) + |\mathbf{S}| (\cos 37°) = 0 \quad \text{and} \quad \textbf{(1)}$$
$$|\mathbf{R}| (\sin 125°) + |\mathbf{S}| (\sin 37°) - 350 = 0. \quad \textbf{(2)}$$

Solving equation (1) for $|R|$, we get

$$|R| = -\frac{|S| (\cos 37°)}{\cos 125°}. \quad \textbf{(3)}$$

Substituting this expression for $|R|$ in equation (2) gives us

$$-\frac{|S| (\cos 37°)}{\cos 125°} (\sin 125°) + |S| (\sin 37°) - 350 = 0.$$

Then solving this equation for $|S|$, we get $|S| \approx 201$, and substituting 201 for $|S|$ in equation (3), we get $|R| \approx 280$. The tensions in the cables are 280 lb and 201 lb.

▶ Now Try Exercise 83.

3.6 EXERCISE SET

Find the component form of the vector given the initial and terminal points. Then find the length of the vector.

1. $\overrightarrow{MN}$; $M(6, -7), N(-3, -2)$

2. $\overrightarrow{CD}$; $C(1, 5), D(5, 7)$

3. $\overrightarrow{FE}$; $E(8, 4), F(11, -2)$

4. $\overrightarrow{BA}$; $A(9, 0), B(9, 7)$

5. $\overrightarrow{KL}$; $K(4, -3), L(8, -3)$

6. $\overrightarrow{GH}$; $G(-6, 10), H(-3, 2)$

7. Find the magnitude of vector $\mathbf{u}$ if $\mathbf{u} = \langle -1, 6 \rangle$.

8. Find the magnitude of vector $\overrightarrow{ST}$ if $\overrightarrow{ST} = \langle -12, 5 \rangle$.

Do the indicated calculations in Exercises 9–26 for the vectors

$$\mathbf{u} = \langle 5, -2 \rangle, \quad \mathbf{v} = \langle -4, 7 \rangle, \quad \text{and} \quad \mathbf{w} = \langle -1, -3 \rangle.$$

9. $\mathbf{u} + \mathbf{w}$

10. $\mathbf{w} + \mathbf{u}$

11. $|3\mathbf{w} - \mathbf{v}|$

12. $6\mathbf{v} + 5\mathbf{u}$

13. $\mathbf{v} - \mathbf{u}$

14. $|2\mathbf{w}|$

15. $5\mathbf{u} - 4\mathbf{v}$

16. $-5\mathbf{v}$

17. $|3\mathbf{u}| - |\mathbf{v}|$

18. $|\mathbf{v}| + |\mathbf{u}|$

19. $\mathbf{v} + \mathbf{u} + 2\mathbf{w}$

20. $\mathbf{w} - (\mathbf{u} + 4\mathbf{v})$

21. $2\mathbf{v} + \mathbf{O}$

22. $10|7\mathbf{w} - 3\mathbf{u}|$

23. $\mathbf{u} \cdot \mathbf{w}$

24. $\mathbf{w} \cdot \mathbf{u}$

25. $\mathbf{u} \cdot \mathbf{v}$

26. $\mathbf{v} \cdot \mathbf{w}$

The vectors $\mathbf{u}$, $\mathbf{v}$, and $\mathbf{w}$ are drawn below. Copy them on a sheet of paper. Then sketch each of the vectors in Exercises 27–30.

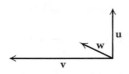

27. $\mathbf{u} + \mathbf{v}$

28. $\mathbf{u} - 2\mathbf{v}$

29. $\mathbf{u} + \mathbf{v} + \mathbf{w}$

30. $\frac{1}{2}\mathbf{u} - \mathbf{w}$

31. Vectors $\mathbf{u}$, $\mathbf{v}$, and $\mathbf{w}$ are determined by the sides of $\triangle ABC$ below.

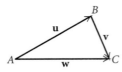

a) Find an expression for $\mathbf{w}$ in terms of $\mathbf{u}$ and $\mathbf{v}$.
b) Find an expression for $\mathbf{v}$ in terms of $\mathbf{u}$ and $\mathbf{w}$.

32. In $\triangle ABC$, vectors $\mathbf{u}$ and $\mathbf{w}$ are determined by the sides shown, where P is the midpoint of side BC. Find an expression for $\mathbf{v}$ in terms of $\mathbf{u}$ and $\mathbf{w}$.

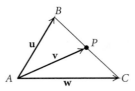

Find a unit vector that has the same direction as the given vector.

33. $\mathbf{v} = \langle -5, 12 \rangle$

34. $\mathbf{u} = \langle 3, 4 \rangle$

35. $\mathbf{w} = \langle 1, -10 \rangle$

36. $\mathbf{a} = \langle 6, -7 \rangle$

37. $\mathbf{r} = \langle -2, -8 \rangle$

38. $\mathbf{t} = \langle -3, -3 \rangle$

Express the vector as a linear combination of the unit vectors $\mathbf{i}$ and $\mathbf{j}$.

39. $\mathbf{w} = \langle -4, 6 \rangle$

40. $\mathbf{r} = \langle -15, 9 \rangle$

41. $\mathbf{s} = \langle 2, 5 \rangle$

42. $\mathbf{u} = \langle 2, -1 \rangle$

Express the vector as a linear combination of $\mathbf{i}$ and $\mathbf{j}$.

43.

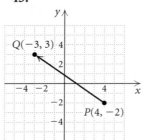

44.

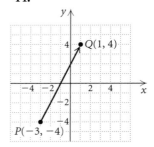

For Exercises 45–48, use the vectors

$\mathbf{u} = 2\mathbf{i} + \mathbf{j}$, $\quad \mathbf{v} = -3\mathbf{i} - 10\mathbf{j}$, $\quad$ and $\quad \mathbf{w} = \mathbf{i} - 5\mathbf{j}$.

Perform the indicated vector operations and state the answer in two forms: **(a)** as a linear combination of $\mathbf{i}$ and $\mathbf{j}$ and **(b)** in component form.

45. $4\mathbf{u} - 5\mathbf{w}$

46. $\mathbf{v} + 3\mathbf{w}$

47. $\mathbf{u} - (\mathbf{v} + \mathbf{w})$

48. $(\mathbf{u} - \mathbf{v}) + \mathbf{w}$

Sketch (include the unit circle) and calculate the unit vector $\mathbf{u} = (\cos \theta)\mathbf{i} + (\sin \theta)\mathbf{j}$ for the given direction angle.

49. $\theta = \dfrac{\pi}{2}$

50. $\theta = \dfrac{\pi}{3}$

51. $\theta = \dfrac{4\pi}{3}$

52. $\theta = \dfrac{3\pi}{2}$

Determine the direction angle θ of the vector, to the nearest degree.

53. $\mathbf{u} = \langle -2, -5 \rangle$

54. $\mathbf{w} = \langle 4, -3 \rangle$

55. $\mathbf{q} = \mathbf{i} + 2\mathbf{j}$

56. $\mathbf{w} = 5\mathbf{i} - \mathbf{j}$

57. $\mathbf{t} = \langle 5, 6 \rangle$

58. $\mathbf{b} = \langle -8, -4 \rangle$

Find the magnitude and the direction angle θ of the vector.

59. $\mathbf{u} = 3[(\cos 45°)\mathbf{i} + (\sin 45°)\mathbf{j}]$

60. $\mathbf{w} = 6[(\cos 150°)\mathbf{i} + (\sin 150°)\mathbf{j}]$

61. $\mathbf{v} = \left\langle -\dfrac{1}{2}, \dfrac{\sqrt{3}}{2} \right\rangle$

62. $\mathbf{u} = -\mathbf{i} - \mathbf{j}$

Find the angle between the given vectors, to the nearest tenth of a degree.

63. $\mathbf{u} = \langle 2, -5 \rangle$, $\mathbf{v} = \langle 1, 4 \rangle$

64. $\mathbf{a} = \langle -3, -3 \rangle$, $\mathbf{b} = \langle -5, 2 \rangle$

65. $\mathbf{w} = \langle 3, 5 \rangle$, $\mathbf{r} = \langle 5, 5 \rangle$

66. $\mathbf{v} = \langle -4, 2 \rangle$, $\mathbf{t} = \langle 1, -4 \rangle$

67. $\mathbf{a} = \mathbf{i} + \mathbf{j}$, $\mathbf{b} = 2\mathbf{i} - 3\mathbf{j}$

68. $\mathbf{u} = 3\mathbf{i} + 2\mathbf{j}$, $\mathbf{v} = -\mathbf{i} + 4\mathbf{j}$

Express each vector in Exercises 69–72 in the form $a\mathbf{i} + b\mathbf{j}$ and sketch each in the coordinate plane.

69. The unit vectors $\mathbf{u} = (\cos \theta)\mathbf{i} + (\sin \theta)\mathbf{j}$ for $\theta = \pi/6$ and $\theta = 3\pi/4$. Include the unit circle $x^2 + y^2 = 1$ in your sketch.

70. The unit vectors $\mathbf{u} = (\cos \theta)\mathbf{i} + (\sin \theta)\mathbf{j}$ for $\theta = -\pi/4$ and $\theta = -3\pi/4$. Include the unit circle $x^2 + y^2 = 1$ in your sketch.

71. The unit vector obtained by rotating $\mathbf{j}$ counterclockwise $3\pi/4$ radians about the origin

72. The unit vector obtained by rotating $\mathbf{j}$ clockwise $2\pi/3$ radians about the origin

For the vectors in Exercises 73 and 74, find the unit vectors $\mathbf{u} = (\cos \theta)\mathbf{i} + (\sin \theta)\mathbf{j}$ in the same direction.

73. $-\mathbf{i} + 3\mathbf{j}$

74. $6\mathbf{i} - 8\mathbf{j}$

For the vectors in Exercises 75 and 76, express each vector in terms of its magnitude and its direction.

75. $2\mathbf{i} - 3\mathbf{j}$

76. $5\mathbf{i} + 12\mathbf{j}$

77. Use a sketch to show that

$$\mathbf{v} = 3\mathbf{i} - 6\mathbf{j} \quad \text{and} \quad \mathbf{u} = -\mathbf{i} + 2\mathbf{j}$$

have opposite directions.

78. Use a sketch to show that

$$\mathbf{v} = 3\mathbf{i} - 6\mathbf{j} \quad \text{and} \quad \mathbf{u} = \tfrac{1}{2}\mathbf{i} - \mathbf{j}$$

have the same direction.

Exercises 79–82 appeared first in Exercise Set 3.5, where we used the law of cosines and the law of sines to solve the applied problems. For this exercise set, solve the problem using the vector form

$$\mathbf{v} = |\mathbf{v}|[(\cos \theta)\mathbf{i} + (\sin \theta)\mathbf{j}].$$

79. *Ship.* A ship sails first N80°E for 120 nautical mi, and then S20°W for 200 nautical mi. How far is the ship, then, from the starting point, and in what direction?

80. *Boat.* A boat heads 35°, propelled by a force of 750 lb. A wind from 320° exerts a force of 150 lb on the boat. How large is the resultant force, and in what direction is the boat moving?

81. *Airplane.* An airplane has an airspeed of 150 km/h. It is to make a flight in a direction of 70° while there is a 25-km/h wind from 340°. What will the airplane's actual heading be?

82. *Airplane.* An airplane flies 32° for 210 mi, and then 280° for 170 mi. How far is the airplane, then, from the starting point, and in what direction?

83. Two cables support a 1000-lb weight, as shown. Find the tension in each cable.

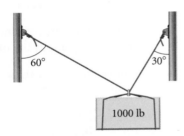

84. A 2500-kg block is suspended by two ropes, as shown. Find the tension in each rope.

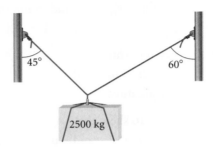

85. A 150-lb sign is hanging from the end of a hinged boom, supported by a cable inclined 42° with the horizontal. Find the tension in the cable and the compression in the boom.

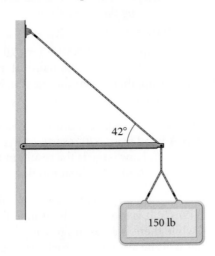

86. A weight of 200 lb is supported by a frame made of two rods and hinged at points A, B, and C. Find the forces exerted by the two rods.

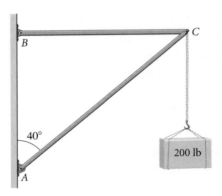

Let $\mathbf{u} = \langle u_1, u_2 \rangle$ and $\mathbf{v} = \langle v_1, v_2 \rangle$. Prove each of the following properties.

87. $\mathbf{u} + \mathbf{v} = \mathbf{v} + \mathbf{u}$

88. $\mathbf{u} \cdot \mathbf{v} = \mathbf{v} \cdot \mathbf{u}$

Collaborative Discussion and Writing

89. Explain how unit vectors are related to the unit circle.

90. Write a vector sum problem for a classmate for which the answer is $\mathbf{v} = 5\mathbf{i} - 8\mathbf{j}$.

Skill Maintenance

Find the slope and the y-intercept of the line with the given equation.

91. $-\frac{1}{5}x - y = 15$ **92.** $y = 7$

Find the zeros of the function.

93. $x^3 - 4x^2 = 0$ **94.** $6x^2 + 7x = 55$

Synthesis

95. If the dot product of two nonzero vectors $\mathbf{u}$ and $\mathbf{v}$ is 0, then the vectors are perpendicular (**orthogonal**). Let $\mathbf{u} = \langle u_1, u_2 \rangle$ and $\mathbf{v} = \langle v_1, v_2 \rangle$.
   a) Prove that if $\mathbf{u} \cdot \mathbf{v} = 0$, then $\mathbf{u}$ and $\mathbf{v}$ are perpendicular.
   b) Give an example of two perpendicular vectors and show that their dot product is 0.

96. If $\overrightarrow{PQ}$ is any vector, what is $\overrightarrow{PQ} + \overrightarrow{QP}$?

97. Find all the unit vectors that are parallel to the vector $\langle 3, -4 \rangle$.

98. Find a vector of length 2 whose direction is the opposite of the direction of the vector $\mathbf{v} = -\mathbf{i} + 2\mathbf{j}$. How many such vectors are there?

99. Given the vector $\overrightarrow{AB} = 3\mathbf{i} - \mathbf{j}$ and A is the point $(2, 9)$, find the point B.

100. Find vector $\mathbf{v}$ from point A to the origin, where $\overrightarrow{AB} = 4\mathbf{i} - 2\mathbf{j}$ and B is the point $(-2, 5)$.

CHAPTER 3 SUMMARY AND REVIEW

Important Properties and Formulas

The Law of Sines

$$\frac{a}{\sin A} = \frac{b}{\sin B} = \frac{c}{\sin C}$$

The Law of Cosines

$$a^2 = b^2 + c^2 - 2bc \cos A,$$
$$b^2 = a^2 + c^2 - 2ac \cos B,$$
$$c^2 = a^2 + b^2 - 2ab \cos C$$

The Area of a Triangle

$$K = \frac{1}{2} bc \sin A = \frac{1}{2} ab \sin C = \frac{1}{2} ac \sin B$$

Complex Numbers

Absolute Value: $|a + bi| = \sqrt{a^2 + b^2}$

Trigonometric Notation: $a + bi = r(\cos \theta + i \sin \theta)$

Multiplication: $r_1(\cos \theta_1 + i \sin \theta_1) \cdot r_2(\cos \theta_2 + i \sin \theta_2) = r_1 r_2[\cos (\theta_1 + \theta_2) + i \sin (\theta_1 + \theta_2)]$

Division: $\dfrac{r_1(\cos \theta_1 + i \sin \theta_1)}{r_2(\cos \theta_2 + i \sin \theta_2)} = \dfrac{r_1}{r_2}[\cos (\theta_1 - \theta_2) + i \sin (\theta_1 - \theta_2)], \quad r_2 \neq 0$

DeMoivre's Theorem

$$[r(\cos\theta + i\sin\theta)]^n = r^n(\cos n\theta + i\sin n\theta)$$

_____

Roots of Complex Numbers

The nth roots of $r(\cos\theta + i\sin\theta)$ are

$$r^{1/n}\left[\cos\left(\frac{\theta}{n} + k\cdot\frac{360°}{n}\right) + i\sin\left(\frac{\theta}{n} + k\cdot\frac{360°}{n}\right)\right], \quad r\neq 0, k = 0, 1, 2,\ldots, n-1.$$

_____

Vectors

If $\mathbf{u} = \langle u_1, u_2\rangle$ and $\mathbf{v} = \langle v_1, v_2\rangle$ and k is a scalar, then:

Length:	$|\mathbf{v}| = \sqrt{v_1^2 + v_2^2}$
Addition:	$\mathbf{u} + \mathbf{v} = \langle u_1 + v_1, u_2 + v_2\rangle$
Subtraction:	$\mathbf{u} - \mathbf{v} = \langle u_1 - v_1, u_2 - v_2\rangle$
Scalar Multiplication:	$k\mathbf{v} = \langle kv_1, kv_2\rangle$
Dot Product:	$\mathbf{u}\cdot\mathbf{v} = u_1v_1 + u_2v_2$
Angle Between Two Vectors:	$\cos\theta = \dfrac{\mathbf{u}\cdot\mathbf{v}}{|\mathbf{u}||\mathbf{v}|}$

REVIEW EXERCISES

Determine whether the statement is true or false.

1. For any point (x, y) on the unit circle, $\langle x, y\rangle$ is a unit vector.   [3.6]

2. The law of sines can be used to solve a triangle when all three sides are known.   [3.1]

3. Two vectors are equivalent if they have the same magnitude and the lines that they are on have the same slope.   [3.5]

4. Vectors $\langle 8, -2\rangle$ and $\langle -8, 2\rangle$ are equivalent. [3.6]

5. Any triangle, right or oblique, can be solved if at least one angle and any other two measures are known.   [3.1]

6. When two angles and an included side of a triangle are known, the triangle cannot be solved using the law of cosines.   [3.2]

Solve $\triangle ABC$, if possible.   [3.1]

7. $a = 23.4$ ft, $b = 15.7$ ft, $c = 8.3$ ft

8. $B = 27°$, $C = 35°$, $b = 19$ in.

9. $A = 133°28'$, $C = 31°42'$, $b = 890$ m

10. $B = 37°$, $b = 4$ yd, $c = 8$ yd

11. Find the area of $\triangle ABC$ if $b = 9.8$ m, $c = 7.3$ m, and $A = 67.3°$.   [3.1]

12. A parallelogram has sides of lengths 3.21 ft and 7.85 ft. One of its angles measures $147°$. Find the area of the parallelogram.   [3.1]

13. _Sandbox._   A child-care center has a triangular-shaped sandbox. Two of the three sides measure 15 ft and 12.5 ft and form an included angle of $42°$. To determine the amount of sand that is needed to fill the box, the director must deter-

mine the area of the floor of the box. Find the area of the floor of the box to the nearest square foot. [3.1]

14. *Flower Garden.* A triangular flower garden has sides of lengths 11 m, 9 m, and 6 m. Find the angles of the garden to the nearest degree. [3.2]

15. In an isosceles triangle, the base angles each measure 52.3° and the base is 513 ft long. Find the lengths of the other two sides to the nearest foot. [3.1]

16. *Airplanes.* Two airplanes leave an airport at the same time. The first flies 175 km/h in a direction of 305.6°. The second flies 220 km/h in a direction of 195.5°. After 2 hr, how far apart are the planes? [3.2]

Graph the complex number and find its absolute value. [3.3]

17. $2 - 5i$

18. 4

19. $2i$

20. $-3 + i$

Find trigonometric notation. [3.3]

21. $1 + i$

22. $-4i$

23. $-5\sqrt{3} + 5i$

24. $\dfrac{3}{4}$

Find standard notation, a + bi. [3.3]

25. $4(\cos 60° + i \sin 60°)$

26. $7(\cos 0° + i \sin 0°)$

27. $5\left(\cos \dfrac{2\pi}{3} + i \sin \dfrac{2\pi}{3}\right)$

28. $2\left[\cos\left(-\dfrac{\pi}{3}\right) + i \sin\left(-\dfrac{\pi}{3}\right)\right]$

Convert to trigonometric notation and then multiply or divide, expressing the answer in standard notation. [3.3]

29. $\left(1 + i\sqrt{3}\right)(1 - i)$

30. $\dfrac{2 - 2i}{2 + 2i}$

31. $\dfrac{2 + 2\sqrt{3}i}{\sqrt{3} - i}$

32. $i\left(3 - 3\sqrt{3}i\right)$

Raise the number to the given power and write trigonometric notation for the answer. [3.3]

33. $[2(\cos 60° + i \sin 60°)]^3$

34. $(1 - i)^4$

Raise the number to the given power and write standard notation for the answer. [3.3]

35. $(1 + i)^6$

36. $\left(\dfrac{1}{2} + \dfrac{\sqrt{3}}{2}i\right)^{10}$

37. Find the square roots of $-1 + i$. [3.3]

38. Find the cube roots of $3\sqrt{3} - 3i$. [3.3]

39. Find and graph the fourth roots of 81. [3.3]

40. Find and graph the fifth roots of 1. [3.3]

Find all the complex solutions of the equation. [3.3]

41. $x^4 - i = 0$

42. $x^3 + 1 = 0$

43. Find the polar coordinates of each of these points. Give three answers for each point. [3.4]

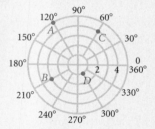

Find the polar coordinates of the point. Express the answer in degrees and then in radians. [3.4]

44. $\left(-4\sqrt{2}, 4\sqrt{2}\right)$

45. $(0, -5)$

Find the rectangular coordinates of the point. [3.4]

46. $\left(3, \dfrac{\pi}{4}\right)$

47. $(-6, -120°)$

Convert to a polar equation. [3.4]

48. $5x - 2y = 6$

49. $y = 3$

50. $x^2 + y^2 = 9$

51. $y^2 - 4x - 16 = 0$

Convert to a rectangular equation.   [3.4]

52. $r = 6$

53. $r + r \sin \theta = 1$

54. $r = \dfrac{3}{1 - \cos \theta}$

55. $r - 2 \cos \theta = 3 \sin \theta$

In Exercises 56–59, match the equation with one of figures (a)–(d), which follow.   [3.4]

a)
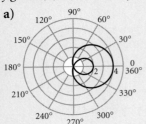

b)
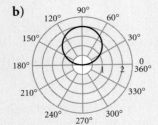

c)
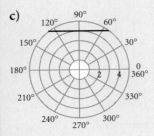

d)
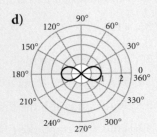

56. $r = 2 \sin \theta$

57. $r^2 = \cos 2\theta$

58. $r = 1 + 3 \cos \theta$

59. $r \sin \theta = 4$

*Magnitudes of vectors **u** and **v** and the angle θ between the vectors are given. Find the magnitude of the sum, **u** + **v**, to the nearest tenth and give the direction by specifying to the nearest degree the angle that it makes with the vector **u**.*   [3.5]

60. $|\mathbf{u}| = 12$, $|\mathbf{v}| = 15$, $\theta = 120°$

61. $|\mathbf{u}| = 41$, $|\mathbf{v}| = 60$, $\theta = 25°$

*The vectors **u**, **v**, and **w** are drawn below. Copy them on a sheet of paper. Then sketch each of the vectors in Exercises 62 and 63.*   [3.5]

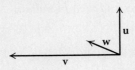

62. $\mathbf{u} - \mathbf{v}$

63. $\mathbf{u} + \frac{1}{2}\mathbf{w}$

64. Forces of 230 N and 500 N act on an object. The angle between the forces is 52°. Find the resultant, giving the angle that it makes with the smaller force.   [3.5]

65. *Wind.* A wind has an easterly component of 15 km/h and a southerly component of 25 km/h. Find the magnitude and the direction of the wind.   [3.5]

66. *Ship.* A ship sails N75°E for 90 nautical mi, and then S10°W for 100 nautical mi. How far is the ship, then, from the starting point, and in what direction?   [3.5]

Find the component form of the vector given the initial and terminal points.   [3.6]

67. $\overrightarrow{AB}$; $A(2, -8)$, $B(-2, -5)$

68. $\overrightarrow{TR}$; $R(0, 7)$, $T(-2, 13)$

69. Find the magnitude of vector **u** if $\mathbf{u} = \langle 5, -6 \rangle$.   [3.6]

Do the calculations in Exercises 70–73 for the vectors

$$\mathbf{u} = \langle 3, -4 \rangle, \quad \mathbf{v} = \langle -3, 9 \rangle \quad \text{and} \quad \mathbf{w} = \langle -2, -5 \rangle.$$
[3.6]

70. $4\mathbf{u} + \mathbf{w}$

71. $2\mathbf{w} - 6\mathbf{v}$

72. $|\mathbf{u}| + |2\mathbf{w}|$

73. $\mathbf{u} \cdot \mathbf{w}$

74. Find a unit vector that has the same direction as $\mathbf{v} = \langle -6, -2 \rangle$.   [3.6]

75. Express the vector $\mathbf{t} = \langle -9, 4 \rangle$ as a linear combination of the unit vectors **i** and **j**.   [3.6]

76. Determine the direction angle θ of the vector $\mathbf{w} = \langle -4, -1 \rangle$ to the nearest degree.   [3.6]

77. Find the magnitude and the direction angle θ of $\mathbf{u} = -5\mathbf{i} - 3\mathbf{j}$.   [3.6]

78. Find the angle between $\mathbf{u} = \langle 3, -7 \rangle$ and $\mathbf{v} = \langle 2, 2 \rangle$ to the nearest tenth of a degree.   [3.6]

79. *Airplane.* An airplane has an airspeed of 160 mph. It is to make a flight in a direction of 80° while there is a 20-mph wind from 310°. What will the airplane's actual heading be?   [3.6]

Do the calculations in Exercises 80–83 for the vectors

$$\mathbf{u} = 2\mathbf{i} + 5\mathbf{j}, \quad \mathbf{v} = -3\mathbf{i} + 10\mathbf{j}, \quad \text{and} \quad \mathbf{w} = 4\mathbf{i} + 7\mathbf{j}.$$
[3.6]

80. $5\mathbf{u} - 8\mathbf{v}$

81. $\mathbf{u} - (\mathbf{v} + \mathbf{w})$

82. $|\mathbf{u} - \mathbf{v}|$

83. $3|\mathbf{w}| + |\mathbf{v}|$

84. Express the vector $\overrightarrow{PQ}$ in the form $a\mathbf{i} + b\mathbf{j}$, if P is the point $(1, -3)$ and Q is the point $(-4, 2)$.   [3.6]

*Express each vector in Exercises 85 and 86 in the form a**i** + b**j** and sketch each in the coordinate plane.* [3.6]

85. The unit vectors **u** = (cos θ)**i** + (sin θ)**j** for θ = π/4 and θ = 5π/4. Include the unit circle $x^2 + y^2 = 1$ in your sketch.

86. The unit vector obtained by rotating **j** counterclockwise 2π/3 radians about the origin.

87. Express the vector 3**i** − **j** as a product of its magnitude and its direction.

88. Determine the trigonometric notation for 1 − i. [3.3]

A. $\sqrt{2}\left(\cos \dfrac{5\pi}{4} + i \sin \dfrac{5\pi}{4}\right)$

B. $\sqrt{2}\left(\cos \dfrac{7\pi}{4} - \sin \dfrac{7\pi}{4}\right)$

C. $\cos \dfrac{7\pi}{4} + i \sin \dfrac{7\pi}{4}$

D. $\sqrt{2}\left(\cos \dfrac{7\pi}{4} + i \sin \dfrac{7\pi}{4}\right)$

89. Convert the polar equation r = 100 to a rectangular equation. [3.4]

A. $x^2 + y^2 = 10{,}000$

B. $x^2 + y^2 = 100$

C. $\sqrt{x^2 + y^2} = 10$

D. $\sqrt{x^2 + y^2} = 1000$

Technology Connection

Use a graphing calculator to convert from rectangular to polar coordinates. Express the answer in degrees and then in radians. [3.4]

90. (−2, 5)

91. (−4.2, $\sqrt{7}$)

Use a graphing calculator to convert from polar to rectangular coordinates. Round the coordinates to the nearest hundredth. [3.4]

92. (2, −15°)

93. $\left(-2.3, \dfrac{\pi}{5}\right)$

Collaborative Discussion and Writing

94. Explain why these statements are not contradictory:

The number 1 has one real cube root.
The number 1 has three complex cube roots.
[3.1]

95. Summarize how you can tell algebraically when solving triangles whether there is no solution, one solution, or two solutions. [3.1], [3.2]

96. *Golf: Distance versus Accuracy.* It is often argued in golf that the farther you hit the ball, the more accurate it must be to stay safe. (Safe means not in the woods, water, or some other hazard.) In his book *Golf and the Spirit* (p. 54), M. Scott Peck asserts "Deviate 5° from your aiming point on a 150-yd shot, and your ball will land approximately 20 yd to the side of where you wanted it to be. Do the same on a 300-yd shot, and it will be 40 yd off target. Twenty yards may well be in the range of safety; 40 yards probably won't. This principle not infrequently allows a mediocre, short-hitting golfer like myself to score better than the long hitter." Check the accuracy of the mathematics in this statement, and comment on Peck's assertion. [3.2]

Synthesis

97. Let **u** = 12**i** + 5**j**. Find a vector that has the same direction as **u** but has length 3. [3.6]

98. A parallelogram has sides of lengths 3.42 and 6.97. Its area is 18.4. Find the sizes of its angles. [3.1]

CHAPTER 3 TEST

Solve △ABC, if possible.

1. $a = 18$ ft, $B = 54°$, $C = 43°$

2. $b = 8$ m, $c = 5$ m, $C = 36°$

3. $a = 16.1$ in., $b = 9.8$ in., $c = 11.2$ in.

4. Find the area of $△ABC$ if $C = 106.4°$, $a = 7$ cm, and $b = 13$ cm.

5. *Distance Across a Lake.*   Points A and B are on opposite sides of a lake. Point C is 52 m from A. The measure of $∠BAC$ is determined to be 108°, and the measure of $∠ACB$ is determined to be 44°. What is the distance from A to B?

6. *Location of Airplanes.*   Two airplanes leave an airport at the same time. The first flies 210 km/h in a direction of 290°. The second flies 180 km/h in a direction of 185°. After 3 hr, how far apart are the planes?

7. Graph: $-4 + i$.

8. Find the absolute value of $2 - 3i$.

9. Find trigonometric notation for $3 - 3i$.

10. Divide and express the result in standard notation $a + bi$:

$$\frac{2\left(\cos \frac{2\pi}{3} + i \sin \frac{2\pi}{3}\right)}{8\left(\cos \frac{\pi}{6} + i \sin \frac{\pi}{6}\right)}.$$

11. Find $(1 - i)^8$ and write standard notation for the answer.

12. Find the polar coordinates of $(-1, \sqrt{3})$. Express the angle in degrees using the smallest possible positive angle.

13. Convert $\left(-1, \frac{2\pi}{3}\right)$ to rectangular coordinates.

14. Convert to a polar equation: $x^2 + y^2 = 10$.

15. Graph: $r = 1 - \cos \theta$.

16. Which of the following is the graph of $r = 3 \cos \theta$?

a)
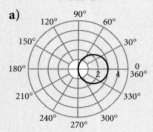

b)
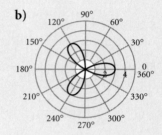

c)
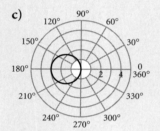

d)
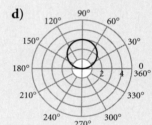

17. For vectors **u** and **v**, $|\mathbf{u}| = 8$, $|\mathbf{v}| = 5$, and the angle between the vectors is 63°. Find $\mathbf{u} + \mathbf{v}$. Give the magnitude to the nearest tenth, and give the direction by specifying the angle that the resultant makes with **u**, to the nearest degree.

18. For $\mathbf{u} = 2\mathbf{i} - 7\mathbf{j}$ and $\mathbf{v} = 5\mathbf{i} + \mathbf{j}$, find $2\mathbf{u} - 3\mathbf{v}$.

19. Find a unit vector in the same direction as $-4\mathbf{i} + 3\mathbf{j}$.

Synthesis

20. A parallelogram has sides of length 15.4 and 9.8. Its area is 72.9. Find the measures of the angles.

Answers

CHAPTER 1

Exercise Set 1.1

1. $\sin \phi = \frac{15}{17}$, $\cos \phi = \frac{8}{17}$, $\tan \phi = \frac{15}{8}$, $\csc \phi = \frac{17}{15}$, $\sec \phi = \frac{17}{8}$, $\cot \phi = \frac{8}{15}$

3. $\sin \alpha = \frac{\sqrt{3}}{2}$, $\cos \alpha = \frac{1}{2}$, $\tan \alpha = \sqrt{3}$, $\csc \alpha = \frac{2\sqrt{3}}{3}$, $\sec \alpha = 2$, $\cot \alpha = \frac{\sqrt{3}}{3}$

5. $\sin \phi = \frac{7\sqrt{65}}{65}$, $\cos \phi = \frac{4\sqrt{65}}{65}$, $\tan \phi = \frac{7}{4}$, $\csc \phi = \frac{\sqrt{65}}{7}$, $\sec \phi = \frac{\sqrt{65}}{4}$, $\cot \phi = \frac{4}{7}$

7. $\csc \alpha = \frac{3}{\sqrt{5}}$, or $\frac{3\sqrt{5}}{5}$; $\sec \alpha = \frac{3}{2}$; $\cot \alpha = \frac{2}{\sqrt{5}}$, or $\frac{2\sqrt{5}}{5}$

9. $\cos \theta = \frac{7}{25}$, $\tan \theta = \frac{24}{7}$, $\csc \theta = \frac{25}{24}$, $\sec \theta = \frac{25}{7}$, $\cot \theta = \frac{7}{24}$

11. $\sin \phi = \frac{2\sqrt{5}}{5}$, $\cos \phi = \frac{\sqrt{5}}{5}$, $\csc \phi = \frac{\sqrt{5}}{2}$, $\sec \phi = \sqrt{5}$, $\cot \phi = \frac{1}{2}$

13. $\sin \theta = \frac{2}{3}$, $\cos \theta = \frac{\sqrt{5}}{3}$, $\tan \theta = \frac{2\sqrt{5}}{5}$, $\sec \theta = \frac{3\sqrt{5}}{5}$, $\cot \theta = \frac{\sqrt{5}}{2}$

15. $\sin \beta = \frac{2\sqrt{5}}{5}$, $\tan \beta = 2$, $\csc \beta = \frac{\sqrt{5}}{2}$, $\sec \beta = \sqrt{5}$, $\cot \beta = \frac{1}{2}$

17. $\frac{\sqrt{2}}{2}$ **19.** 2 **21.** $\frac{\sqrt{3}}{3}$ **23.** $\frac{1}{2}$ **25.** 1 **27.** 2

29. 62.4 m   **31.** 9.72°   **33.** 35.01°   **35.** 3.03°

37. 49.65°   **39.** 0.25°   **41.** 5.01°   **43.** 17°36′

45. 83°1′30″   **47.** 11°45′   **49.** 47°49′36″   **51.** 0°54′

53. 39°27′   **55.** 0.6293   **57.** 0.0737   **59.** 1.2765

61. 0.7621   **63.** 0.9336   **65.** 12.4288   **67.** 1.0000

69. 1.7032   **71.** 30.8°   **73.** 12.5°   **75.** 64.4°

77. 46.5°   **79.** 25.2°   **81.** 38.6°   **83.** 45°   **85.** 60°

87. 45°   **89.** 60°   **91.** 30°

93. $\cos 20° = \sin 70° = \dfrac{1}{\sec 20°}$

95. $\tan 52° = \cot 38° = \dfrac{1}{\cot 52°}$

97. $\sin 25° \approx 0.4226$, $\cos 25° \approx 0.9063$, $\tan 25° \approx 0.4663$, $\csc 25° \approx 2.3662$, $\sec 25° \approx 1.1034$, $\cot 25° \approx 2.1445$

99. $\sin 18°49′55″ \approx 0.3228$, $\cos 18°49′55″ \approx 0.9465$, $\tan 18°49′55″ \approx 0.3411$, $\csc 18°49′55″ \approx 3.0979$, $\sec 18°49′55″ \approx 1.0565$, $\cot 18°49′55″ \approx 2.9317$

101. $\sin 8° = q$, $\cos 8° = p$, $\tan 8° = \dfrac{1}{r}$, $\csc 8° = \dfrac{1}{q}$, $\sec 8° = \dfrac{1}{p}$, $\cot 8° = r$ **103.** Discussion and Writing

105.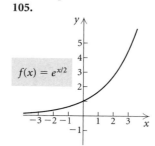

106.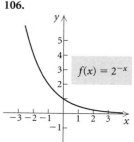

107.   **108.**

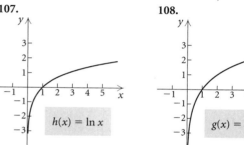

109. 4   **110.** 9.21   **111.** 343

112. $\frac{101}{97}$ **113.** 0.6534

115. Area $= \frac{1}{2}ab$. But $a = c \sin A$, so Area $= \frac{1}{2}bc \sin A$.

Exercise Set 1.2

1. $F = 60°, d = 3, f \approx 5.2$
3. $A = 22.7°, a \approx 52.7, c \approx 136.6$
5. $P = 47°38', n \approx 34.4, p \approx 25.4$
7. $B = 2°17', b \approx 0.39, c = 9.74$
9. $A \approx 77.2°, B \approx 12.8°, a \approx 439$
11. $B = 42.42°, a \approx 35.7, b \approx 32.6$
13. $B = 55°, a \approx 28.0, c \approx 48.8$
15. $A \approx 62.4°, B \approx 27.6°, a \approx 3.56$ **17.** Approximately 34°
19. About 62.2 ft   **21.** 110 ft   **23.** 750 ft
25. About 92.9 cm   **27.** About 599 ft
29. Radius: 9.15 in.; length: 73.20 in.; width: 54.90 in.
31. 17.9 ft   **33.** About 8 km   **35.** About 19.5 mi
37. About 24 km   **39.** Discussion and Writing
40. $3\sqrt{10}$, or about 9.487
41. $10\sqrt{2}$, or about 14.142
42. $\ln t = 4$ **43.** $10^{-3} = 0.001$ **45.** 3.3
47. Cut so that $\theta = 79.38°$ **49.** $\theta \approx 27°$

Exercise Set 1.3

1. III   **3.** III   **5.** I   **7.** III   **9.** II   **11.** II
13. $434°, 794°, -286°, -646°$
15. $475.3°, 835.3°, -244.7°, -604.7°$
17. $180°, 540°, -540°, -900°$ **19.** $72.89°, 162.89°$
21. $77°56'46'', 167°56'46''$ **23.** $44.8°, 134.8°$
25. $\sin \beta = \frac{5}{13}, \cos \beta = -\frac{12}{13}, \tan \beta = -\frac{5}{12}, \csc \beta = \frac{13}{5},$
$\sec \beta = -\frac{13}{12}, \cot \beta = -\frac{12}{5}$
27. $\sin \phi = -\frac{2\sqrt{7}}{7}, \cos \phi = -\frac{\sqrt{21}}{7}, \tan \phi = \frac{2\sqrt{3}}{3},$
$\csc \phi = -\frac{\sqrt{7}}{2}, \sec \phi = -\frac{\sqrt{21}}{3}, \cot \phi = \frac{\sqrt{3}}{2}$
29. $\sin \theta = -\frac{2\sqrt{13}}{13}, \cos \theta = \frac{3\sqrt{13}}{13}, \tan \theta = -\frac{2}{3}$
31. $\sin \theta = \frac{5\sqrt{41}}{41}, \cos \theta = \frac{4\sqrt{41}}{41}, \tan \theta = \frac{5}{4}$
33. $\cos \theta = -\frac{2\sqrt{2}}{3}, \tan \theta = \frac{\sqrt{2}}{4}, \csc \theta = -3,$
$\sec \theta = -\frac{3\sqrt{2}}{4}, \cot \theta = 2\sqrt{2}$
35. $\sin \theta = -\frac{\sqrt{5}}{5}, \cos \theta = \frac{2\sqrt{5}}{5}, \tan \theta = -\frac{1}{2},$
$\csc \theta = -\sqrt{5}, \sec \theta = \frac{\sqrt{5}}{2}$
37. $\sin \phi = -\frac{4}{5}, \tan \phi = -\frac{4}{3}, \csc \phi = -\frac{5}{4}, \sec \phi = \frac{5}{3},$
$\cot \phi = -\frac{3}{4}$

39. $30°; -\frac{\sqrt{3}}{2}$ **41.** $45°; 1$ **43.** 0   **45.** $45°; -\frac{\sqrt{2}}{2}$
47. $30°; 2$ **49.** $30°; \sqrt{3}$ **51.** $30°; -\frac{\sqrt{3}}{3}$
53. Not defined   **55.** -1 **57.** $60°; \sqrt{3}$
59. $45°; \frac{\sqrt{2}}{2}$ **61.** $45°; -\sqrt{2}$ **63.** 1   **65.** 0   **67.** 0
69. 0   **71.** Positive: cos, sec; negative: sin, csc, tan, cot
73. Positive: tan, cot; negative: sin, csc, cos, sec
75. Positive: sin, csc; negative: cos, sec, tan, cot
77. Positive: all   **79.** $\sin 319° = -0.6561,$
$\cos 319° = 0.7547, \tan 319° = -0.8693, \csc 319° \approx -1.5242,$
$\sec 319° \approx 1.3250, \cot 319° \approx -1.1504$
81. $\sin 115° = 0.9063, \cos 115° = -0.4226,$
$\tan 115° = -2.1445, \csc 115° \approx 1.1034, \sec 115° \approx -2.3663,$
$\cot 115° \approx -0.4663$ **83.** East: about 130 km; south: 75 km
85. About 223 km   **87.** -1.1585 **89.** -1.4910
91. 0.8771   **93.** 0.4352   **95.** 0.9563   **97.** 2.9238
99. 275.4°   **101.** 200.1°   **103.** 288.1°   **105.** 72.6°
107. Discussion and Writing
109.

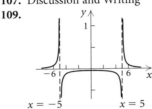

$$f(x) = \frac{1}{x^2 - 25}$$

110.

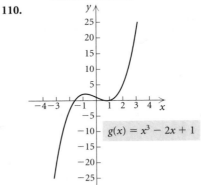

$g(x) = x^3 - 2x + 1$

111. Domain: $\{x | x \neq -2\}$; range: $\{x | x \neq 1\}$
112. Domain: $\{x | x \neq -\frac{3}{2} \ and \ x \neq 5\}$;
range: all real numbers
113. 12   **114.** $-2, 3$ **115.** $(12, 0)$
116. $(-2, 0), (3, 0)$ **117.** 19.625 in.

Exercise Set 1.4

1.

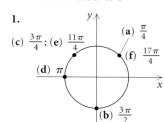

(c) $\frac{3\pi}{4}$; (e) $\frac{11\pi}{4}$ (a) $\frac{\pi}{4}$ (f) $\frac{17\pi}{4}$ (d) π (b) $\frac{3\pi}{2}$

3.

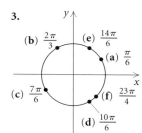

(b) $\frac{2\pi}{3}$ (e) $\frac{14\pi}{6}$ (a) $\frac{\pi}{6}$ (c) $\frac{7\pi}{6}$ (f) $\frac{23\pi}{4}$ (d) $\frac{10\pi}{6}$

5. $M: \frac{2\pi}{3}, -\frac{4\pi}{3}$; $N: \frac{3\pi}{2}, -\frac{\pi}{2}$; $P: \frac{5\pi}{4}, -\frac{3\pi}{4}$; $Q: \frac{11\pi}{6}, -\frac{\pi}{6}$

7.

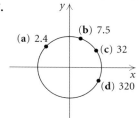

(a) 2.4   (b) 7.5   (c) 32   (d) 320

9. $\frac{9\pi}{4}, -\frac{7\pi}{4}$ **11.** $\frac{19\pi}{6}, -\frac{5\pi}{6}$ **13.** $\frac{4\pi}{3}, -\frac{8\pi}{3}$

15. Complement: $\frac{\pi}{6}$; supplement: $\frac{2\pi}{3}$

17. Complement: $\frac{\pi}{8}$; supplement: $\frac{5\pi}{8}$

19. Complement: $\frac{5\pi}{12}$; supplement: $\frac{11\pi}{12}$

21. $\frac{5\pi}{12}$ **23.** $\frac{10\pi}{9}$ **25.** $-\frac{214.6\pi}{180}$, or $-\frac{1073\pi}{900}$ **27.** $-\pi$

29. $\frac{12.5\pi}{180}$, or $\frac{5\pi}{72}$ **31.** $-\frac{17\pi}{9}$ **33.** 4.19   **35.** -1.05

37. 2.06   **39.** 0.02   **41.** 6.02   **43.** 1.66   **45.** $-135°$

47. 1440°   **49.** 57.30°   **51.** 134.47°   **53.** 225°

55. $-5156.62°$ **57.** 51.43°

59. $0° = 0$ radians, $30° = \frac{\pi}{6}$, $45° = \frac{\pi}{4}$, $60° = \frac{\pi}{3}$, $90° = \frac{\pi}{2}$, $135° = \frac{3\pi}{4}$, $180° = \pi$, $225° = \frac{5\pi}{4}$, $270° = \frac{3\pi}{2}$, $315° = \frac{7\pi}{4}$, $360° = 2\pi$

61. 2.29   **63.** 5.50 in.   **65.** 1.1; 63°   **67.** 3.2 yd

69. $\frac{5\pi}{3}$, or about 5.24   **71.** 3150 $\frac{\text{cm}}{\text{min}}$

73. About 12,003 revolutions per hour   **75.** 1047 mph

77. 19,205 revolutions/hr   **79.** About 202

81. Left to the student   **83.** Discussion and Writing

85. One-to-one   **86.** [1.1] Cosine of θ

87. Exponential function   **88.** Horizontal asymptote

89. Odd function   **90.** Natural

91. Horizontal line; inverse   **92.** Logarithm

93. 111.7 km; 69.8 mi   **95.** (a) 5°37′30″; (b) 19°41′15″

97. 1.676 radians/sec   **99.** 1.46 nautical miles

Exercise Set 1.5

1. (a) $\left(-\frac{3}{4}, -\frac{\sqrt{7}}{4}\right)$; (b) $\left(\frac{3}{4}, \frac{\sqrt{7}}{4}\right)$; (c) $\left(\frac{3}{4}, -\frac{\sqrt{7}}{4}\right)$

3. (a) $\left(\frac{2}{5}, \frac{\sqrt{21}}{5}\right)$; (b) $\left(-\frac{2}{5}, -\frac{\sqrt{21}}{5}\right)$; (c) $\left(-\frac{2}{5}, \frac{\sqrt{21}}{5}\right)$

5. $\left(\frac{\sqrt{2}}{2}, -\frac{\sqrt{2}}{2}\right)$ **7.** 0   **9.** $\sqrt{3}$ **11.** 0   **13.** $-\frac{\sqrt{3}}{2}$

15. Not defined   **17.** $\frac{\sqrt{3}}{2}$ **19.** $-\frac{\sqrt{2}}{2}$ **21.** 0   **23.** 0

25. 0.4816   **27.** 1.3065   **29.** -2.1599 **31.** 1

33. -1.1747 **35.** -1 **37.** -0.7071 **39.** 0

41. 0.8391

43. (a)

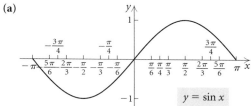

$y = \sin x$

(b)

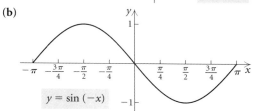

$y = \sin(-x)$

(c) same as (b); (d) the same

45. (a) See Exercise 43(a);

(b)

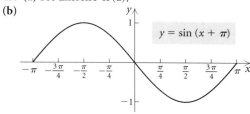

$y = \sin(x + \pi)$

(c) same as (b); (d) the same

47. (a)

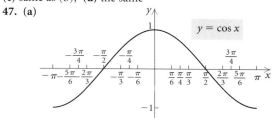

$y = \cos x$

(b)

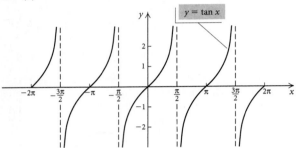

(c) same as (b); **(d)** the same

49. (a)

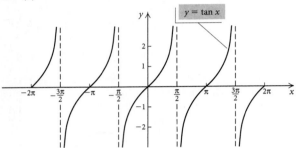

(b)

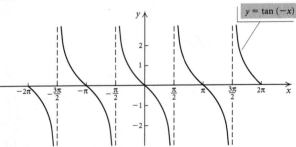

(c) same as (b); **(d)** the same

51. Even: cosine, secant; odd: sine, tangent, cosecant, cotangent   **53.** Positive: I, III; negative: II, IV

55. Positive: I, IV; negative: II, III

57. Domain: $(-\infty, \infty)$; range: $[0, 1]$; period: π; amplitude: $\frac{1}{2}$ **59.** 1   **61.** Discussion and Writing

63.

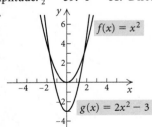

Stretch the graph of f vertically, then shift it down 3 units.

64.

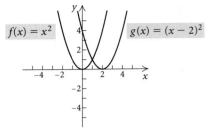

Shift the graph of f right 2 units.

65.

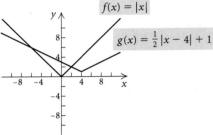

Shift the graph of f to the right 4 units, shrink it vertically, then shift it up 1 unit.

66.

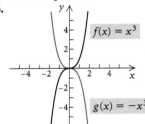

Reflect the graph of f across the x-axis.

67. $y = -(x - 2)^3 - 1$ **68.** $y = \dfrac{1}{4x} + 3$

69. $\cos x$ **71.** $\sin x$ **73.** $\sin x$ **75.** $-\cos x$

77. $-\sin x$ **79. (a)** $\dfrac{\pi}{2} + 2k\pi, k \in \mathbb{Z}$; **(b)** $\pi + 2k\pi,$

$k \in \mathbb{Z}$; **(c)** $k\pi, k \in \mathbb{Z}$

81. $\left[-\dfrac{\pi}{2} + 2k\pi, \dfrac{\pi}{2} + 2k\pi\right], k \in \mathbb{Z}$

83. $\left\{x \mid x \neq \dfrac{\pi}{2} + k\pi, k \in \mathbb{Z}\right\}$

85.

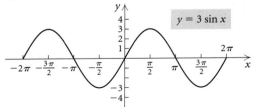

87.

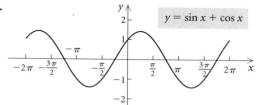

$y = \sin x + \cos x$

89. (a) $\triangle OPA \sim \triangle ODB$;

Thus, $\dfrac{AP}{OA} = \dfrac{BD}{OB}$

$\dfrac{\sin \theta}{\cos \theta} = \dfrac{BD}{1}$

$\tan \theta = BD$

(b) $\triangle OPA \sim \triangle ODB$;

$\dfrac{OD}{OP} = \dfrac{OB}{OA}$

$\dfrac{OD}{1} = \dfrac{1}{\cos \theta}$

$OD = \sec \theta$

(c) $\triangle OAP \sim \triangle ECO$;

$\dfrac{OE}{PO} = \dfrac{CO}{AP}$

$\dfrac{OE}{1} = \dfrac{1}{\sin \theta}$

$OE = \csc \theta$

(d) $\triangle OAP \sim \triangle ECO$

$\dfrac{CE}{AO} = \dfrac{CO}{AP}$

$\dfrac{CE}{\cos \theta} = \dfrac{1}{\sin \theta}$

$CE = \dfrac{\cos \theta}{\sin \theta}$

$CE = \cot \theta$

Visualizing the Graph

1. J   **2.** H   **3.** E   **4.** F   **5.** B   **6.** D   **7.** G
8. A   **9.** C   **10.** I

Exercise Set 1.6

1. Amplitude: 1; period: 2π; phase shift: 0

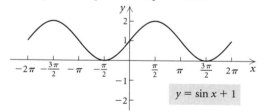

$y = \sin x + 1$

3. Amplitude: 3; period: 2π; phase shift: 0

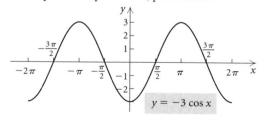

$y = -3 \cos x$

5. Amplitude: $\frac{1}{2}$; period: 2π; phase shift: 0

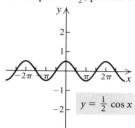

$y = \frac{1}{2} \cos x$

7. Amplitude: 1; period: π; phase shift: 0

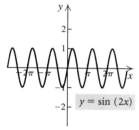

$y = \sin (2x)$

9. Amplitude: 2; period: 4π; phase shift: 0

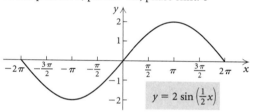

$y = 2 \sin \left(\frac{1}{2}x\right)$

11. Amplitude: $\dfrac{1}{2}$; period: 2π; phase shift: $-\dfrac{\pi}{2}$

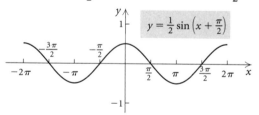

$y = \frac{1}{2} \sin \left(x + \frac{\pi}{2}\right)$

13. Amplitude: 3; period: 2π; phase shift: π

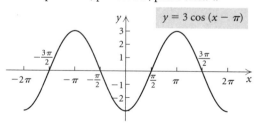

$y = 3 \cos (x - \pi)$

15. Amplitude: $\frac{1}{3}$; period: 2π; phase shift: 0

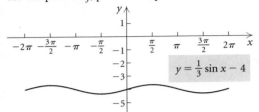

17. Amplitude: 1; period: 2π; phase shift: 0

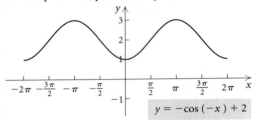

19. Amplitude: 2; period: 4π; phase shift: π

21. Amplitude: $\dfrac{1}{2}$; period: π; phase shift: $-\dfrac{\pi}{4}$

23. Amplitude: 3; period: 2; phase shift: $\dfrac{3}{\pi}$

25. Amplitude: $\frac{1}{2}$; period: 1; phase shift: 0
27. Amplitude: 1; period: 4π; phase shift: π
29. Amplitude: 1; period: 1; phase shift: 0

31. Amplitude: $\dfrac{1}{4}$; period: 2; phase shift: $\dfrac{4}{\pi}$

33. (b)   **35.** (h)   **37.** (a)   **39.** (f)

41. $y = \frac{1}{2}\cos x + 1$ **43.** $y = \cos\left(x + \dfrac{\pi}{2}\right) - 2$

45.

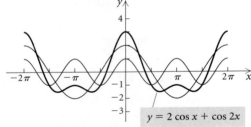

47.

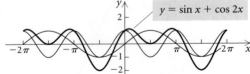

49.

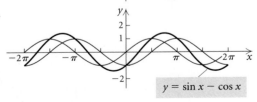

51.

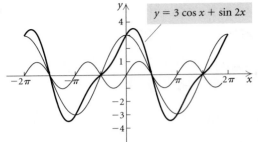

53.

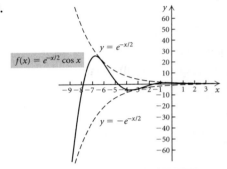

55.

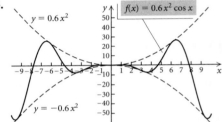

57.

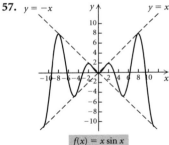

59.

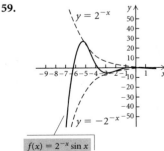

61.

$y = x + \sin x$

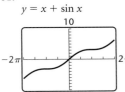

63.

$y = \cos x - x$

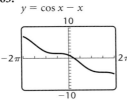

65.

$y = \cos 2x + 2x$

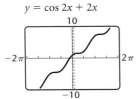

67.

$y = 4 \cos 2x - 2 \sin x$

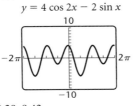

69. $-9.42, -6.28, -3.14, 3.14, 6.28, 9.42$

71. $-3.14, 0, 3.14$

73. (a) $y = 101.6 + 3 \sin\left(\frac{\pi}{8}x\right)$ (b) $104.6°, 98.6°$

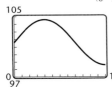

75. Discussion and Writing

77. Rational   **78.** Logarithmic

79. Quartic   **80.** Linear

81. [1.6] Trigonometric   **82.** Exponential

83. Linear   **84.** [1.6] Trigonometric

85. Cubic   **86.** Exponential

87. Maximum: 8; minimum: 4

89.

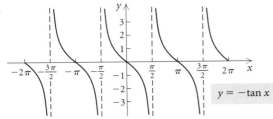

$y = -\tan x$

91.

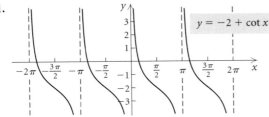

$y = -2 + \cot x$

93.

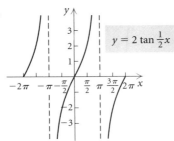

$y = 2 \tan \frac{1}{2}x$

95.

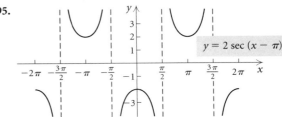

$y = 2 \sec (x - \pi)$

97.

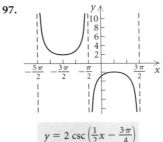

$y = 2 \csc\left(\frac{1}{2}x - \frac{3\pi}{4}\right)$

99. Amplitude: 3000; period: 90; phase shift: 10   **101.** 4 in.

Review Exercises: Chapter 1

1. False   **2.** True   **3.** True   **4.** True   **5.** False

6. False   **7.** $\sin \theta = \frac{3\sqrt{73}}{73}$, $\cos \theta = \frac{8\sqrt{73}}{73}$, $\tan \theta = \frac{3}{8}$,

$\csc \theta = \frac{\sqrt{73}}{3}$, $\sec \theta = \frac{\sqrt{73}}{8}$, $\cot \theta = \frac{8}{3}$

8. $\cos \beta = \frac{3}{10}$, $\tan \beta = \frac{\sqrt{91}}{3}$, $\csc \beta = \frac{10\sqrt{91}}{91}$,

$\sec \beta = \frac{10}{3}$, $\cot \beta = \frac{3\sqrt{91}}{91}$ **9.** $\frac{\sqrt{2}}{2}$ **10.** $\frac{\sqrt{3}}{3}$

11. $-\frac{\sqrt{2}}{2}$ **12.** $\frac{1}{2}$ **13.** Not defined   **14.** $-\sqrt{3}$

15. $\frac{2\sqrt{3}}{3}$ **16.** -1 **17.** $22°16'12''$ **18.** $47.56°$

19. 0.4452 **20.** 1.1315 **21.** 0.9498 **22.** -0.9092

23. -1.5282 **24.** -0.2778 **25.** $205.3°$

26. $47.2°$ **27.** $60°$ **28.** $60°$ **29.** $45°$ **30.** $30°$

31. sin 30.9° ≈ 0.5135, cos 30.9° ≈ 0.8581,
tan 30.9° ≈ 0.5985, csc 30.9° ≈ 1.9474, sec 30.9° ≈ 1.1654,
cot 30.9° ≈ 1.6709   **32.** $b ≈ 4.5$, $A ≈ 58.1°$, $B ≈ 31.9°$
33. $A = 38.83°$, $b ≈ 37.9$, $c ≈ 48.6$ **34.** 1748 m
35. 14 ft   **36.** II   **37.** I   **38.** IV   **39.** 425°, −295°

40. $\dfrac{\pi}{3}, -\dfrac{5\pi}{3}$ **41.** Complement: 76.6°; supplement: 166.6°

42. Complement: $\dfrac{\pi}{3}$; supplement: $\dfrac{5\pi}{6}$

43. $\sin\theta = \dfrac{3\sqrt{13}}{13}$, $\cos\theta = \dfrac{-2\sqrt{13}}{13}$, $\tan\theta = -\dfrac{3}{2}$,

$\csc\theta = \dfrac{\sqrt{13}}{3}$, $\sec\theta = -\dfrac{\sqrt{13}}{2}$, $\cot\theta = -\dfrac{2}{3}$

44. $\sin\theta = -\dfrac{2}{3}$, $\cos\theta = -\dfrac{\sqrt{5}}{3}$, $\cot\theta = \dfrac{\sqrt{5}}{2}$,

$\sec\theta = -\dfrac{3\sqrt{5}}{5}$, $\csc\theta = -\dfrac{3}{2}$ **45.** About 1743 mi

46.

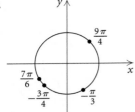

47. $\frac{121}{150}\pi$, 2.53   **48.** $-\dfrac{\pi}{6}$, −0.52   **49.** 270°   **50.** 171.89°

51. −257.83°   **52.** 1980°   **53.** $\dfrac{7\pi}{4}$, or 5.5 cm

54. 2.25, 129°   **55.** About 37.7 ft/min
56. 497,829 radians/hr   **57.** $\left(\frac{3}{5}, \frac{4}{5}\right), \left(-\frac{3}{5}, -\frac{4}{5}\right), \left(-\frac{3}{5}, \frac{4}{5}\right)$

58. −1   **59.** 1   **60.** $-\dfrac{\sqrt{3}}{2}$ **61.** $\frac{1}{2}$ **62.** $\dfrac{\sqrt{3}}{3}$ **63.** −1

64. −0.9056   **65.** 0.9218   **66.** Not defined   **67.** 4.3813
68. −6.1685   **69.** 0.8090

70.

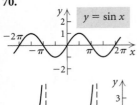
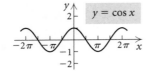
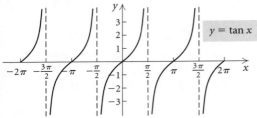

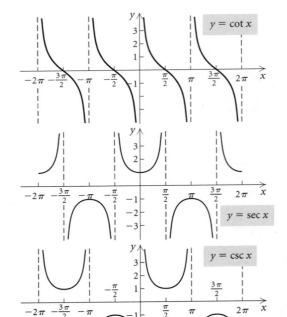

71. Period of sin, cos, sec, csc: 2π; period of tan, cot: π
72.

FUNCTION	DOMAIN	RANGE
Sine	$(-\infty, \infty)$	$[-1, 1]$
Cosine	$(-\infty, \infty)$	$[-1, 1]$
Tangent	$\left\{ x \,\middle|\, x \neq \dfrac{\pi}{2} + k\pi, k \in \mathbb{Z} \right\}$	$(-\infty, \infty)$

73.

FUNCTION	I	II	III	IV
Sine	+	+	−	−
Cosine	+	−	−	+
Tangent	+	−	+	−

74. Amplitude: 1; period: 2π; phase shift: $-\dfrac{\pi}{2}$

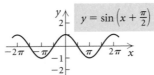

75. Amplitude: $\frac{1}{2}$; period: π; phase shift: $\frac{\pi}{4}$

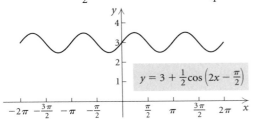

$$y = 3 + \frac{1}{2}\cos\left(2x - \frac{\pi}{2}\right)$$

76. (d)   **77.** (a)   **78.** (c)   **79.** (b)

80.

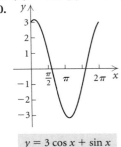

$$y = 3\cos x + \sin x$$

81.

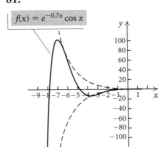

$$f(x) = e^{-0.7x}\cos x$$

82. C   **83.** B

84. Discussion and Writing: Both degrees and radians are units of angle measure. A degree is defined to be $\frac{1}{360}$ of one complete positive revolution. Degree notation has been in use since Babylonian times. Radians are defined in terms of intercepted arc length on a circle, with one radian being the measure of the angle for which the arc length equals the radius. There are 2π radians in one complete revolution.

85. Discussion and Writing: The graph of the cosine function is shaped like a continuous wave, with "high" points at $y = 1$ and "low" points at $y = -1$. The maximum value of the cosine function is 1, and it occurs at all points where $x = 2k\pi$, $k \in \mathbb{Z}$.

86. Discussion and Writing: No; $\sin x$ is never greater than 1.

87. Domain: $(-\infty, \infty)$; range: $[-3, 3]$; period 4π

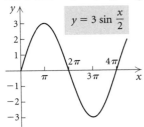

$$y = 3\sin\frac{x}{2}$$

88. $y_2 = 2\sin\left(x + \frac{\pi}{2}\right) - 2$

89. The domain consists of the intervals
$$\left(-\frac{\pi}{2} + 2k\pi, \frac{\pi}{2} + 2k\pi\right), k \in \mathbb{Z}.$$

90. $\cos x = -0.7890$, $\tan x = -0.7787$, $\cot x = -1.2842$, $\sec x = -1.2674$, $\csc x = 1.6276$

Test: Chapter 1

1. [1.1] $\sin\theta = \frac{4}{\sqrt{65}}$, or $\frac{4\sqrt{65}}{65}$; $\cos\theta = \frac{7}{\sqrt{65}}$, or $\frac{7\sqrt{65}}{65}$; $\tan\theta = \frac{4}{7}$; $\csc\theta = \frac{\sqrt{65}}{4}$; $\sec\theta = \frac{\sqrt{65}}{7}$; $\cot\theta = \frac{7}{4}$

2. [1.3] $\frac{\sqrt{3}}{2}$ **3.** [1.3] -1 **4.** [1.4] -1 **5.** [1.4] $-\sqrt{2}$

6. [1.1] $38.47°$ **7.** [1.3] -0.2419 **8.** [1.3] -0.2079

9. [1.4] -5.7588 **10.** [1.4] 0.7827 **11.** [1.1] $30°$

12. [1.1] $\sin 61.6° \approx 0.8796$; $\cos 61.6° \approx 0.4756$; $\tan 61.6° \approx 1.8495$; $\csc 61.6° \approx 1.1369$; $\sec 61.6° \approx 2.1026$; $\cot 61.6° \approx 0.5407$ **13.** [1.2] $B = 54.1°$, $a \approx 32.6$, $c \approx 55.7$

14. [1.3] Answers may vary; $472°$, $-248°$ **15.** [1.4] $\frac{\pi}{6}$

16. [1.3] $\cos\theta = \frac{5}{\sqrt{41}}$; $\tan\theta = -\frac{4}{5}$; $\csc\theta = -\frac{\sqrt{41}}{4}$; $\sec\theta = \frac{\sqrt{41}}{5}$; $\cot\theta = -\frac{5}{4}$ **17.** [1.4] $\frac{7\pi}{6}$ **18.** [1.4] $135°$

19. [1.4] $\frac{16\pi}{3} \approx 16.755$ cm   **20.** [1.5] 1   **21.** [1.5] 2π

22. [1.5] $\frac{\pi}{2}$ **23.** [1.6] (c)   **24.** [1.2] About 444 ft

25. [1.2] About 272 mi   **26.** [1.4] $18\pi \approx 56.55$ m/min

27. [1.6]

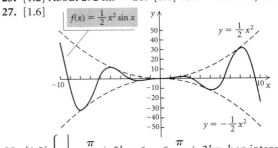

$$f(x) = \frac{1}{2}x^2\sin x$$
$$y = \frac{1}{2}x^2$$
$$y = -\frac{1}{2}x^2$$

28. [1.5] $\left\{x \middle| -\frac{\pi}{2} + 2k\pi < x < \frac{\pi}{2} + 2k\pi, k \text{ an integer}\right\}$

CHAPTER 2

Exercise Set 2.1

1. $\sin^2 x - \cos^2 x$ **3.** $\sin y + \cos y$ **5.** $1 - 2\sin\phi\cos\phi$

7. $\sin^3 x + \csc^3 x$ **9.** $\cos x(\sin x + \cos x)$

11. $(\sin x + \cos x)(\sin x - \cos x)$

13. $(2\cos x + 3)(\cos x - 1)$

15. $(\sin x + 3)(\sin^2 x - 3\sin x + 9)$ **17.** $\tan x$

19. $\sin x + 1$ **21.** $\frac{2\tan t + 1}{3\tan t + 1}$ **23.** 1

25. $\frac{5\cot\phi}{\sin\phi + \cos\phi}$ **27.** $\frac{1 + 2\sin s + 2\cos s}{\sin^2 s - \cos^2 s}$

29. $\frac{5(\sin\theta - 3)}{3}$ **31.** $\sin x\cos x$

33. $\sqrt{\cos \alpha}\,(\sin \alpha - \cos \alpha)$ **35.** $1 - \sin y$

37. $\dfrac{\sqrt{\sin x \cos x}}{\cos x}$ **39.** $\dfrac{\sqrt{2}\cot y}{2}$ **41.** $\dfrac{\cos x}{\sqrt{\sin x \cos x}}$

43. $\dfrac{1 + \sin y}{\cos y}$ **45.** $\cos \theta = \dfrac{\sqrt{a^2 - x^2}}{a},\ \tan \theta = \dfrac{x}{\sqrt{a^2 - x^2}}$

47. $\sin \theta = \dfrac{\sqrt{x^2 - 9}}{x},\ \cos \theta = \dfrac{3}{x}$ **49.** $\sin \theta \tan \theta$

51. $\dfrac{\sqrt{6} - \sqrt{2}}{4}$ **53.** $\dfrac{\sqrt{3} + 1}{1 - \sqrt{3}}$, or $-2 - \sqrt{3}$

55. $\dfrac{\sqrt{6} + \sqrt{2}}{4}$ **57.** $\sin 59° \approx 0.8572$

59. $\cos 24° \approx 0.9135$ **61.** $\tan 52° \approx 1.2799$

63. $\tan(\mu + \nu) = \dfrac{\sin(\mu + \nu)}{\cos(\mu + \nu)}$

$= \dfrac{\sin \mu \cos \nu + \cos \mu \sin \nu}{\cos \mu \cos \nu - \sin \mu \sin \nu}$

$= \dfrac{\sin \mu \cos \nu + \cos \mu \sin \nu}{\cos \mu \cos \nu - \sin \mu \sin \nu} \cdot \dfrac{\frac{1}{\cos \mu \cos \nu}}{\frac{1}{\cos \mu \cos \nu}}$

$= \dfrac{\dfrac{\sin \mu}{\cos \mu} + \dfrac{\sin \nu}{\cos \nu}}{1 - \dfrac{\sin \mu \sin \nu}{\cos \mu \cos \nu}}$

$= \dfrac{\tan \mu + \tan \nu}{1 - \tan \mu \tan \nu}$

65. 0 **67.** $-\frac{7}{25}$ **69.** -1.5789 **71.** 0.7071

73. $2 \sin \alpha \cos \beta$ **75.** $\cos u$ **77.** Left to the student

79. Discussion and Writing **81.** All real numbers

82. No solution **83.** [1.1] 1.9417 **84.** [1.1] 1.6645

85. $0°$; the lines are parallel **87.** $\dfrac{3\pi}{4}$, or $135°$ **89.** $4.57°$

91. $\dfrac{\cos(x + h) - \cos x}{h}$

$= \dfrac{\cos x \cos h - \sin x \sin h - \cos x}{h}$

$= \dfrac{\cos x \cos h - \cos x}{h} - \dfrac{\sin x \sin h}{h}$

$= \cos x \left(\dfrac{\cos h - 1}{h}\right) - \sin x \left(\dfrac{\sin h}{h}\right)$

93. Let $x = \dfrac{\pi}{5}$. Then $\dfrac{\sin 5x}{x} = \dfrac{\sin \pi}{\pi/5} = 0 \neq \sin 5$. Answers may vary.

95. Let $\alpha = \dfrac{\pi}{4}$. Then $\cos(2\alpha) = \cos \dfrac{\pi}{2} = 0$, but

$2 \cos \alpha = 2 \cos \dfrac{\pi}{4} = \sqrt{2}$. Answers may vary.

97. Let $x = \dfrac{\pi}{6}$. Then $\dfrac{\cos 6x}{\cos x} = \dfrac{\cos \pi}{\cos \dfrac{\pi}{6}} = \dfrac{-1}{\sqrt{3}/2} \neq 6$.

Answers may vary.

99. $\dfrac{6 - 3\sqrt{3}}{9 + 2\sqrt{3}} \approx 0.0645$

101. $168.7°$ **103.** $\cos 2\theta = \cos^2 \theta - \sin^2 \theta$, or $1 - 2\sin^2 \theta$, or $2\cos^2 \theta - 1$

105. $\tan\left(x + \dfrac{\pi}{4}\right) = \dfrac{\tan x + \tan \dfrac{\pi}{4}}{1 - \tan x \tan \dfrac{\pi}{4}} = \dfrac{1 + \tan x}{1 - \tan x}$

107. $\sin(\alpha + \beta) + \sin(\alpha - \beta) = \sin \alpha \cos \beta + \cos \alpha \sin \beta + \sin \alpha \cos \beta - \cos \alpha \sin \beta = 2 \sin \alpha \cos \beta$

Exercise Set 2.2

1. (a) $\tan \dfrac{3\pi}{10} \approx 1.3763$, $\csc \dfrac{3\pi}{10} \approx 1.2361$, $\sec \dfrac{3\pi}{10} \approx 1.7013$, $\cot \dfrac{3\pi}{10} \approx 0.7266$; **(b)** $\sin \dfrac{\pi}{5} \approx 0.5878$, $\cos \dfrac{\pi}{5} \approx 0.8090$, $\tan \dfrac{\pi}{5} \approx 0.7266$, $\csc \dfrac{\pi}{5} \approx 1.7013$, $\sec \dfrac{\pi}{5} \approx 1.2361$, $\cot \dfrac{\pi}{5} \approx 1.3763$

3. (a) $\cos \theta = -\dfrac{2\sqrt{2}}{3}$, $\tan \theta = -\dfrac{\sqrt{2}}{4}$, $\csc \theta = 3$, $\sec \theta = -\dfrac{3\sqrt{2}}{4}$, $\cot \theta = -2\sqrt{2}$;

(b) $\sin\left(\dfrac{\pi}{2} - \theta\right) = -\dfrac{2\sqrt{2}}{3}$, $\cos\left(\dfrac{\pi}{2} - \theta\right) = \dfrac{1}{3}$, $\tan\left(\dfrac{\pi}{2} - \theta\right) = -2\sqrt{2}$, $\csc\left(\dfrac{\pi}{2} - \theta\right) = -\dfrac{3\sqrt{2}}{4}$, $\sec\left(\dfrac{\pi}{2} - \theta\right) = 3$, $\cot\left(\dfrac{\pi}{2} - \theta\right) = -\dfrac{\sqrt{2}}{4}$;

(c) $\sin\left(\theta - \dfrac{\pi}{2}\right) = \dfrac{2\sqrt{2}}{3}$, $\cos\left(\theta - \dfrac{\pi}{2}\right) = \dfrac{1}{3}$, $\tan\left(\theta - \dfrac{\pi}{2}\right) = 2\sqrt{2}$, $\csc\left(\theta - \dfrac{\pi}{2}\right) = \dfrac{3\sqrt{2}}{4}$, $\sec\left(\theta - \dfrac{\pi}{2}\right) = 3$, $\cot\left(\theta - \dfrac{\pi}{2}\right) = \dfrac{\sqrt{2}}{4}$

5. $\sec\left(x + \dfrac{\pi}{2}\right) = -\csc x$ **7.** $\tan\left(x - \dfrac{\pi}{2}\right) = -\cot x$

9. $\sin 2\theta = \frac{24}{25}$, $\cos 2\theta = -\frac{7}{25}$, $\tan 2\theta = -\frac{24}{7}$; II

11. $\sin 2\theta = \frac{24}{25}$, $\cos 2\theta = -\frac{7}{25}$, $\tan 2\theta = -\frac{24}{7}$; II

13. $\sin 2\theta = -\frac{120}{169}$, $\cos 2\theta = \frac{119}{169}$, $\tan 2\theta = -\frac{120}{119}$; IV

15. $\cos 4x = 1 - 8\sin^2 x \cos^2 x$, or $\cos^4 x - 6\sin^2 x \cos^2 x + \sin^4 x$, or $8\cos^4 x - 8\cos^2 x + 1$

17. $\dfrac{\sqrt{2+\sqrt{3}}}{2}$ **19.** $\dfrac{\sqrt{2+\sqrt{2}}}{2}$ **21.** $2+\sqrt{3}$

23. 0.6421 **25.** 0.1735 **27.** $\cos x$ **29.** 1

31. $\cos 2x$ **33.** 8

35. (d);
$$\dfrac{\cos 2x}{\cos x - \sin x} = \dfrac{\cos^2 x - \sin^2 x}{\cos x - \sin x}$$
$$= \dfrac{(\cos x + \sin x)(\cos x - \sin x)}{\cos x - \sin x}$$
$$= \cos x + \sin x$$
$$= \dfrac{\sin x}{\sin x}(\cos x + \sin x)$$
$$= \sin x\left(\dfrac{\cos x}{\sin x} + \dfrac{\sin x}{\sin x}\right)$$
$$= \sin x\,(\cot x + 1)$$

37. (d); $\dfrac{\sin 2x}{2\cos x} = \dfrac{2\sin x \cos x}{2\cos x} = \sin x$

39. Discussion and Writing   **41.** [2.1] $\sin^2 x$ **42.** [2.1] 1

43. [2.1] $-\cos^2 x$ **44.** [2.1] $\csc^2 x$ **45.** [2.1] 1

46. [2.1] $\sec^2 x$ **47.** [2.1] $\cos^2 x$ **48.** [2.1] $\tan^2 x$

49. [1.5] (a), (e)   **50.** [1.5] (b), (c), (f)   **51.** [1.5] (d)

52. [1.5] (e)   **53.** $\sin 141° \approx 0.6293$, $\cos 141° \approx -0.7772$, $\tan 141° \approx -0.8097$, $\csc 141° \approx 1.5891$, $\sec 141° \approx -1.2867$, $\cot 141° \approx -1.2350$

55. $-\cos x\,(1 + \cot x)$ **57.** $\cot^2 y$

59. $\sin\theta = -\dfrac{15}{17}$, $\cos\theta = -\dfrac{8}{17}$, $\tan\theta = \dfrac{15}{8}$

61. (a) 9.80359 m/sec^2; (b) 9.80180 m/sec^2;
(c) $g = 9.78049(1 + 0.005264\sin^2\phi + 0.000024\sin^4\phi)$

Exercise Set 2.3

1.

$\sec x - \sin x \tan x$	$\cos x$
$\dfrac{1}{\cos x} - \sin x \cdot \dfrac{\sin x}{\cos x}$	
$\dfrac{1 - \sin^2 x}{\cos x}$	
$\dfrac{\cos^2 x}{\cos x}$	
$\cos x$	$\cos x$

3.

$\dfrac{1 - \cos x}{\sin x}$	$\dfrac{\sin x}{1 + \cos x}$
	$\dfrac{\sin x}{1 + \cos x} \cdot \dfrac{1 - \cos x}{1 - \cos x}$
	$\dfrac{\sin x\,(1 - \cos x)}{1 - \cos^2 x}$
	$\dfrac{\sin x\,(1 - \cos x)}{\sin^2 x}$
	$\dfrac{1 - \cos x}{\sin x}$

5.

$\dfrac{1 + \tan\theta}{1 - \tan\theta} + \dfrac{1 + \cot\theta}{1 - \cot\theta}$	0
$\dfrac{1 + \dfrac{\sin\theta}{\cos\theta}}{1 - \dfrac{\sin\theta}{\cos\theta}} + \dfrac{1 + \dfrac{\cos\theta}{\sin\theta}}{1 - \dfrac{\cos\theta}{\sin\theta}}$	
$\dfrac{\dfrac{\cos\theta + \sin\theta}{\cos\theta}}{\dfrac{\cos\theta - \sin\theta}{\cos\theta}} + \dfrac{\dfrac{\sin\theta + \cos\theta}{\sin\theta}}{\dfrac{\sin\theta - \cos\theta}{\sin\theta}}$	
$\dfrac{\cos\theta + \sin\theta}{\cos\theta} \cdot \dfrac{\cos\theta}{\cos\theta - \sin\theta} + $	
$\dfrac{\sin\theta + \cos\theta}{\sin\theta} \cdot \dfrac{\sin\theta}{\sin\theta - \cos\theta}$	
$\dfrac{\cos\theta + \sin\theta}{\cos\theta - \sin\theta} + \dfrac{\sin\theta + \cos\theta}{\sin\theta - \cos\theta}$	
$\dfrac{\cos\theta + \sin\theta}{\cos\theta - \sin\theta} - \dfrac{\cos\theta + \sin\theta}{\cos\theta - \sin\theta}$	
0	

7.

$\dfrac{\cos^2\alpha + \cot\alpha}{\cos^2\alpha - \cot\alpha}$	$\dfrac{\cos^2\alpha \tan\alpha + 1}{\cos^2\alpha \tan\alpha - 1}$
$\dfrac{\cos^2\alpha + \dfrac{\cos\alpha}{\sin\alpha}}{\cos^2\alpha - \dfrac{\cos\alpha}{\sin\alpha}}$	$\dfrac{\cos^2\alpha \dfrac{\sin\alpha}{\cos\alpha} + 1}{\cos^2\alpha \dfrac{\sin\alpha}{\cos\alpha} - 1}$
$\dfrac{\cos\alpha\left(\cos\alpha + \dfrac{1}{\sin\alpha}\right)}{\cos\alpha\left(\cos\alpha - \dfrac{1}{\sin\alpha}\right)}$	$\dfrac{\sin\alpha\cos\alpha + 1}{\sin\alpha\cos\alpha - 1}$
$\dfrac{\cos\alpha + \dfrac{1}{\sin\alpha}}{\cos\alpha - \dfrac{1}{\sin\alpha}}$	
$\dfrac{\sin\alpha\cos\alpha + 1}{\sin\alpha}$	
$\dfrac{\sin\alpha\cos\alpha - 1}{\sin\alpha}$	
$\dfrac{\sin\alpha\cos\alpha + 1}{\sin\alpha\cos\alpha - 1}$	

9.

$\dfrac{2\tan\theta}{1 + \tan^2\theta}$	$\sin 2\theta$
$\dfrac{2\tan\theta}{\sec^2\theta}$	$2\sin\theta\cos\theta$
$\dfrac{2\sin\theta}{\cos\theta} \cdot \dfrac{\cos^2\theta}{1}$	
$2\sin\theta\cos\theta$	

11.

$1 - \cos 5\theta \cos 3\theta - \sin 5\theta \sin 3\theta$	$2 \sin^2 \theta$
$1 - [\cos 5\theta \cos 3\theta + \sin 5\theta \sin 3\theta]$	$1 - \cos 2\theta$
$1 - \cos (5\theta - 3\theta)$	
$1 - \cos 2\theta$	

13.

$2 \sin \theta \cos^3 \theta + 2 \sin^3 \theta \cos \theta$	$\sin 2\theta$
$2 \sin \theta \cos \theta (\cos^2 \theta + \sin^2 \theta)$	$2 \sin \theta \cos \theta$
$2 \sin \theta \cos \theta$	

15.

$\dfrac{\tan x - \sin x}{2 \tan x}$	$\sin^2 \dfrac{x}{2}$
$\dfrac{1}{2}\left[\dfrac{\dfrac{\sin x}{\cos x} - \sin x}{\dfrac{\sin x}{\cos x}}\right]$	$\dfrac{1 - \cos x}{2}$
$\dfrac{1}{2} \dfrac{\sin x - \sin x \cos x}{\cos x} \cdot \dfrac{\cos x}{\sin x}$	
$\dfrac{1 - \cos x}{2}$	

17.

$\sin (\alpha + \beta) \sin (\alpha - \beta)$	$\sin^2 \alpha - \sin^2 \beta$
$\left(\begin{matrix}\sin \alpha \cos \beta + \\ \cos \alpha \sin \beta\end{matrix}\right)\left(\begin{matrix}\sin \alpha \cos \beta - \\ \cos \alpha \sin \beta\end{matrix}\right)$	$\begin{matrix}1 - \cos^2 \alpha - \\ (1 - \cos^2 \beta)\end{matrix}$
$\sin^2 \alpha \cos^2 \beta - \cos^2 \alpha \sin^2 \beta$	$\cos^2 \beta - \cos^2 \alpha$
$\begin{matrix}\cos^2 \beta (1 - \cos^2 \alpha) - \\ \cos^2 \alpha (1 - \cos^2 \beta)\end{matrix}$	
$\begin{matrix}\cos^2 \beta - \cos^2 \alpha \cos^2 \beta - \\ \cos^2 \alpha + \cos^2 \alpha \cos^2 \beta\end{matrix}$	
$\cos^2 \beta - \cos^2 \alpha$	

19.

$\tan \theta (\tan \theta + \cot \theta)$	$\sec^2 \theta$
$\tan^2 \theta + \tan \theta \cot \theta$	
$\tan^2 \theta + 1$	
$\sec^2 \theta$	

21.

$\dfrac{1 + \cos^2 x}{\sin^2 x}$	$2 \csc^2 x - 1$
$\dfrac{1}{\sin^2 x} + \dfrac{\cos^2 x}{\sin^2 x}$	
$\csc^2 x + \cot^2 x$	
$\csc^2 x + \csc^2 x - 1$	
$2 \csc^2 x - 1$	

23.

$\dfrac{1 + \sin x}{1 - \sin x} + \dfrac{\sin x - 1}{1 + \sin x}$	$4 \sec x \tan x$
$\dfrac{(1 + \sin x)^2 - (1 - \sin x)^2}{1 - \sin^2 x}$	$4 \cdot \dfrac{1}{\cos x} \cdot \dfrac{\sin x}{\cos x}$
$\dfrac{(1 + 2 \sin x + \sin^2 x) - (1 - 2 \sin x + \sin^2 x)}{\cos^2 x}$	$\dfrac{4 \sin x}{\cos^2 x}$
$\dfrac{4 \sin x}{\cos^2 x}$	

25.

$\cos^2 \alpha \cot^2 \alpha$	$\cot^2 \alpha - \cos^2 \alpha$
$(1 - \sin^2 \alpha) \cot^2 \alpha$	
$\cot^2 \alpha - \sin^2 \alpha \cdot \dfrac{\cos^2 \alpha}{\sin^2 \alpha}$	
$\cot^2 \alpha - \cos^2 \alpha$	

27.

$2 \sin^2 \theta \cos^2 \theta + \cos^4 \theta$	$1 - \sin^4 \theta$
$\cos^2 \theta (2 \sin^2 \theta + \cos^2 \theta)$	$(1 + \sin^2 \theta)(1 - \sin^2 \theta)$
$\cos^2 \theta (\sin^2 \theta + \sin^2 \theta + \cos^2 \theta)$	$(1 + \sin^2 \theta)(\cos^2 \theta)$
$\cos^2 \theta (\sin^2 \theta + 1)$	

29.

$\dfrac{1 + \sin x}{1 - \sin x}$	$(\sec x + \tan x)^2$
$\dfrac{1 + \sin x}{1 - \sin x} \cdot \dfrac{1 + \sin x}{1 + \sin x}$	$\left(\dfrac{1}{\cos x} + \dfrac{\sin x}{\cos x}\right)^2$
$\dfrac{(1 + \sin x)^2}{1 - \sin^2 x}$	$\dfrac{(1 + \sin x)^2}{\cos^2 x}$
$\dfrac{(1 + \sin x)^2}{\cos^2 x}$	

31. Sine sum and difference identities:

$$\sin (x + y) = \sin x \cos y + \cos x \sin y,$$
$$\sin (x - y) = \sin x \cos y - \cos x \sin y.$$

Add the sum and difference identities:

$$\sin (x + y) + \sin (x - y) = 2 \sin x \cos y$$
$$\tfrac{1}{2}[\sin (x + y) + \sin (x - y)] = \sin x \cos y. \quad (3)$$

Subtract the difference identity from the sum identity:

$$\sin (x + y) - \sin (x - y) = 2 \cos x \sin y$$
$$\tfrac{1}{2}[\sin (x + y) - \sin (x - y)] = \cos x \sin y. \quad (4)$$

33. $\sin 3\theta - \sin 5\theta = 2\cos \dfrac{8\theta}{2} \sin \dfrac{-2\theta}{2} = -2 \cos 4\theta \sin\theta$

35. $\sin 8\theta + \sin 5\theta = 2 \sin \dfrac{13\theta}{2} \cos \dfrac{3\theta}{2}$

37. $\sin 7u \sin 5u = \tfrac{1}{2}(\cos 2u - \cos 12u)$

39. $7 \cos \theta \sin 7\theta = \tfrac{7}{2}[\sin 8\theta - \sin (-6\theta)]$
$\qquad\qquad\qquad = \tfrac{7}{2}(\sin 8\theta + \sin 6\theta)$

41. $\cos 55° \sin 25° = \tfrac{1}{2}(\sin 80° - \sin 30°) = \tfrac{1}{2} \sin 80° - \tfrac{1}{4}$

43.

$\sin 4\theta + \sin 6\theta$	$\cot \theta (\cos 4\theta - \cos 6\theta)$
$2 \sin \dfrac{10\theta}{2} \cos \dfrac{-2\theta}{2}$	$\dfrac{\cos \theta}{\sin \theta}\left(2 \sin \dfrac{10\theta}{2} \sin \dfrac{2\theta}{2}\right)$
$2 \sin 5\theta \cos(-\theta)$	$\dfrac{\cos \theta}{\sin \theta}(2 \sin 5\theta \sin \theta)$
$2 \sin 5\theta \cos \theta$	$2 \sin 5\theta \cos \theta$

45.

$\cot 4x\,(\sin x + \sin 4x + \sin 7x)$	$\cos x + \cos 4x + \cos 7x$
$\dfrac{\cos 4x}{\sin 4x}\left(\sin 4x + 2\sin\dfrac{8x}{2}\cos\dfrac{-6x}{2}\right)$	$\cos 4x + 2\cos\dfrac{8x}{2}\cdot\cos\dfrac{6x}{2}$
$\dfrac{\cos 4x}{\sin 4x}\,(\sin 4x + 2\sin 4x\cos 3x)$	$\cos 4x + 2\cos 4x\cdot\cos 3x$
$\cos 4x\,(1 + 2\cos 3x)$	$\cos 4x\,(1 + 2\cos 3x)$

47.

$\cot\dfrac{x+y}{2}$	$\dfrac{\sin y - \sin x}{\cos x - \cos y}$
$\dfrac{\cos\dfrac{x+y}{2}}{\sin\dfrac{x+y}{2}}$	$\dfrac{2\cos\dfrac{x+y}{2}\sin\dfrac{y-x}{2}}{2\sin\dfrac{x+y}{2}\sin\dfrac{y-x}{2}}$
	$\dfrac{\cos\dfrac{x+y}{2}}{\sin\dfrac{x+y}{2}}$

49.

$\tan\dfrac{\theta+\phi}{2}\,(\sin\theta - \sin\phi)$	$\tan\dfrac{\theta-\phi}{2}\,(\sin\theta + \sin\phi)$
$\dfrac{\sin\dfrac{\theta+\phi}{2}}{\cos\dfrac{\theta+\phi}{2}}\left(2\cos\dfrac{\theta+\phi}{2}\sin\dfrac{\theta-\phi}{2}\right)$	$\dfrac{\sin\dfrac{\theta-\phi}{2}}{\cos\dfrac{\theta-\phi}{2}}\left(2\sin\dfrac{\theta+\phi}{2}\cos\dfrac{\theta-\phi}{2}\right)$
$2\sin\dfrac{\theta+\phi}{2}\cdot\sin\dfrac{\theta-\phi}{2}$	$2\sin\dfrac{\theta+\phi}{2}\cdot\sin\dfrac{\theta-\phi}{2}$

51. B;

$\dfrac{\cos x + \cot x}{1 + \csc x}$	$\cos x$
$\dfrac{\dfrac{\cos x}{1} + \dfrac{\cos x}{\sin x}}{1 + \dfrac{1}{\sin x}}$	
$\dfrac{\sin x\cos x + \cos x}{\sin x}\cdot\dfrac{\sin x}{\sin x + 1}$	
$\dfrac{\cos x\,(\sin x + 1)}{\sin x + 1}$	
$\cos x$	

53. A;

$\sin x\cos x + 1$	$\dfrac{\sin^3 x - \cos^3 x}{\sin x - \cos x}$
	$\dfrac{(\sin x - \cos x)(\sin^2 x + \sin x\cos x + \cos^2 x)}{\sin x - \cos x}$
	$\sin^2 x + \sin x\cos x + \cos^2 x$
	$\sin x\cos x + 1$

55. C;

$\dfrac{1}{\cot x\,\sin^2 x}$	$\tan x + \cot x$
$\dfrac{1}{\dfrac{\cos x}{\sin x}\cdot\sin^2 x}$	$\dfrac{\sin x}{\cos x} + \dfrac{\cos x}{\sin x}$
$\dfrac{1}{\cos x\,\sin x}$	$\dfrac{\sin^2 x + \cos^2 x}{\cos x\,\sin x}$
	$\dfrac{1}{\cos x\,\sin x}$

57. Discussion and Writing

59. (a), (d)

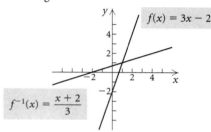

$f^{-1}(x) = \dfrac{x+2}{3}$

(b) yes; (c) $f^{-1}(x) = \dfrac{x+2}{3}$

60. (a), (d)

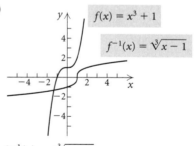

(b) yes; (c) $f^{-1}(x) = \sqrt[3]{x-1}$

61. (a), (d)

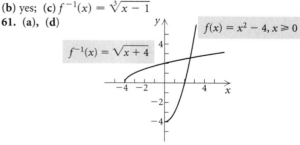

(b) yes; (c) $f^{-1}(x) = \sqrt{x+4}$

62. (a), (d)

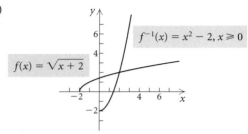

$f(x) = \sqrt{x+2}$

$f^{-1}(x) = x^2 - 2,\ x \geq 0$

(b) yes; **(c)** $f^{-1}(x) = x^2 - 2,\ x \geq 0$

63. $0, \frac{5}{2}$ **64.** $-4, \frac{7}{3}$ **65.** $\pm 2, \pm 3i$

66. $5 \pm 2\sqrt{6}$ **67.** 27   **68.** 9

69.

$\ln |\tan x|$	$-\ln |\cot x|$
$\ln \left| \dfrac{1}{\cot x} \right|$	
$\ln |1| - \ln |\cot x|$	
$0 - \ln |\cot x|$	
$-\ln |\cot x|$	

71. $\log(\cos x - \sin x) + \log(\cos x + \sin x)$
$= \log[(\cos x - \sin x)(\cos x + \sin x)]$
$= \log(\cos^2 x - \sin^2 x) = \log \cos 2x$

73. $\dfrac{1}{\omega C(\tan \theta + \tan \phi)} = \dfrac{1}{\omega C\left(\dfrac{\sin \theta}{\cos \theta} + \dfrac{\sin \phi}{\cos \phi}\right)}$

$= \dfrac{1}{\omega C\left(\dfrac{\sin \theta \cos \phi + \sin \phi \cos \theta}{\cos \theta \cos \phi}\right)}$

$= \dfrac{\cos \theta \cos \phi}{\omega C \sin(\theta + \phi)}$

Exercise Set 2.4

1. $-\dfrac{\pi}{3}, -60°$ **3.** $\dfrac{\pi}{4}, 45°$ **5.** $\dfrac{\pi}{4}, 45°$ **7.** $0, 0°$

9. $\dfrac{\pi}{6}, 30°$ **11.** $\dfrac{\pi}{6}, 30°$ **13.** $-\dfrac{\pi}{6}, -30°$

15. $-\dfrac{\pi}{6}, -30°$ **17.** $\dfrac{\pi}{2}, 90°$ **19.** $\dfrac{\pi}{3}, 60°$

21. $0.3520, 20.2°$ **23.** $1.2917, 74.0°$ **25.** $2.9463, 168.8°$

27. $-0.1600, -9.2°$ **29.** $0.8289, 47.5°$

31. $-0.9600, -55.0°$

33. $\sin^{-1}: [-1, 1];\ \cos^{-1}: [-1, 1];\ \tan^{-1}: (-\infty, \infty)$

35. $\theta = \sin^{-1}\left(\dfrac{2000}{d}\right)$ **37.** 0.3 **39.** $\dfrac{\pi}{4}$ **41.** $\dfrac{\pi}{5}$

43. $-\dfrac{\pi}{3}$ **45.** $\dfrac{1}{2}$ **47.** 1 **49.** $\dfrac{\pi}{3}$ **51.** $\dfrac{\sqrt{11}}{33}$

53. $-\dfrac{\pi}{6}$ **55.** $\dfrac{a}{\sqrt{a^2 + 9}}$ **57.** $\dfrac{\sqrt{q^2 - p^2}}{p}$ **59.** $\dfrac{p}{3}$

61. $\dfrac{\sqrt{3}}{2}$ **63.** $-\dfrac{\sqrt{2}}{10}$ **65.** $xy + \sqrt{(1 - x^2)(1 - y^2)}$

67. 0.9861   **69.** Discussion and Writing

71. Discussion and Writing   **72.** [1.5] Periodic

73. [1.4] Radian measure   **74.** [1.1] Similar

75. [1.2] Angle of depression   **76.** [3.4] Angular speed

77. [1.3] Supplementary   **78.** [1.5] Amplitude

79. [1.1] Acute   **80.** [1.5] Circular

81.

$\sin^{-1} x + \cos^{-1} x$	$\dfrac{\pi}{2}$
$\sin(\sin^{-1} x + \cos^{-1} x)$	$\sin \dfrac{\pi}{2}$
$[\sin(\sin^{-1} x)][\cos(\cos^{-1} x)] +$	
$\quad [\cos(\sin^{-1} x)][\sin(\cos^{-1} x)]$	1
$x \cdot x + \sqrt{1 - x^2} \cdot \sqrt{1 - x^2}$	
$x^2 + 1 - x^2$	
1	

83.

$\sin^{-1} x$	$\tan^{-1} \dfrac{x}{\sqrt{1 - x^2}}$
$\sin(\sin^{-1} x)$	$\sin\left(\tan^{-1} \dfrac{x}{\sqrt{1 - x^2}}\right)$
x	x

85.

$\sin^{-1} x$	$\cos^{-1} \sqrt{1 - x^2}$
$\sin(\sin^{-1} x)$	$\sin\left(\cos^{-1} \sqrt{1 - x^2}\right)$
x	x

87. $\theta = \tan^{-1} \dfrac{y + h}{x} - \tan^{-1} \dfrac{y}{x};\ 38.7°$

Visualizing the Graph

1. D   **2.** G   **3.** C   **4.** H   **5.** I   **6.** A   **7.** E
8. J   **9.** F   **10.** B

Exercise Set 2.5

1. $\dfrac{\pi}{6} + 2k\pi, \dfrac{11\pi}{6} + 2k\pi$, or $30° + k \cdot 360°, 330° + k \cdot 360°$

3. $\dfrac{2\pi}{3} + k\pi$, or $120° + k \cdot 180°$

5. $\dfrac{\pi}{6} + 2k\pi, \dfrac{5\pi}{6} + 2k\pi$, or $30° + k \cdot 360°, 150° + k \cdot 360°$

7. $\dfrac{3\pi}{4} + 2k\pi, \dfrac{5\pi}{4} + 2k\pi$, or $135° + k \cdot 360°, 225° + k \cdot 360°$

9. $98.09°, 261.91°$ **11.** $\dfrac{4\pi}{3}, \dfrac{5\pi}{3}$ **13.** $\dfrac{\pi}{4}, \dfrac{3\pi}{4}, \dfrac{5\pi}{4}, \dfrac{7\pi}{4}$

15. $\dfrac{\pi}{6}, \dfrac{5\pi}{6}, \dfrac{3\pi}{2}$ **17.** $\dfrac{\pi}{6}, \dfrac{\pi}{2}, \dfrac{3\pi}{2}, \dfrac{11\pi}{6}$

19. $109.47°, 120°, 240°, 250.53°$

21. $0, \dfrac{\pi}{4}, \dfrac{3\pi}{4}, \pi, \dfrac{5\pi}{4}, \dfrac{7\pi}{4}$ **23.** $139.81°, 220.19°$

25. $37.22°, 169.35°, 217.22°, 349.35°$ **27.** $0, \pi, \dfrac{7\pi}{6}, \dfrac{11\pi}{6}$

29. $0, \dfrac{\pi}{2}, \pi, \dfrac{3\pi}{2}$ **31.** $0, \pi$ **33.** $\dfrac{3\pi}{4}, \dfrac{7\pi}{4}$

35. $\dfrac{2\pi}{3}, \dfrac{4\pi}{3}, \dfrac{3\pi}{2}$ **37.** $\dfrac{\pi}{4}, \dfrac{3\pi}{4}, \dfrac{5\pi}{4}, \dfrac{7\pi}{4}$ **39.** $\dfrac{\pi}{12}, \dfrac{5\pi}{12}$

41. $0.967, 1.853, 4.109, 4.994$ **43.** $\dfrac{2\pi}{3}, \dfrac{4\pi}{3}$

45. Left to the student   **47.** $1.114, 2.773$

49. 0.515 **51.** $0.422, 1.756$

53. **(a)** $y = 7 \sin(-2.6180x + 0.5236) + 7$;
(b) $\$10{,}500, \$13{,}062$

55. Discussion and Writing

57. [1.2] $B = 35°, b \approx 140.7, c \approx 245.4$

58. [1.2] $R \approx 15.5°, T \approx 74.5°, t \approx 13.7$ **59.** 36

60. 14 **61.** $\dfrac{\pi}{3}, \dfrac{2\pi}{3}, \dfrac{4\pi}{3}, \dfrac{5\pi}{3}$ **63.** $\dfrac{\pi}{3}, \dfrac{4\pi}{3}$

65. 0 **67.** $e^{3\pi/2 + 2k\pi}$, where k (an integer) ≤ -1

69. 1.24 days, 6.76 days   **71.** $16.5°$N   **73.** 1 **75.** 0.1923

Review Exercises: Chapter 2

1. True   **2.** True   **3.** True   **4.** False   **5.** False

6. $\csc^2 x$ **7.** 1 **8.** $\tan^2 y - \cot^2 y$ **9.** $\dfrac{(\cos^2 x + 1)^2}{\cos^2 x}$

10. $\csc x (\sec x - \csc x)$ **11.** $(3 \sin y + 5)(\sin y - 4)$

12. $(10 - \cos u)(100 + 10 \cos u + \cos^2 u)$ **13.** 1

14. $\frac{1}{2} \sec x$ **15.** $\dfrac{3 \tan x}{\sin x - \cos x}$ **16.** $\dfrac{3 \cos y + 3 \sin y + 2}{\cos^2 y - \sin^2 y}$

17. 1 **18.** $\frac{1}{4} \cot x$ **19.** $\sin x + \cos x$ **20.** $\dfrac{\cos x}{1 - \sin x}$

21. $\dfrac{\cos x}{\sqrt{\sin x}}$ **22.** $3 \sec \theta$ **23.** $\cos x \cos \dfrac{3\pi}{2} - \sin x \sin \dfrac{3\pi}{2}$

24. $\dfrac{\tan 45° - \tan 30°}{1 + \tan 45° \tan 30°}$ **25.** $\cos(27° - 16°)$, or $\cos 11°$

26. $\dfrac{-\sqrt{6} - \sqrt{2}}{4}$ **27.** $2 - \sqrt{3}$ **28.** -0.3745

29. $-\sin x$ **30.** $\sin x$ **31.** $-\cos x$

32. **(a)** $\sin \alpha = -\dfrac{4}{5}, \tan \alpha = \dfrac{4}{3}, \cot \alpha = \dfrac{3}{4},$

$\sec \alpha = -\dfrac{5}{3}, \csc \alpha = -\dfrac{5}{4}$; **(b)** $\sin\left(\dfrac{\pi}{2} - \alpha\right) = -\dfrac{3}{5},$

$\cos\left(\dfrac{\pi}{2} - \alpha\right) = -\dfrac{4}{5}, \tan\left(\dfrac{\pi}{2} - \alpha\right) = \dfrac{3}{4},$

$\cot\left(\dfrac{\pi}{2} - \alpha\right) = \dfrac{4}{3}, \sec\left(\dfrac{\pi}{2} - \alpha\right) = -\dfrac{5}{4},$

$\csc\left(\dfrac{\pi}{2} - \alpha\right) = -\dfrac{5}{3}$; **(c)** $\sin\left(\alpha + \dfrac{\pi}{2}\right) = -\dfrac{3}{5},$

$\cos\left(\alpha + \dfrac{\pi}{2}\right) = \dfrac{4}{5}, \tan\left(\alpha + \dfrac{\pi}{2}\right) = -\dfrac{3}{4},$

$\cot\left(\alpha + \dfrac{\pi}{2}\right) = -\dfrac{4}{3}, \sec\left(\alpha + \dfrac{\pi}{2}\right) = \dfrac{5}{4},$

$\csc\left(\alpha + \dfrac{\pi}{2}\right) = -\dfrac{5}{3}$ **33.** $-\sec x$

34. $\tan 2\theta = \dfrac{24}{7}, \cos 2\theta = \dfrac{7}{25}, \sin 2\theta = \dfrac{24}{25}$; I

35. $\dfrac{\sqrt{2 - \sqrt{2}}}{2}$

36. $\sin 2\beta = 0.4261, \cos \dfrac{\beta}{2} = 0.9940, \cos 4\beta = 0.6369$

37. $\cos x$ **38.** 1 **39.** $\sin 2x$ **40.** $\tan 2x$

41.

$\dfrac{1 - \sin x}{\cos x}$	$\dfrac{\cos x}{1 + \sin x}$
$\dfrac{1 - \sin x}{\cos x} \cdot \dfrac{\cos x}{\cos x}$	$\dfrac{\cos x}{1 + \sin x} \cdot \dfrac{1 - \sin x}{1 - \sin x}$
$\dfrac{\cos x - \sin x \cos x}{\cos^2 x}$	$\dfrac{\cos x - \sin x \cos x}{1 - \sin^2 x}$
	$\dfrac{\cos x - \sin x \cos x}{\cos^2 x}$

42.

$\dfrac{1 + \cos 2\theta}{\sin 2\theta}$	$\cot \theta$
$\dfrac{1 + 2\cos^2 \theta - 1}{2 \sin \theta \cos \theta}$	$\dfrac{\cos \theta}{\sin \theta}$
$\dfrac{\cos \theta}{\sin \theta}$	

43.

$\dfrac{\tan y + \sin y}{2 \tan y}$	$\cos^2 \dfrac{y}{2}$
$\dfrac{1}{2}\left[\dfrac{\dfrac{\sin y + \sin y \cos y}{\cos y}}{\dfrac{\sin y}{\cos y}}\right]$	$\dfrac{1 + \cos y}{2}$
$\dfrac{1}{2}\left[\dfrac{\sin y (1 + \cos y)}{\cos y} \cdot \dfrac{\cos y}{\sin y}\right]$	
$\dfrac{1 + \cos y}{2}$	

44.

$\dfrac{\sin x - \cos x}{\cos^2 x}$	$\dfrac{\tan^2 x - 1}{\sin x + \cos x}$
	$\dfrac{\dfrac{\sin^2 x}{\cos^2 x} - 1}{\sin x + \cos x}$
	$\dfrac{\dfrac{\sin^2 x - \cos^2 x}{\cos^2 x}}{\sin x + \cos x} \cdot \dfrac{1}{\sin x + \cos x}$
	$\dfrac{\sin x - \cos x}{\cos^2 x}$

Additional fragments near problems 27 and 33 (top right):

$\cot\left(\alpha + \dfrac{\pi}{2}\right) = -\dfrac{4}{3}, \sec\left(\alpha + \dfrac{\pi}{2}\right) = \dfrac{5}{4},$

$\csc\left(\alpha + \dfrac{\pi}{2}\right) = -\dfrac{5}{3}$ **33.** $-\sec x$

45. $3 \cos 2\theta \sin \theta = \frac{3}{2}(\sin 3\theta - \sin \theta)$

46. $\sin \theta - \sin 4\theta = -2 \cos \dfrac{5\theta}{2} \sin \dfrac{3\theta}{2}$

47. $-\dfrac{\pi}{6}, -30°$ **48.** $\dfrac{\pi}{6}, 30°$ **49.** $\dfrac{\pi}{4}, 45°$ **50.** $0, 0°$

51. $1.7920, 102.7°$ **52.** $0.3976, 22.8°$ **53.** $\frac{1}{2}$ **54.** $\dfrac{\sqrt{3}}{3}$

55. $\dfrac{\pi}{7}$ **56.** $\dfrac{\sqrt{2}}{2}$ **57.** $\dfrac{3}{\sqrt{b^2 + 9}}$ **58.** $-\frac{7}{25}$

59. $\dfrac{3\pi}{4} + 2k\pi, \dfrac{5\pi}{4} + 2k\pi$, or $135° + k \cdot 360°$, $225° + k \cdot 360°$

60. $\dfrac{\pi}{3} + k\pi$, or $60° + k \cdot 180°$ **61.** $\dfrac{\pi}{6}, \dfrac{5\pi}{6}, \dfrac{7\pi}{6}, \dfrac{11\pi}{6}$

62. $\dfrac{\pi}{4}, \dfrac{\pi}{2}, \dfrac{3\pi}{4}, \dfrac{5\pi}{4}, \dfrac{3\pi}{2}, \dfrac{7\pi}{4}$ **63.** $\dfrac{2\pi}{3}, \pi, \dfrac{4\pi}{3}$ **64.** $0, \pi$

65. $\dfrac{\pi}{4}, \dfrac{3\pi}{4}, \dfrac{5\pi}{4}, \dfrac{7\pi}{4}$ **66.** $0, \dfrac{\pi}{2}, \pi, \dfrac{3\pi}{2}$ **67.** $\dfrac{7\pi}{12}, \dfrac{23\pi}{12}$

68. $0.864, 2.972, 4.006, 6.114$ **69.** B **70.** A

71. B;

$\csc x - \cos x \cot x$	$\sin x$
$\dfrac{1}{\sin x} - \cos x \cdot \dfrac{\cos x}{\sin x}$	
$\dfrac{1 - \cos^2 x}{\sin x}$	
$\dfrac{\sin^2 x}{\sin x}$	
$\sin x$	

72. D;

$\dfrac{1}{\sin x \cos x} - \dfrac{\cos x}{\sin x}$	$\dfrac{\sin x \cos x}{1 - \sin^2 x}$
$\dfrac{1}{\sin x \cos x} - \dfrac{\cos^2 x}{\sin x \cos x}$	$\dfrac{\sin x \cos x}{\cos^2 x}$
$\dfrac{1 - \cos^2 x}{\sin x \cos x}$	$\dfrac{\sin x}{\cos x}$
$\dfrac{\sin^2 x}{\sin x \cos x}$	
$\dfrac{\sin x}{\cos x}$	

73. A;

$\dfrac{\cot x - 1}{1 - \tan x}$	$\dfrac{\csc x}{\sec x}$
$\dfrac{\dfrac{\cos x}{\sin x} - \dfrac{\sin x}{\sin x}}{\dfrac{\cos x}{\cos x} - \dfrac{\sin x}{\cos x}}$	$\dfrac{\dfrac{1}{\sin x}}{\dfrac{1}{\cos x}}$
$\dfrac{\cos x - \sin x}{\sin x} \cdot \dfrac{\cos x}{\cos x - \sin x}$	$\dfrac{1}{\sin x} \cdot \dfrac{\cos x}{1}$
	$\dfrac{\cos x}{\sin x}$

74. C;

$\dfrac{\cos x + 1}{\sin x} + \dfrac{\sin x}{\cos x + 1}$	$\dfrac{2}{\sin x}$
$\dfrac{(\cos x + 1)^2 + \sin^2 x}{\sin x\,(\cos x + 1)}$	
$\dfrac{\cos^2 x + 2 \cos x + 1 + \sin^2 x}{\sin x\,(\cos x + 1)}$	
$\dfrac{2 \cos x + 2}{\sin x\,(\cos x + 1)}$	
$\dfrac{2(\cos x + 1)}{\sin x\,(\cos x + 1)}$	
$\dfrac{2}{\sin x}$	

75. 4.917 **76.** No solution in $[0, 2\pi)$

77. Discussion and Writing

(a) $2 \cos^2 x - 1 = \cos 2x = \cos^2 x - \sin^2 x$

$= 1 \cdot (\cos^2 x - \sin^2 x)$

$= (\cos^2 x + \sin^2 x)(\cos^2 x - \sin^2 x)$

$= \cos^4 x - \sin^4 x;$

(b) $\cos^4 x - \sin^4 x = (\cos^2 x + \sin^2 x)(\cos^2 x - \sin^2 x)$

$= 1 \cdot (\cos^2 x - \sin^2 x)$

$= \cos^2 x - \sin^2 x = \cos 2x$

$= 2 \cos^2 x - 1;$

(c)

$2 \cos^2 x - 1$	$\cos^4 x - \sin^4 x$
$\cos 2x$	$(\cos^2 x + \sin^2 x)(\cos^2 x - \sin^2 x)$
	$1 \cdot (\cos^2 x - \sin^2 x)$
	$\cos^2 x - \sin^2 x$
	$\cos 2x$

Answers may vary. Method 2 may be the more efficient because it involves straightforward factorization and simplification. Method 1(a) requires a "trick" such as multiplying by a particular expression equivalent to 1.

78. Discussion and Writing: The ranges of the inverse trigonometric functions are restricted in order that they might be functions.

79. $108.4°$

80.

$\cos (u + v) = \cos u \cos v - \sin u \sin v$

$= \cos u \cos v - \cos \left(\dfrac{\pi}{2} - u\right) \cos \left(\dfrac{\pi}{2} - v\right)$

81. $\cos^2 x$

82. $\sin \theta = \sqrt{\dfrac{1}{2} + \dfrac{\sqrt{6}}{5}}$; $\cos \theta = \sqrt{\dfrac{1}{2} - \dfrac{\sqrt{6}}{5}}$;

$\tan \theta = \sqrt{\dfrac{5 + 2\sqrt{6}}{5 - 2\sqrt{6}}}$ **83.** $\ln e^{\sin t} = \log_e e^{\sin t} = \sin t$

84.

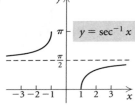

85. Let $x = \dfrac{\sqrt{2}}{2}$. Then $\tan^{-1}\dfrac{\sqrt{2}}{2} \approx 0.6155$ and

$\dfrac{\sin^{-1}\dfrac{\sqrt{2}}{2}}{\cos^{-1}\dfrac{\sqrt{2}}{2}} = \dfrac{\dfrac{\pi}{4}}{\dfrac{\pi}{4}} = 1.$ **86.** $\dfrac{\pi}{2}, \dfrac{3\pi}{2}$

Test: Chapter 2

1. [2.1] $2\cos x + 1$ **2.** [2.1] 1 **3.** [2.1] $\dfrac{\cos\theta}{1 + \sin\theta}$

4. [2.1] $2\cos\theta$ **5.** [2.1] $\dfrac{\sqrt{2} + \sqrt{6}}{4}$ **6.** [2.1] $\dfrac{3 - \sqrt{3}}{3 + \sqrt{3}}$

7. [2.1] $\frac{120}{169}$ **8.** [2.2] $\dfrac{\sqrt{5}}{3}$ **9.** [2.2] $\frac{24}{25}$, II

10. [2.2] $\dfrac{\sqrt{2 + \sqrt{3}}}{2}$ **11.** [2.2] 0.9304 **12.** [2.2] $3\sin 2x$

13. [2.3]

$\csc x - \cos x \cot x$	$\sin x$
$\dfrac{1}{\sin x} - \cos x \cdot \dfrac{\cos x}{\sin x}$	
$\dfrac{1 - \cos^2 x}{\sin x}$	
$\dfrac{\sin^2 x}{\sin x}$	
$\sin x$	

14. [2.3]

$(\sin x + \cos x)^2$	$1 + \sin 2x$
$\sin^2 x + 2\sin x \cos x + \cos^2 x$	
$1 + 2\sin x \cos x$	
$1 + \sin 2x$	

15. [2.3]

$(\csc\beta + \cot\beta)^2$	$\dfrac{1 + \cos\beta}{1 - \cos\beta}$
$\left(\dfrac{1}{\sin\beta} + \dfrac{\cos\beta}{\sin\beta}\right)^2$	$\dfrac{1 + \cos\beta}{1 - \cos\beta} \cdot \dfrac{1 + \cos\beta}{1 + \cos\beta}$
$\left(\dfrac{1 + \cos\beta}{\sin\beta}\right)^2$	$\dfrac{(1 + \cos\beta)^2}{1 - \cos^2\beta}$
$\dfrac{(1 + \cos\beta)^2}{\sin^2\beta}$	$\dfrac{(1 + \cos\beta)^2}{\sin^2\beta}$

16. [2.3]

$\dfrac{1 + \sin\alpha}{1 + \csc\alpha}$	$\dfrac{\tan\alpha}{\sec\alpha}$
	$\dfrac{\sin\alpha}{\cos\alpha}$
$\dfrac{1 + \sin\alpha}{1 + \dfrac{1}{\sin\alpha}}$	$\dfrac{1}{\cos\alpha}$
$\dfrac{1 + \sin\alpha}{\dfrac{\sin\alpha + 1}{\sin\alpha}}$	$\sin\alpha$
$\sin\alpha$	

17. [2.4] $\cos 8\alpha - \cos\alpha = -2\sin\dfrac{9\alpha}{2}\sin\dfrac{7\alpha}{2}$

18. [2.4] $4\sin\beta\cos 3\beta = 2(\sin 4\beta - \sin 2\beta)$

19. [2.4] $-45°$ **20.** [2.4] $\dfrac{\pi}{3}$ **21.** [2.4] 2.3072

22. [2.4] $\dfrac{\sqrt{3}}{2}$ **23.** [2.4] $\dfrac{5}{\sqrt{x^2 - 25}}$ **24.** [2.4] 0

25. [2.5] $\dfrac{\pi}{6}, \dfrac{5\pi}{6}, \dfrac{7\pi}{6}, \dfrac{11\pi}{6}$ **26.** [2.5] $0, \dfrac{\pi}{4}, \dfrac{3\pi}{4}, \pi$

27. [2.5] $\dfrac{\pi}{2}, \dfrac{11\pi}{6}$ **28.** [2.2] $\sqrt{\frac{11}{12}}$

CHAPTER 3

Exercise Set 3.1

1. $A = 121°, a \approx 33, c \approx 14$ **3.** $B \approx 57.4°, C \approx 86.1°,$
$c \approx 40$, or $B \approx 122.6°, C \approx 20.9°, c \approx 14$ **5.** $B \approx 44°24',$
$A \approx 74°26', a \approx 33.3$ **7.** $A = 110.36°, a \approx 5$ mi, $b \approx 3$ mi
9. $B \approx 83.78°, A \approx 12.44°, a \approx 12.30$ yd   **11.** $B \approx 14.7°,$
$C \approx 135.0°, c \approx 28.04$ cm   **13.** No solution
15. $B = 125.27°, b \approx 302$ m, $c \approx 138$ m   **17.** 8.2 ft^2
19. 12 yd^2 **21.** 596.98 ft^2 **23.** 76.3 m   **25.** 787 ft^2
27. About 51 ft   **29.** From A: about 35 mi; from B:
about 66 mi   **31.** About 22 mi   **33.** Discussion and
Writing   **35.** [1.1] $1.348, 77.2°$ **36.** [1.1] No angle
37. [1.1] $18.24°$ **38.** [1.1] $125.06°$ **39.** 5
40. [1.3] $\dfrac{\sqrt{3}}{2}$ **41.** [1.3] $\dfrac{\sqrt{2}}{2}$ **42.** [1.3] $-\dfrac{\sqrt{3}}{2}$
43. [1.3] $-\dfrac{1}{2}$ **44.** 2 **45.** Use the formula for the area of
a triangle and the law of sines.

$$K = \frac{1}{2}bc\sin A \quad \text{and} \quad b = \frac{c\sin B}{\sin C},$$
$$\text{so} \quad K = \frac{c^2\sin A\sin B}{2\sin C}.$$
$$K = \frac{1}{2}ab\sin C \quad \text{and} \quad b = \frac{a\sin B}{\sin A},$$
$$\text{so} \quad K = \frac{a^2\sin B\sin C}{2\sin A}.$$
$$K = \frac{1}{2}bc\sin A \quad \text{and} \quad c = \frac{b\sin C}{\sin B},$$
$$\text{so} \quad K = \frac{b^2\sin A\sin C}{2\sin B}.$$

47.

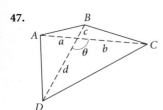

For the quadrilateral $ABCD$, we have

$$\text{Area} = \frac{1}{2}bd\sin\theta + \frac{1}{2}ac\sin\theta$$

$$+ \frac{1}{2}ad(\sin 180° - \theta) + \frac{1}{2}bc\sin(180° - \theta)$$

Note: $\sin\theta = \sin(180° - \theta)$.

$$= \frac{1}{2}(bd + ac + ad + bc)\sin\theta$$

$$= \frac{1}{2}(a + b)(c + d)\sin\theta$$

$$= \frac{1}{2}d_1 d_2 \sin\theta,$$

where $d_1 = a + b$ and $d_2 = c + d$.
49. 44.1 " from wall 1 and 104.3 " from wall 4

Exercise Set 3.2

1. $a \approx 15$, $B \approx 24°$, $C \approx 126°$ **3.** $A \approx 36.18°$, $B \approx 43.53°$, $C \approx 100.29°$ **5.** $b \approx 75$ m, $A \approx 94°51'$, $C \approx 12°29'$
7. $A \approx 24.15°$, $B \approx 30.75°$, $C \approx 125.10°$ **9.** No solution
11. $A \approx 79.93°$, $B \approx 53.55°$, $C \approx 46.52°$
13. $c \approx 45.17$ mi, $A \approx 89.3°$, $B \approx 42.0°$ **15.** $a \approx 13.9$ in., $B \approx 36.127°$, $C \approx 90.417°$ **17.** Law of sines; $C = 98°$, $a \approx 96.7$, $c \approx 101.9$ **19.** Law of cosines; $A \approx 73.71°$, $B \approx 51.75°$, $C \approx 54.54°$ **21.** Cannot be solved
23. Law of cosines; $A \approx 33.71°$, $B \approx 107.08°$, $C \approx 39.21°$
25. About 367 ft   **27.** About 1.5 mi
29. $S \approx 112.5°$, $T \approx 27.2°$, $U \approx 40.3°$ **31.** About 912 km
33. (a) About 16 ft; **(b)** about 122 ft²   **35.** About 4.7 cm
37. Discussion and Writing   **39.** Quartic
40. Linear   **41.** [1.5] Trigonometric
42. Exponential   **43.** Rational
44. Cubic   **45.** Exponential
46. Logarithmic   **47.** [1.5] Trigonometric
48. Quadratic   **49.** About 9386 ft
51. $A = \frac{1}{2}a^2 \sin\theta$; when $\theta = 90°$

Exercise Set 3.3

1. 5;   **3.** 1;

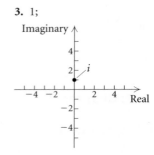

5. $\sqrt{17}$;   **7.** 3;

9. $3 - 3i$; $3\sqrt{2}\left(\cos\dfrac{7\pi}{4} + i\sin\dfrac{7\pi}{4}\right)$, or $3\sqrt{2}(\cos 315° + i\sin 315°)$

11. $4i$; $4\left(\cos\dfrac{\pi}{2} + i\sin\dfrac{\pi}{2}\right)$, or $4(\cos 90° + i\sin 90°)$

13. $\sqrt{2}\left(\cos\dfrac{7\pi}{4} + i\sin\dfrac{7\pi}{4}\right)$, or $\sqrt{2}(\cos 315° + i\sin 315°)$

15. $3\left(\cos\dfrac{3\pi}{2} + i\sin\dfrac{3\pi}{2}\right)$, or $3(\cos 270° + i\sin 270°)$

17. $2\left(\cos\dfrac{\pi}{6} + i\sin\dfrac{\pi}{6}\right)$, or $2(\cos 30° + i\sin 30°)$

19. $\dfrac{2}{5}(\cos 0 + i\sin 0)$, or $\dfrac{2}{5}(\cos 0° + i\sin 0°)$

21. $6\left(\cos\dfrac{5\pi}{4} + i\sin\dfrac{5\pi}{4}\right)$, or $6(\cos 225° + i\sin 225°)$

23. $\dfrac{3\sqrt{3}}{2} + \dfrac{3}{2}i$ **25.** $-10i$ **27.** $2 + 2i$ **29.** $2i$

31. $\dfrac{\sqrt{2}}{2} - \dfrac{\sqrt{6}}{2}i$ **33.** $4(\cos 42° + i\sin 42°)$

35. $11.25(\cos 56° + i\sin 56°)$ **37.** 4

39. $-i$ **41.** $6 + 6\sqrt{3}i$ **43.** $-2i$

45. $8(\cos \pi + i \sin \pi)$ **47.** $8\left(\cos \dfrac{3\pi}{2} + i \sin \dfrac{3\pi}{2}\right)$

49. $\dfrac{27}{2} + \dfrac{27\sqrt{3}}{2}i$ **51.** $-4 + 4i$ **53.** -1

55. $-\dfrac{\sqrt{2}}{2} + \dfrac{\sqrt{2}}{2}i, \dfrac{\sqrt{2}}{2} - \dfrac{\sqrt{2}}{2}i$

57. $2(\cos 157.5° + i \sin 157.5°), 2(\cos 337.5° + i \sin 337.5°)$

59. $\dfrac{\sqrt{3}}{2} + \dfrac{1}{2}i, -\dfrac{\sqrt{3}}{2} + \dfrac{1}{2}i, -i$

61. $\sqrt[3]{4}(\cos 110° + i \sin 110°), \sqrt[3]{4}(\cos 230° + i \sin 230°),$
$\sqrt[3]{4}(\cos 350° + i \sin 350°)$

63. $2, 2i, -2, -2i;$

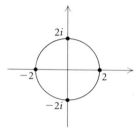

65. $\cos 36° + i \sin 36°,$
$\cos 108° + i \sin 108°, -1,$
$\cos 252° + i \sin 252°,$
$\cos 324° + i \sin 324°;$

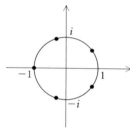

67. $\sqrt[10]{8}, \sqrt[10]{8}(\cos 36° + i \sin 36°), \sqrt[10]{8}(\cos 72° + i \sin 72°),$
$\sqrt[10]{8}(\cos 108° + i \sin 108°), \sqrt[10]{8}(\cos 144° + i \sin 144°), -\sqrt[10]{8},$
$\sqrt[10]{8}(\cos 216° + i \sin 216°), \sqrt[10]{8}(\cos 252° + i \sin 252°),$
$\sqrt[10]{8}(\cos 288° + i \sin 288°), \sqrt[10]{8}(\cos 324° + i \sin 324°)$

69. $\dfrac{\sqrt{3}}{2} + \dfrac{1}{2}i, i, -\dfrac{\sqrt{3}}{2} + \dfrac{1}{2}i, -\dfrac{\sqrt{3}}{2} - \dfrac{1}{2}i, -i, \dfrac{\sqrt{3}}{2} - \dfrac{1}{2}i$

71. $1, -\dfrac{1}{2} + \dfrac{\sqrt{3}}{2}i, -\dfrac{1}{2} - \dfrac{\sqrt{3}}{2}i$

73. $\cos 67.5° + i \sin 67.5°, \cos 157.5° + i \sin 157.5°,$
$\cos 247.5° + i \sin 247.5°, \cos 337.5° + i \sin 337.5°$

75. $\sqrt{3} + i, 2i, -\sqrt{3} + i, -\sqrt{3} - i, -2i, \sqrt{3} - i$

77. Left to the student   **79.** Discussion and Writing

81. [1.4] $15°$ **82.** [1.4] $540°$ **83.** [1.4] $\dfrac{11\pi}{6}$

84. [1.4] $-\dfrac{5\pi}{4}$ **85.** $3\sqrt{5}$

86.

(graph with points (0, 3), (2, −1), $\left(-\dfrac{1}{2}, -4\right)$)

87. [1.5] $\dfrac{\sqrt{3}}{2}$

88. [1.5] $\dfrac{\sqrt{3}}{2}$ **89.** [1.5] $\dfrac{\sqrt{2}}{2}$ **90.** [1.5] $\dfrac{1}{2}$

91. $-\dfrac{1+\sqrt{3}}{2} + \dfrac{1+\sqrt{3}}{2}i, -\dfrac{1-\sqrt{3}}{2} + \dfrac{1-\sqrt{3}}{2}i$

93. $\cos \theta - i \sin \theta$

95. $z = a + bi, |z| = \sqrt{a^2 + b^2}; \bar{z} = a - bi,$
$|\bar{z}| = \sqrt{a^2 + (-b)^2} = \sqrt{a^2 + b^2}, \therefore |z| = |\bar{z}|$

97. $|(a + bi)^2| = |a^2 - b^2 + 2abi| = \sqrt{(a^2 - b^2)^2 + 4a^2b^2}$
$= \sqrt{a^4 + 2a^2b^2 + b^4} = a^2 + b^2,$
$|a + bi|^2 = \left(\sqrt{a^2 + b^2}\right)^2 = a^2 + b^2$

99. $\dfrac{z}{w} = \dfrac{r_1(\cos \theta_1 + i \sin \theta_1)}{r_2(\cos \theta_2 + i \sin \theta_2)}$
$= \dfrac{r_1}{r_2}(\cos(\theta_1 - \theta_2) + i \sin(\theta_1 - \theta_2)),$
$\left|\dfrac{z}{w}\right| = \sqrt{\left[\dfrac{r_1}{r_2} \cos(\theta_1 - \theta_2)\right]^2 + \left[\dfrac{r_1}{r_2} \sin(\theta_1 - \theta_2)\right]^2}$
$= \sqrt{\dfrac{r_1^2}{r_2^2}} = \dfrac{|r_1|}{|r_2|};$
$|z| = \sqrt{(r_1 \cos \theta_1)^2 + (r_1 \sin \theta_1)^2} = \sqrt{r_1^2} = |r_1|;$
$|w| = \sqrt{(r_2 \cos \theta_2)^2 + (r_2 \sin \theta_2)^2} = \sqrt{r_2^2} = |r_2|;$
Then $\left|\dfrac{z}{w}\right| = \dfrac{|r_1|}{|r_2|} = \dfrac{|z|}{|w|}.$

101.

(graph with line $z + \bar{z} = 3$ at $\dfrac{3}{2}$)

Visualizing the Graph
1. J   **2.** C   **3.** E   **4.** H   **5.** I   **6.** A   **7.** D
8. G   **9.** B   **10.** F

Exercise Set 3.4

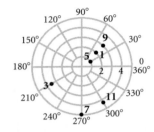

13. A: $(4, 30°)$, $(4, 390°)$, $(-4, 210°)$; B: $(5, 300°)$, $(5, -60°)$, $(-5, 120°)$; C: $(2, 150°)$, $(2, 510°)$, $(-2, 330°)$; D: $(3, 225°)$, $(3, -135°)$, $(-3, 45°)$; answers may vary

15. $(3, 270°)$, $\left(3, \dfrac{3\pi}{2}\right)$ **17.** $(6, 300°)$, $\left(6, \dfrac{5\pi}{3}\right)$

19. $(8, 330°)$, $\left(8, \dfrac{11\pi}{6}\right)$ **21.** $(2, 225°)$, $\left(2, \dfrac{5\pi}{4}\right)$

23. $(2, 60°)$, $\left(2, \dfrac{\pi}{3}\right)$ **25.** $(5, 315°)$, $\left(5, \dfrac{7\pi}{4}\right)$

27. $\left(\dfrac{5}{2}, \dfrac{5\sqrt{3}}{2}\right)$ **29.** $\left(-\dfrac{3\sqrt{2}}{2}, -\dfrac{3\sqrt{2}}{2}\right)$

31. $\left(-\dfrac{3}{2}, -\dfrac{3\sqrt{3}}{2}\right)$ **33.** $(-1, \sqrt{3})$ **35.** $(-\sqrt{3}, -1)$

37. $(3\sqrt{3}, -3)$ **39.** $r(3\cos\theta + 4\sin\theta) = 5$
41. $r\cos\theta = 5$ **43.** $r = 6$ **45.** $r^2\cos^2\theta = 25r\sin\theta$
47. $r^2\sin^2\theta - 5r\cos\theta - 25 = 0$ **49.** $r^2 = 2r\cos\theta$
51. $x^2 + y^2 = 25$ **53.** $y = 2$ **55.** $y^2 = -6x + 9$
57. $x^2 - 9x + y^2 - 7y = 0$ **59.** $x = 5$ **61.** $y = -\sqrt{3}x$
63.   **65.**

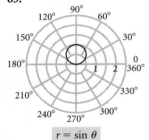

$r = \sin\theta$

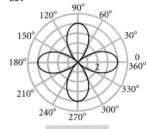

$r = 4\cos 2\theta$

67.   **69.**

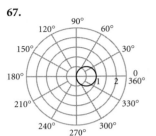

$r = \cos\theta$

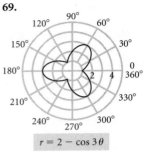

$r = 2 - \cos 3\theta$

71. $(7.616, 66.8°)$, $(7.616, 1.166)$
73. $(4.643, 132.9°)$, $(4.643, 2.320)$ **75.** $(2.19, -2.05)$
77. $(1.30, -3.99)$ **79.** (d)   **81.** (g)   **83.** (j)   **85.** (b)
87. (e)   **89.** (k)

91.   **93.**

$r = \sin\theta\tan\theta$

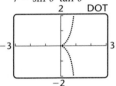

$r = e^{\theta/10}$

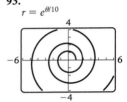

95.   **97.**

$r = \cos 2\theta\sec\theta$

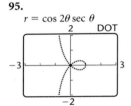

$r = \frac{1}{4}\tan^2\theta\sec\theta$

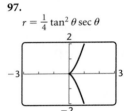

99. Discussion and Writing   **101.** 12   **102.** $\frac{1}{5}$
103.   **104.**

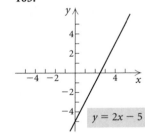

$y = 2x - 5$

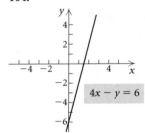

$4x - y = 6$

105.

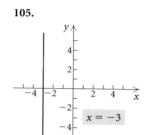

106.

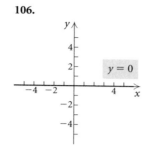

107. $y^2 = -4x + 4$

Exercise Set 3.5

1. Yes **3.** No **5.** Yes **7.** No **9.** No **11.** Yes
13. 55 N, 55° **15.** 929 N, 19° **17.** 57.0, 38°
19. 18.4, 37° **21.** 20.9, 58° **23.** 68.3, 18°
25. 11 ft/sec, 63° **27.** 726 lb, 47° **29.** 60°
31. 70.7 east; 70.7 south
33. Horizontal: 215.17 mph forward; vertical: 65.78 mph up
35. Horizontal: 390 lb forward; vertical: 675.5 lb up
37. Northerly: 115 km/h; westerly: 164 km/h
39. Perpendicular: 90.6 lb; parallel: 42.3 lb **41.** 48.1 lb
43. Discussion and Writing **45.** Natural
46. [2.2] Half-angle **47.** [1.4] Linear speed
48. [1.1] Cosine **49.** [2.1] Identity
50. [1.1] Cotangent of θ **51.** [1.3] Coterminal
52. [3.1] Sines **53.** Horizontal line; inverse
54. [1.3] Reference angle; acute
55. **(a)** (4.950, 4.950); **(b)** (0.950, −1.978)

Exercise Set 3.6

1. $\langle -9, 5 \rangle$; $\sqrt{106}$ **3.** $\langle -3, 6 \rangle$; $3\sqrt{5}$ **5.** $\langle 4, 0 \rangle$; 4
7. $\sqrt{37}$ **9.** $\langle 4, -5 \rangle$ **11.** $\sqrt{257}$ **13.** $\langle -9, 9 \rangle$
15. $\langle 41, -38 \rangle$ **17.** $\sqrt{261} - \sqrt{65}$ **19.** $\langle -1, -1 \rangle$
21. $\langle -8, 14 \rangle$ **23.** 1 **25.** −34
27.

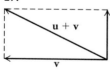

29.

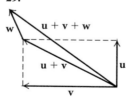

31. **(a)** $w = u + v$; **(b)** $v = w - u$
33. $\left\langle -\frac{5}{13}, \frac{12}{13} \right\rangle$ **35.** $\left\langle \frac{1}{\sqrt{101}}, -\frac{10}{\sqrt{101}} \right\rangle$
37. $\left\langle -\frac{1}{\sqrt{17}}, -\frac{4}{\sqrt{17}} \right\rangle$ **39.** $w = -4i + 6j$
41. $s = 2i + 5j$ **43.** $-7i + 5j$
45. **(a)** $3i + 29j$; **(b)** $\langle 3, 29 \rangle$ **47.** **(a)** $4i + 16j$; **(b)** $\langle 4, 16 \rangle$

49. j, or $\langle 0, 1 \rangle$ **51.** $-\frac{1}{2}i - \frac{\sqrt{3}}{2}j$, or $\left\langle -\frac{1}{2}, -\frac{\sqrt{3}}{2} \right\rangle$
53. 248° **55.** 63° **57.** 50° **59.** $|u| = 3$; $\theta = 45°$
61. 1; 120° **63.** 144.2° **65.** 14.0° **67.** 101.3°
69.

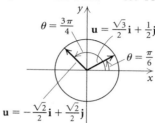

71. $u = -\frac{\sqrt{2}}{2}i - \frac{\sqrt{2}}{2}j$ **73.** $u = -\frac{\sqrt{10}}{10}i + \frac{3\sqrt{10}}{10}j$

75. $\sqrt{13}\left(\frac{2\sqrt{13}}{13}i - \frac{3\sqrt{13}}{13}j \right)$

77.

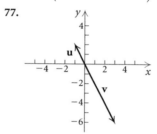

79. 174 nautical mi, S17°E **81.** 60°
83. 500 lb on left, 866 lb on right
85. Cable: 224-lb tension; boom: 167-lb compression
87. $u + v = \langle u_1, u_2 \rangle + \langle v_1, v_2 \rangle$
　　　$= \langle u_1 + v_1, u_2 + v_2 \rangle$
　　　$= \langle v_1 + u_1, v_2 + u_2 \rangle$
　　　$= \langle v_1, v_2 \rangle + \langle u_1, u_2 \rangle$
　　　$= v + u$
89. Discussion and Writing
91. $-\frac{1}{5}$; $(0, -15)$ **92.** 0; $(0, 7)$
93. 0, 4 **94.** $-\frac{11}{3}, \frac{5}{2}$
95. **(a)** $\cos \theta = \dfrac{u \cdot v}{|u| \, |v|} = \dfrac{0}{|u| \, |v|}$, $\therefore \cos \theta = 0$ and $\theta = 90°$.
(b) Answers may vary. $u = \langle 2, -3 \rangle$ and $v = \langle -3, -2 \rangle$;
$u \cdot v = 2(-3) + (-3)(-2) = 0$
97. $\frac{3}{5}i - \frac{4}{5}j$, $-\frac{3}{5}i + \frac{4}{5}j$ **99.** (5, 8)

Review Exercises: Chapter 3

1. True **2.** False **3.** False **4.** False **5.** False **6.** True
7. $A \approx 153°, B \approx 18°, C \approx 9°$
8. $A = 118°, a \approx 37$ in., $c \approx 24$ in.
9. $B = 14°50', a \approx 2523$ m, $c \approx 1827$ m
10. No solution **11.** 33 m^2 **12.** 13.72 ft^2 **13.** 63 ft^2
14. 92°, 33°, 55° **15.** 419 ft **16.** About 650 km

17. $\sqrt{29}$;

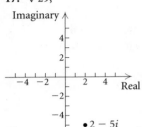

18. 4;

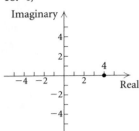

19. 2;

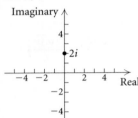

20. $\sqrt{10}$;

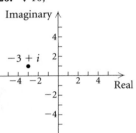

21. $\sqrt{2}\left(\cos\dfrac{\pi}{4} + i\sin\dfrac{\pi}{4}\right)$, or $\sqrt{2}(\cos 45° + i\sin 45°)$

22. $4\left(\cos\dfrac{3\pi}{2} + i\sin\dfrac{3\pi}{2}\right)$, or $4(\cos 270° + i\sin 270°)$

23. $10\left(\cos\dfrac{5\pi}{6} + i\sin\dfrac{5\pi}{6}\right)$, or $10(\cos 150° + i\sin 150°)$

24. $\frac{3}{4}(\cos 0 + i\sin 0)$, or $\frac{3}{4}(\cos 0° + i\sin 0°)$

25. $2 + 2\sqrt{3}i$　**26.** 7　**27.** $-\dfrac{5}{2} + \dfrac{5\sqrt{3}}{2}i$

28. $1 - \sqrt{3}i$　**29.** $1 + \sqrt{3} + (-1 + \sqrt{3})i$

30. $-i$　**31.** $2i$　**32.** $3\sqrt{3} + 3i$

33. $8(\cos 180° + i\sin 180°)$

34. $4(\cos 7\pi + i\sin 7\pi)$　**35.** $-8i$

36. $-\dfrac{1}{2} - \dfrac{\sqrt{3}}{2}i$

37. $\sqrt[4]{2}\left(\cos\dfrac{3\pi}{8} + i\sin\dfrac{3\pi}{8}\right)$,

$\sqrt[4]{2}\left(\cos\dfrac{11\pi}{8} + i\sin\dfrac{11\pi}{8}\right)$

38. $\sqrt[3]{6}(\cos 110° + i\sin 110°)$,

$\sqrt[3]{6}(\cos 230° + i\sin 230°)$, $\sqrt[3]{6}(\cos 350° + i\sin 350°)$

39. $3, 3i, -3, -3i$

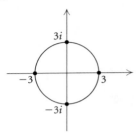

40. $1, \cos 72° + i\sin 72°, \cos 144° + i\sin 144°$,
$\cos 216° + i\sin 216°, \cos 288° + i\sin 288°$

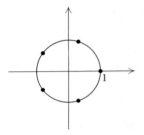

41. $\cos 22.5° + i\sin 22.5°, \cos 112.5° + i\sin 112.5°$,
$\cos 202.5° + i\sin 202.5°, \cos 292.5° + i\sin 292.5°$

42. $\dfrac{1}{2} + \dfrac{\sqrt{3}}{2}i, -1, \dfrac{1}{2} - \dfrac{\sqrt{3}}{2}i$

43. A: $(5, 120°), (5, 480°), (-5, 300°)$; B: $(3, 210°)$,
$(-3, 30°), (-3, 390°)$; C: $(4, 60°), (4, 420°), (-4, 240°)$;
D: $(1, 300°), (1, -60°), (-1, 120°)$; answers may vary

44. $(8, 135°), \left(8, \dfrac{3\pi}{4}\right)$　**45.** $(5, 270°), \left(5, \dfrac{3\pi}{2}\right)$

46. $\left(\dfrac{3\sqrt{2}}{2}, \dfrac{3\sqrt{2}}{2}\right)$　**47.** $\left(3, 3\sqrt{3}\right)$

48. $r(5\cos\theta - 2\sin\theta) = 6$　**49.** $r\sin\theta = 3$

50. $r = 3$　**51.** $r^2\sin^2\theta - 4r\cos\theta - 16 = 0$

52. $x^2 + y^2 = 36$　**53.** $x^2 + 2y = 1$

54. $y^2 - 6x = 9$　**55.** $x^2 - 2x + y^2 - 3y = 0$

56. (b)　**57.** (d)　**58.** (a)　**59.** (c)

60. $13.7, 71°$　**61.** $98.7, 15°$

62.

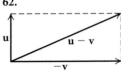

63.

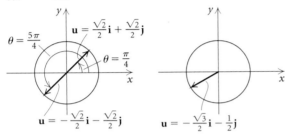

64. 666.7 N, 36°   **65.** 29 km/h, 149°
66. 102.4 nautical mi, S43°E   **67.** $\langle -4, 3 \rangle$
68. $\langle 2, -6 \rangle$ **69.** $\sqrt{61}$ **70.** $\langle 10, -21 \rangle$
71. $\langle 14, -64 \rangle$ **72.** $5 + \sqrt{116}$ **73.** 14
74. $\left\langle -\dfrac{3}{\sqrt{10}}, -\dfrac{1}{\sqrt{10}} \right\rangle$ **75.** $-9\mathbf{i} + 4\mathbf{j}$
76. 194.0°   **77.** $\sqrt{34}$; $\theta = 211.0°$
78. 111.8°   **79.** 85.1°   **80.** $34\mathbf{i} - 55\mathbf{j}$
81. $\mathbf{i} - 12\mathbf{j}$ **82.** $5\sqrt{2}$
83. $3\sqrt{65} + \sqrt{109}$ **84.** $-5\mathbf{i} + 5\mathbf{j}$
85.   **86.**

87. $\sqrt{10}\left(\dfrac{3\sqrt{10}}{10}\mathbf{i} - \dfrac{\sqrt{10}}{10}\mathbf{j} \right)$
88. D   **89.** A
90. (5.385, 111.8°), (5.385, 1.951)
91. (4.964, 147.8°), (4.964, 2.579)   **92.** (1.93, −0.52)
93. (−1.86, −1.35)
94. Discussion and Writing: A nonzero complex number has n different complex nth roots. Thus, 1 has three different complex cube roots, one of which is the real number 1. The other two are complex conjugates. Since the set of reals is a subset of the set of complex numbers, the real cube root of 1 is also a complex root of 1.
95. Discussion and Writing: A triangle has no solution when a sine or cosine value found is less than −1 or greater than 1. A triangle also has no solution if the sum of the angle measures calculated is greater than 180°. A triangle has only one solution if only one possible answer is found, or if one of the possible answers has an angle sum greater than 180°. A triangle has two solutions when two possible answers are found and neither results in an angle sum greater than 180°.
96. Discussion and Writing: For 150-yd shot, about 13.1 yd; for 300-yd shot, about 26.2 yd
97. $\frac{36}{13}\mathbf{i} + \frac{15}{13}\mathbf{j}$ **98.** 50.52°, 129.48°

7. [3.3]

8. [3.3] $\sqrt{13}$ **9.** [3.3] $3\sqrt{2}(\cos 315° + i \sin 315°)$
10. [3.3] $\frac{1}{4}i$ **11.** [3.3] 16
12. [3.4] $2(\cos 120° + i \sin 120°)$
13. [3.4] $\left(\dfrac{1}{2}, -\dfrac{\sqrt{3}}{2} \right)$ **14.** [3.4] $r = \sqrt{10}$
15. [3.4]

$r = 1 - \cos \theta$

16. [3.4] (a)   **17.** [3.5] Magnitude: 11.2; direction: 23.4°
18. [3.6] $-11\mathbf{i} - 17\mathbf{j}$ **19.** [3.6] $-\frac{4}{5}\mathbf{i} + \frac{3}{5}\mathbf{j}$
20. [3.1] 28.9°, 151.1°

Test: Chapter 3

1. [3.1] $A = 83°$, $b \approx 14.7$ ft, $c \approx 12.4$ ft
2. [3.1] $A \approx 73.9°$, $B \approx 70.1°$, $a \approx 8.2$ m, or $A \approx 34.1°$, $B \approx 109.9°$, $a \approx 4.8$ m
3. [3.2] $A \approx 99.9°$, $B \approx 36.8°$, $C \approx 43.3°$
4. [3.1] About 43.6 cm^2 **5.** [3.1] About 77 m
6. [3.5] About 930 km

Trigonometry

Trigonometric Functions

Acute Angles

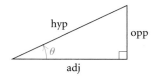

Any Angle

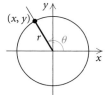

Real Numbers

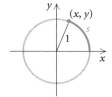

$$\sin \theta = \frac{\text{opp}}{\text{hyp}}, \quad \csc \theta = \frac{\text{hyp}}{\text{opp}},$$
$$\cos \theta = \frac{\text{adj}}{\text{hyp}}, \quad \sec \theta = \frac{\text{hyp}}{\text{adj}},$$
$$\tan \theta = \frac{\text{opp}}{\text{adj}}, \quad \cot \theta = \frac{\text{adj}}{\text{opp}}$$

$$\sin \theta = \frac{y}{r}, \quad \csc \theta = \frac{r}{y},$$
$$\cos \theta = \frac{x}{r}, \quad \sec \theta = \frac{r}{x},$$
$$\tan \theta = \frac{y}{x}, \quad \cot \theta = \frac{x}{y}$$

$$\sin s = y, \quad \csc s = \frac{1}{y},$$
$$\cos s = x, \quad \sec s = \frac{1}{x},$$
$$\tan s = \frac{y}{x}, \quad \cot s = \frac{x}{y}$$

Basic Trigonometric Identities

$$\sin (-x) = -\sin x,$$
$$\cos (-x) = \cos x,$$
$$\tan (-x) = -\tan x,$$

$$\tan x = \frac{\sin x}{\cos x},$$
$$\cot x = \frac{\cos x}{\sin x},$$

$$\csc x = \frac{1}{\sin x},$$
$$\sec x = \frac{1}{\cos x},$$
$$\cot x = \frac{1}{\tan x}$$

Pythagorean Identities

$$\sin^2 x + \cos^2 x = 1,$$
$$1 + \cot^2 x = \csc^2 x,$$
$$1 + \tan^2 x = \sec^2 x$$

Identities Involving $\pi/2$

$$\sin (\pi/2 - x) = \cos x,$$
$$\cos (\pi/2 - x) = \sin x, \quad \sin (x \pm \pi/2) = \pm\cos x,$$
$$\tan (\pi/2 - x) = \cot x, \quad \cos (x \pm \pi/2) = \mp\sin x$$

Sum and Difference Identities

$$\sin (u \pm v) = \sin u \cos v \pm \cos u \sin v,$$
$$\cos (u \pm v) = \cos u \cos v \mp \sin u \sin v,$$
$$\tan (u \pm v) = \frac{\tan u \pm \tan v}{1 \mp \tan u \tan v}$$

Double-Angle Identities

$$\sin 2x = 2 \sin x \cos x,$$
$$\cos 2x = \cos^2 x - \sin^2 x$$
$$= 1 - 2 \sin^2 x$$
$$= 2 \cos^2 x - 1,$$
$$\tan 2x = \frac{2 \tan x}{1 - \tan^2 x}$$

Half-Angle Identities

$$\sin \frac{x}{2} = \pm\sqrt{\frac{1 - \cos x}{2}}, \quad \cos \frac{x}{2} = \pm\sqrt{\frac{1 + \cos x}{2}},$$
$$\tan \frac{x}{2} = \pm\sqrt{\frac{1 - \cos x}{1 + \cos x}} = \frac{\sin x}{1 + \cos x} = \frac{1 - \cos x}{\sin x}$$

(*continued*)

Trigonometry *(continued)*

The Law of Sines

In any $\triangle ABC$,

$$\frac{a}{\sin A} = \frac{b}{\sin B} = \frac{c}{\sin C}.$$

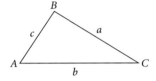

The Law of Cosines

In any $\triangle ABC$,

$$a^2 = b^2 + c^2 - 2bc \cos A,$$
$$b^2 = a^2 + c^2 - 2ac \cos B,$$
$$c^2 = a^2 + b^2 - 2ab \cos C.$$

Trigonometric Function Values of Special Angles

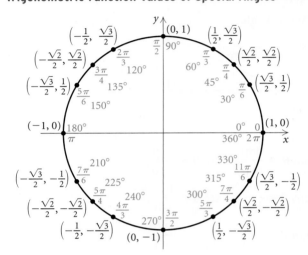

Graphs of Trigonometric Functions

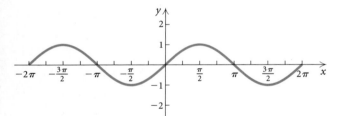

The sine function: $f(x) = \sin x$

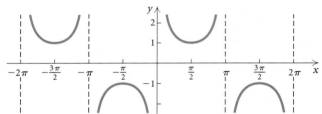

The cosecant function: $f(x) = \csc x$

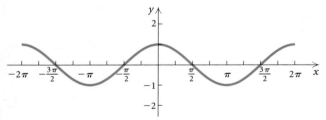

The cosine function: $f(x) = \cos x$

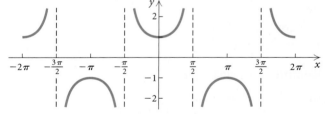

The secant function: $f(x) = \sec x$

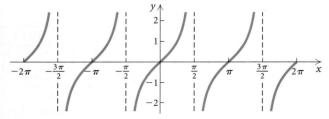

The tangent function: $f(x) = \tan x$

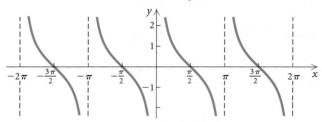

The cotangent function: $f(x) = \cot x$